ENCYCLOPEDIA OF PLANT PHYSIOLOGY

EDITED BY

W. RUHLAND

COEDITORS

E. ASHBY · J. BONNER · M. GEIGER-HUBER · W. O. JAMES
A. LANG · D. MÜLLER · M. G. STÅLFELT

VOLUME IX

THE METABOLISM OF SULFUR- AND PHOSPHORUS-CONTAINING COMPOUNDS

CONTRIBUTORS

H. G. ALBAUM · B. AXELROD · R. S. BANDURSKI · TH. BERSIN · K. HASSE · A. KJÆR
J. A. LOVERN · B. J. D. MEEUSE · J. R. P. O'BRIEN · W. SCHWARTZ · P. SCHWARZE
TE MAY CHING · M. D. THOMAS · W. W. UMBREIT · J. M. WIAME

SUBEDITOR

P. SCHWARZE

WITH 30 FIGURES

Springer-Verlag
Berlin Heidelberg GmbH
1958

HANDBUCH DER PFLANZENPHYSIOLOGIE

HERAUSGEGEBEN VON

W. RUHLAND

IN GEMEINSCHAFT MIT

E. ASHBY · J. BONNER · M. GEIGER-HUBER · W. O. JAMES
A. LANG · D. MÜLLER · M. G. STÅLFELT

BAND IX

DER STOFFWECHSEL DER SCHWEFEL- UND PHOSPHORHALTIGEN VERBINDUNGEN

BEARBEITET VON

H. G. ALBAUM · B. AXELROD · R. S. BANDURSKI · TH. BERSIN · K. HASSE · A. KJÆR
J. A. LOVERN · B. J. D. MEEUSE · J. R. P. O'BRIEN · W. SCHWARTZ · P. SCHWARZE
TE MAY CHING · M. D. THOMAS · W. W. UMBREIT · J. M. WIAME

REDIGIERT VON

P. SCHWARZE

MIT 30 ABBILDUNGEN

Springer-Verlag
Berlin Heidelberg GmbH
1958

ISBN 978-3-662-21768-9   ISBN 978-3-662-21766-5 (eBook)
DOI 10.1007/978-3-662-21766-5

© by Springer-Verlag Berlin Heidelberg 1958
Ursprünglich erschienen bei Springer-Verlag oHG. Berlin · Göttingen · Heidelberg 1958.
Softcover reprint of the hardcover 1st edition 1958

Inhaltsverzeichnis. — Contents.

I. Einführende Bemerkungen.

II. Der Stoffwechsel der S-haltigen Verbindungen.

Mitarbeiter von Band IX. — Contributors to volume IX.

Dr. HARRY G. ALBAUM, Professor of Biology, Biology Department, Brooklyn College, Brooklyn 10, New York (USA).

Dr. BERNARD AXELROD, Associate Professor of Biochemistry, Department of Biochemistry, Purdue University, West Lafayette, Indiana (USA).

Dr. ROBERT S. BANDURSKI, Associate Professor, Department of Botany and Plant Pathology, Michigan State University, East Lansing, Michigan (USA).

Professor Dr. phil. THEODOR BERSIN, Laboratorien Hausmann A.G., St. Gallen (Schweiz).

Professor Dr. phil. KURT HASSE, Institut für organische Chemie, Technische Hochschule Karlsruhe.

Professor Dr. phil. ANDERS KJÆR, Royal Veterinary and Agricultural College, Copenhagen (Denmark).

Dr. JOHN ARNOLD LOVERN, Department of Scientific and Industrial Research, Torry Research Station, Aberdeen (Great Britain).

Dr. BASTIAAN JACOB DIRK MEEUSE, Associate-Professor of Botany, Botany Department, University of Washington, Seattle 5, Wash. (USA).

Dr. J. R. P. O'BRIEN, Director of Department of Clinical Biochemistry, Reader in Clinical Biochemistry, Department of Clinical Biochemistry, University of Oxford, Oxford (Great Britain).

Professor Dr. phil. habil. W. SCHWARTZ, Direktor des Institutes für Mikrobiologie der Ernst-Moritz-Arndt-Universität, Greifswald.

Dr. PAUL SCHWARZE, Max-Planck-Institut für Züchtungsforschung, Köln-Vogelsang.

TE MAY CHING, Ph. D., Assistant Professor in Seed Technology, Oregon State College, Corvallis, Ore. (USA).

MOYER D. THOMAS, D. Sc. (Oxan), Senior Scientist, Stanford Research Institute, Menlo Park, California (USA).

Dr. WAYNE W. UMBREIT, Associate Director, Merck Institute for Therapeutic Research, Rahway, New Jersey (USA).

Dr. J. M. WIAME, Professeur à l'Université de Bruxelles et Directeur du Service de Recherches du C.E.R.I.A., Bruxelles. Service de Recherches, C.E.R.I.A., 1 Avenue E. Gryson, Anderlecht-Bruxelles (Belgique).

Einführende Bemerkungen.

Von

P. Schwarze.

Man kennt heute eine große Zahl von S- und P-haltigen Pflanzenstoffen, deren Herkunft und Bedeutung im Stoffwechsel vielfach untersucht worden ist. Der Schwefel ist ein Bestandteil des Eiweißes und damit ein Baustein der den gesamten Stoffwechsel beherrschenden Enzyme. Bauelement vieler Enzyme ist auch die Phosphorsäure. Diese steht außerdem im Dienste der Energieübertragung und spielt damit eine äußerst wichtige Rolle bei der Synthese der verschiedensten Pflanzenstoffe.

Die Funktionen, die S- und P-haltige Stoffe bei den spezifischen Stoffwechselprozessen zu erfüllen haben, machen es erforderlich, diese Verbindungen jeweils im Zusammenhang mit diesen zu behandeln, und dadurch ergibt es sich, daß die zahlreichen Einzelbefunde über den gesamten Stoffwechsel verteilt sind, in diesem Werk also auf insgesamt 9 Bände (IV—XII). Diese Sachlage ließ es wünschenswert erscheinen, die bisher vorliegenden Erkenntnisse über den S- und P-Stoffwechsel außerdem möglichst geschlossen darzustellen. Es wurde dabei Sorge getragen, Wiederholungen, die sich innerhalb des groß angelegten Werkes nicht vermeiden lassen, in tragbaren Grenzen zu halten, indem im vorliegenden Band ein Teilgebiet dann nur kurz und überblicksmäßig behandelt wurde, wenn dessen ausführliche Darstellung an anderem Ort unvermeidlich war. Das gilt in besonderem Maße für die P-Verbindungen und die an ihren Umsetzungen beteiligten Vorgänge. Meist wurde jedoch der umgekehrte Weg eingeschlagen und auf eine ausführliche Darstellung an anderer Stelle zugunsten einer solchen in diesem Band verzichtet.

Nicht behandelt sind im vorliegenden Band die S-haltigen Aminosäuren und die Eiweiße, zu deren obligaten Bestandteilen diese S-Verbindungen gehören; von der Gruppe der P-haltigen Stoffe wurden die Nucleinsäuren ausgeschlossen. Die S-haltigen Aminosäuren nehmen eine Schlüsselstellung im N-Umsatz ein, und die Eiweiße sind die Gipfelleistung dieses Stoffwechselzweiges. Die Behandlung beider Typen von Schwefelverbindungen mußte deshalb dem Band VIII vorbehalten bleiben. Bei den Nucleinsäuren ist es ebenfalls weniger der N-Gehalt, der ihre Aufnahme in den Band VIII erforderlich machte, als vielmehr ihre Funktion, die sie als Träger und Vermittler der Spezifität bei der Eiweißsynthese erfüllen.

Der Stoffwechsel der S-haltigen Verbindungen. Übersicht.

Von

P. Schwarze.

Seit LIEBIG weiß man, daß die Pflanze neben einer Reihe anderer Stoffe dem Boden Schwefel entnimmt, und SACHS und LIEBIG konnten experimentell einwandfrei nachweisen, daß dieses Element unentbehrlich ist. Es gehört zu jener Gruppe von Mineralstoffen, die, wie Stickstoff, Phosphor, Kalium, Calcium, Magnesium und Eisen, im Gegensatz zu den „Spurenelementen" in relativ großer Menge für die normale Entwicklung gebraucht werden. Über seine Funktion im Stoffwechsel, auf welche die Unentbehrlichkeit hinweist, war zu PFEFFERS Zeiten noch nichts bekannt. „Der Schwefel gehört zur Constitution der meisten Proteinstoffe, die durchschnittlich nur 0,2—2,4 Proc., also weniger als P enthalten. Zwar kommt außerdem der Schwefel im Senföl und gewissen anderen Verbindungen vor, indes läßt sich aus diesen Erfahrungen nicht entnehmen, ob er noch anderweitige generelle Bedeutung im Stoffwechsel hat" (Pflanzenphysiologie 1897, S. 423). „Da in den höheren Pflanzen der Schwefel meist in Form des Sulfations zur Aufnahme gelangt und in den Proteinstoffen nur SH-Gruppen vorkommen, so muß allgemein auch hier eine Sulfatreduktion stattfinden. Über diese Prozesse, die sich natürlich auch in den nicht chlorophyllhaltigen Pflanzen abspielen müssen, ist noch nicht das mindeste bekannt." Mit diesen Worten äußert sich CZAPEK 1925 (Biochemie der Pflanzen, Band III, S. 171) über den wichtigsten Teil des Schwefelstoffwechsels der höheren Pflanze. KOSTYTSCHEW stellt 1926 in seinem Lehrbuch der chemischen Physiologie als bemerkenswert heraus, daß S von der höheren Pflanze nur in Form von Sulfat assimiliert werden kann, für Pilze bis zu einem gewissen Grad aber Sulfite und Thiosulfat als S-Quelle in Frage kommen. Die chemische Seite der S-Assimilation und des S-Stoffwechsels überhaupt ist aber noch völlig ungeklärt. KOSTYTSCHEW weist damals auf die Notwendigkeit hin, Intermediärprodukte des S-Stoffwechsels abzufangen.

Aufbauend auf diesem relativ schwachen Fundament haben sich Physiologie und Biochemie der seither vergangenen Jahrzehnte intensiv mit dem S-Stoffwechsel der Pflanzen beschäftigt und dank neuer analytischer Methoden unser Wissen auf diesem Teilgebiet der Physiologie beträchtlich vermehrt. Den Untersuchungen an der höheren Pflanze gingen oft solche an Mikroorganismen voraus, und häufig kamen der Pflanzenphysiologie die Arbeiten über den Stoffwechsel von Mensch und Tier zugute, die von der medizinischen Physiologie und Biologie vorangetrieben wurden.

Wir kennen heute eine stattliche Zahl S-haltiger Pflanzenstoffe und haben auch Einblick in ihren Stoffwechsel gewonnen. Neben den am Grundstoffwechsel beteiligten S-Verbindungen hat man eine Anzahl sekundärer Pflanzenstoffe isolieren und charakterisieren können. Eingehend untersucht wurden die S-Spezialisten unter den Mikroorganismen, aber auch bei Vertretern mit „normalem" S-Stoffwechsel hat man wichtige Informationen über die allen Organismen gemeinsamen Züge des S-Stoffwechsels erhalten. Die vielen Einzelbefunde lassen sich zu einem Bild des S-Kreislaufes in der Natur zusammenfügen.

Am Stoffwechsel beteiligte S-haltige Verbindungen (Abschnitt II B). Schwefelverbindungen sind die Aminosäuren Cystein, Homocystein und Methionin, die beide frei, hauptsächlich aber im Eiweiß gebunden in den Organismen vorkommen. Sie sind als N-Verbindungen im vorangehenden Band VIII ausführlich besprochen. Dort wird auch ihre Fähigkeit zur Ausbildung von Schwefelbrücken und deren Bedeutung für den Zusammenhalt und die Gestalt des Eiweißmoleküls behandelt.

Als Eiweißbaustein ist der Schwefel Bestandteil aller Enzyme und bei einer Gruppe von ihnen, den Thiolenzymen (BARRON 1955), Voraussetzung für ihre Aktivität. Nur wenn eine bestimmte Anzahl von freien SH-Gruppen in ihrem Molekül enthalten ist, entfalten diese Enzyme ihre Wirkung. Bei dem verbreiteten und gut bekannten Thiolenzym Urease scheint nach BERSIN (1940) jedes sechste S-Atom im Molekül (Molekulargewicht 483000) in Form von SH-Gruppen vorhanden zu sein. Thiolenzyme finden sich unter den Hydrolasen (Urease, Papain, bestimmte bakterielle Proteasen, Thiaminasen), Redoxasen (Flavin-, Pyridin- und Pyridoxalenzyme) und Transferasen. Sie sind unter anderem daran zu erkennen, daß sie sich durch Oxydationsmittel und mercaptidbildende Fermentinhibitoren inaktivieren und durch Thiolverbindungen wieder reaktivieren lassen.

Bei einer Reihe von Enzymen ist der Schwefel auch Baustein des Coenzyms. Gut bekannt ist das Coenzym der Carboxylase, jenes verbreiteten Enzyms, das Pyruvat in CO_2 und Acetaldehyd spaltet. Cocarboxylase ist der Pyrophosphatester des auch Aneurin genannten, zur Gruppe der B-Vitamine gehörenden Thiamins (s. Beitrag O'BRIEN in diesem Band und VENNESLAND in Band XII).

Cocarboxylase = Thiaminpyrophosphat

Seine Unentbehrlichkeit für viele tierische Organismen beruht darauf, daß diese bei der Synthese der Cocarboxylase auf das Thiazolderivat Thiamin zurückgreifen müssen, da ihnen der Mechanismus zu seiner Bildung fehlt. Thiaminpyrophosphat ist auch das Coenzym der Transketolase und der Pyruvatoxydase. Die Transketolase spaltet Ribulose-5-phosphat in 3-Glycerinaldehyd und „aktivierten Glykolaldehyd" (Glykolaldehyd-Thiaminpyrophosphat), von denen sich letzterer mit Ribose-5-phosphat zu Sedoheptulose-7-phosphat verbindet (HORECKER und SMYRNOTIS 1953; s. auch ALBAUM in diesem Band). Die Pyruvatoxydase decarboxyliert Brenztraubensäure oxydativ zu Essigsäure (Näheres s. Beitrag MILLERD in Band XII). Nach REED und DE BUSK (1952) sollte nicht Thiaminpyrophosphat selbst das Coenzym dieser Oxydase sein, sondern dessen Kondensationsprodukt mit Liponsäure, das Lipothiamidpyrophosphat. Neuere Befunde sprechen aber gegen eine covalente Bindung zwischen Thiaminpyrophosphat und Liponsäure (GUNSALUS et al. 1954). Die Liponsäure scheint jedoch in diesem System die Übertragung des Acetylrestes, wahrscheinlich aus der Verbindung mit Thiaminpyrophosphat, auf das Coenzym A zu vermitteln.

Liponsäure oder 6,8-Thioctansäure, 1951 erstmalig aus Leber isoliert (Reed, de Busk, Gunsalus und Hornberger), leitet sich formal von der n-Caprinsäure ab. Werden je ein Wasserstoffatom des ε- und η-Kohlenstoffatoms durch die SH-Gruppe ersetzt, entsteht die Dihydroliponsäure, und diese geht durch Dehydrierung unter Ausbildung eines fünfgliedrigen Ringes mit einer S—S-Brücke in die Liponsäure über.

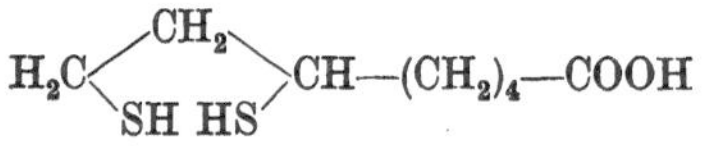

Dihydro-Liponsäure 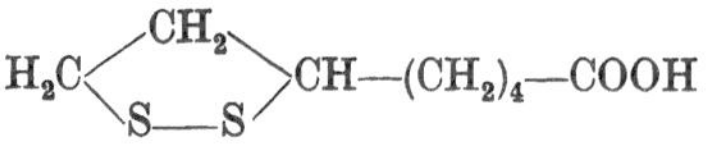   Liponsäure

Als Sulfoxyd der α-Liponsäure erwies sich die β-Liponsäure, die zweite Form dieser physiologisch bedeutsamen Substanz. Neben den schwach bis stark sauren Liponsäuren gibt es eine neutrale Form, einen Carboxylester der α-Liponsäure, der in den Chloroplasten lokalisiert und mit der Hill-Reaktion in Zusammenhang stehen soll. Die wasserlöslichen Formen der Liponsäure werden für Proteinkomplexe gehalten. Die Liponsäuren selbst sind, worauf der Name hinweist, in Fettlösungsmitteln leicht löslich. Eine Zusammenfassung der Arbeiten über die Chemie und Biochemie der α-Liponsäure wurde unlängst von Grisebach (1956) gegeben.

Als Thiol, und zwar als Pyrophosphorsäureester des Adenyl-pantothenyl-thioäthanolamins (Gregory, Novelli und Lipmann 1952), erwies sich auch das von Lipmann (1945) aufgefundene Coenzym A, das sich mit Acetat zum Acetyl-Coenzym A, dem „aktiven Acetat", verbindet und dieses reaktionsfähig macht (Näheres s. Albaum in diesem Band und Decker und Lynen, Band XII). Aus Acetat werden mittels Coenzym A Fettsäuren aufgebaut, und den gleichen Mechanismus setzt die Pflanze bei der Synthese der Carotine, der Phytosterine, des Kautschuks und vielleicht anderer sekundärer Pflanzenstoffe ein (s. die Beiträge von Stumpf, Band VII, und Goodwin, Heusner und Arreguin, Band X). Über die weite Verbreitung von Coenzym A in höheren Pflanzen hat Seifter (1954) berichtet.

Es ist wahrscheinlich, daß auch das einen reduzierten Thiophenring enthaltende Biotin die Funktion eines Coenzyms in einer Reihe von Enzymreaktionen erfüllt (Lichstein 1951).

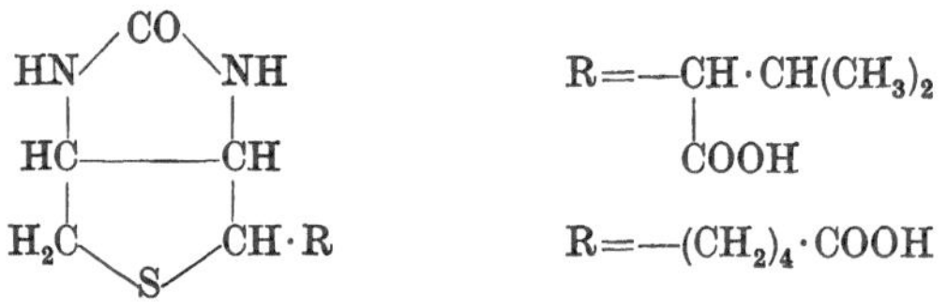

Seit langem kennt man das in der Thiol- und in der Disulfidform auftretende Glutathion, das als Peptid im vorhergehenden Band VIII behandelt wird. Es ist wie das Cystein/Cystin-System eine wesentliche Komponente biologischer Redoxmechanismen und übernimmt die Rolle eines Cosubstrates bei der Einwirkung der Ketonaldehydmutase auf Ketonaldehyd (Lohmann 1932).

Diese Beispiele zeigen, daß eine Reihe von S-Verbindungen Schlüsselstellungen im Stoffwechsel einnimmt und ihre spezifische physiologische Funktion häufig an den Schwefel im Molekül gebunden ist. Darin unterscheiden sie sich von den sekundären S-Verbindungen wie den Senfölen, die nur sporadisch vorkommen und denen eine katalytische oder andersartige Funktion im Stoffwechsel nicht zuerkannt werden kann, mindestens nicht den häufig anzutreffenden großen Mengen. Es ist damit zu rechnen, daß sich mit den zur Zeit bekannten S-haltigen

Verbindungen ihre Zahl nicht erschöpft. Nur jene hat man bisher aufgefunden, die relativ stabil sind oder leicht feststellbare Wirkungen entfalten, während empfindliche oder nur in Spuren vorkommende Vertreter, die aber trotzdem unentbehrlich sein können, sich dem Nachweis entzogen haben.

Die Assimilation des Schwefels (Abschnitt II C). Die höhere Pflanze nimmt den Schwefel als Sulfat auf und reduziert ihn zu den physiologisch wirksamen Thiolen. Daß diese Reduktion stattfinden muß, weiß man, seit die verschiedenen phytochemischen Formen des Schwefels bekannt sind, doch hat man bei der höheren Pflanze den Mechanismus der Reduktion bisher nicht aufklären können. Es handelt sich um einen endothermen Prozeß, der nicht photochemisch vor sich geht, sondern mit Hilfe von Energie, die dem Atmungsprozeß entnommen wird. In *Neurospora* scheint sich die Reduktion auf zweierlei Art zu vollziehen (s. Horowitz 1950): Sulfat wird zu Sulfit reduziert, dieses in Thiosulfat umgewandelt, das wiederum mit einer 3-C-Verbindung Cystein ergibt. Mit diesem Mechanismus soll ein zweiter im Gleichgewicht stehen, nach dem Sulfat und eine 3-C-Verbindung unter Bildung von Cysteinsäure reagieren, die über Cysteinsulfinsäure in Cystein übergeht. Untersuchungen an Mutanten von *Escherichia coli* (Lampen und Mitarbeiter 1947) sprechen dafür, daß Sulfit, Hyposulfit, Thiosulfat, Sulfid und Thioglykolat Glieder der Reaktionskette zwischen Sulfat und Cystein sind und Cystein über Cystathionin und Homocystein in Methionin umgewandelt wird. Sulfid wird dabei nicht frei, sondern im „statu nascendi" von einem 3-C-Körper übernommen. Ähnliche Ergebnisse, wenigstens im anorganischen Teil der Kette, wurden mit Mutanten von *Ophiostoma multiannulatum* (Fries 1945) erhalten, während Befunde an Mutanten von *Aspergillus nidulans* dagegen sprechen, daß Sulfid als Glied der Kette auftritt. Es scheint vielmehr, daß Thiosulfat-Serin die erste organische S-Verbindung ist (Hockenhull 1950). Nicht jede der Wachstum ermöglichenden S-Verbindungen muß jedoch ein Zwischenglied im normalen Gang der Sulfatreduktion sein; die Versuche geben aber einen Begriff von der Fähigkeit der betreffenden Organismen, S-Verbindungen umzusetzen.

Von den bei Mikroorganismen vermuteten Zwischenstufen ist bei den höheren Pflanzen mit Sicherheit nur das Sulfit nachgewiesen. Es tritt kurz nach der Überführung von S-Mangelpflanzen in ein S-haltiges Milieu auf. Die Klarlegung des Weges der Sulfatreduktion bei der höheren Pflanze ist eines der wesentlichen Probleme der Stoffwechselphysiologie.

Sulfat ist das normale Ausgangsmaterial für den S-Stoffwechsel der höheren Pflanze. Weniger wirksam, aber doch verwertbar ist SO_2, wenn es in sublethaler Konzentration in der Luft vorkommt (z. B. Thomas, Hendricks und Hill 1944). Selbst elementarer (radioaktiver) Schwefel, der auf Blätter und Früchte aufgestäubt wurde, ließ sich nach einiger Zeit im Eiweiß und anderen organischen Verbindungen nachweisen (Turrell und Chervenak 1949, Turrell 1950, Turrell und Weber 1955). Den höheren Pflanzen in Form von einfachen Sulfhydrylen zugeführter Schwefel wird wahrscheinlich vollständig zu Sulfat oxidiert, wenn nicht gerade Bedarf an diesen Stoffen besteht (Mothes 1939, Barrien und Wood 1939, Miller 1947).

Sulfat wird von den Pflanzen in der gleichen Weise wie andere Anionen absorbiert. Die S-Aufnahme und die S-Quellen des Bodens und der Gewässer werden in Band IV behandelt (Robertson: Aufnahme; Wiklander und Gessner: Mineralstoffquellen; Baumeister und Stiles: spezielle Nährstoffe).

Der Schwefel wandert in der Hauptsache als Sulfat (z. B. Mothes und Specht 1934), doch können offenbar auch die reduzierten organischen Formen

Methionin, Cystein, Cystin und Glutathion von älteren zu jüngeren Teilen transportiert werden, wenn Schwefelmangel in der Pflanze herrscht. Die Wanderung von der Wurzel nach den oberirdischen Organen läßt sich mit radioaktivem Schwefel gut verfolgen. Wanderungswege sind die Gefäße, von denen aus die Verteilung auf die Nachbargewebe stattfindet. Mit Hilfe von Radioautogrammen hat man zeigen können, daß z.B. in den Blättern die Aktivität zunächst auf die Gefäße beschränkt, später aber auch auf die Intercostalfelder diffus verteilt ist. Die Anwendung von radioaktivem Schwefel, der sich dank seiner günstigen Halbwertszeit und der noch relativ intensiven β-Strahlung für diesen Zweck gut eignet, hat neue Erkenntnisse über den Schwefelumsatz, die Verteilung und die Ablagerungsorte gebracht. Beim Weizen und der Tomate z.B. (HARRISON 1944, THOMAS 1948) hat man eine Konzentrierung des Schwefels im wachsenden Embryo feststellen und dessen Verteilung auf die einzelnen Gewebsschichten ermitteln können. Junge Gewebe sind reich an reduziertem Schwefel, Eiweißschwefel und einfachen Sulfhydrylkörpern (NIGHTINGALE und Mitarbeiter 1932, HEISERICH 1935, MOTHES 1939). Mit zunehmendem Alter steigt der Sulfatgehalt stark an, wohl weil die Sulfataufnahme zwangsläufig weitergeht, auch wenn kein Schwefel mehr reduziert wird und die Gewebe bereits mit S-Verbindungen „gesättigt" sind (MOTHES 1939). Die Bedeutung des Schwefels als Eiweißbestandteil kommt unter anderem in der großen Ähnlichkeit der Symptome des S- und N-Mangels zum Ausdruck, worauf z.B. von NIGHTINGALE und Mitarbeitern (1932) hingewiesen wurde.

Von großem physiologischem Interesse ist die Hypothese, daß S-Verbindungen an der Photosynthese beteiligt sind. Nach CALVIN und BARLTROP (1952) sollten die Spaltprodukte des Wassers von der Disulfidform der Liponsäure aufgenommen und weitergeleitet werden. Die Steigerung der Hill-Reaktion und das nach neueren Versuchen wahrscheinliche Vorkommen der Liponsäure in Chloroplasten (s. GRISEBACH 1956) stützen zwar diese Vorstellung, doch beweisen sie noch nicht, daß die Liponsäure der natürliche Überträger ist. CALVIN (1956) schränkt neuerdings seine Hypothese ein, indem er annimmt, daß die Liponsäure lediglich den Photowasserstoff überträgt. ARONOFF (1957) geht auf das Problem in einem kürzlich erschienenen Sammelbericht über die Photosynthese näher ein. In diesem Handbuch werden die Untersuchungen, die sich mit der Rolle der Liponsäure bei der Photosynthese befassen, in Band V, Abschnitt „Photosynthese der grünen Pflanze", ausführlich behandelt.

„Sekundäre" Schwefelverbindungen (Abschnitt II D). Viele der bisher behandelten S-Verbindungen sind unentbehrliche Bestandteile der Zelle, und ihre in den Grundzügen umrissenen Umsetzungen und Funktionen gehören dem elementaren Bestand des Zellstoffwechsels an. Die Differenzierung während der Ontogenese ist mit Verschiebungen im stofflichen Aufbau und im Ablauf der einzelnen Stoffwechselvorgänge verbunden, doch bleibt die elementare physiologische Ausrüstung im wesentlichen wohl dieselbe.

Neben den am Grundstoffwechsel beteiligten S-Verbindungen gibt es eine Reihe „sekundärer" S-haltiger Pflanzenstoffe. Die Bezeichnung „sekundär" bringt zum Ausdruck, daß diese Stoffe und Reaktionen, die zu ihrer Bildung und Speicherung führen, nicht jeder Pflanze eigen sind. Eine scharfe Abgrenzung zwischen diesen beiden Stoffgruppen ist allerdings nicht möglich, da es sich oft nicht entscheiden läßt, ob ein Stoff, der in großer Menge ohne Bedeutung für die Pflanze ist, in Spuren nicht doch eine wichtige Rolle spielt.

Sekundäre S-Verbindungen enthalten die Vertreter der nahe verwandten Familien der Cruciferen, Resedaceen und Capparidaceen sowie der weiter entfernt stehenden Familien der Tropaeolaceen und Limnanthaceen. Sporadisches

Vorkommen wird für die Familien der Salvadoraceen, der Caricaceen und Euphorbiaceen angegeben. Seit langem bekannt sind die durch einen charakteristischen Geruch gekennzeichneten Senföle. Senföle sind die Ester der Isothiocyansäure (RNCS), Kohlenwasserstoffe, die außer dem Schwefel auch noch Stickstoff im Molekül enthalten. Mit ihnen vergesellschaftet kommen häufig N-freie sekundäre S-Verbindungen vor, die einfachen Thiole oder Mercaptane (RSH), Sulfide oder Thioäther (RSR$_1$), Disulfide (RS—SR$_1$) und Polysulfide sowie die Oxydationsprodukte von Sulfiden, die Sulfoxyde (R—SO—R$_1$) und Sulfone (R—SO$_2$—R$_1$). Da sich Senföle und Lauchöle an ihrem typischen Geruch leicht erkennen lassen, ist nicht damit zu rechnen, daß noch wesentliche neue Vorkommen entdeckt werden, doch sind die in den bekannten Arten enthaltenen sekundären S-Verbindungen erst unvollkommen erforscht.

Das Thiol des Methylalkohols, das einfachste Thiol, hat man z. B. in den Wurzeln von *Raphanus sativus* L. (NAKAMURA 1925) und in den Blättern von *Lasianthus* (KOOLHAAS 1931) nachgewiesen. Der holzzerstörende Pilz *Scopulariopsis brevicaulis* bildet diese Verbindung aus Methionin und δ-Methylcystein (CHALLENGER 1951). Auf ähnliche Weise entsteht dieses Thiol wahrscheinlich auch in der höheren Pflanze. Zu einer Anhäufung von Alkylthiolen kann es deshalb nicht kommen, weil diese Stoffe sehr reaktionsfähig sind und leicht in Disulfide und ähnliche Verbindungen übergehen.

Disulfide und Polysulfide bilden neben einfachen Sulfiden die Komponenten des bei verschiedenen Arten der Gattung *Allium* vorkommenden Lauchöles. Divinylsulfid ist die Hauptkomponente des Bärlauchöles *(Allium ursinum)*, Allylsulfid der wesentliche Bestandteil des Öles aus *Allium cepa*. Bei der fraktionierten Destillation des Knoblauchöles wurden verschiedene Sulfide, Allylpropyldisulfid, Diallyldisulfid und Diallyltrisulfid festgestellt. Diese Sulfide kommen wahrscheinlich nicht präformiert in der Pflanze vor, sondern entstehen erst bei der Destillation aus einem noch unbekannten Sulfid. Dimethylsulfid hat man als Bestandteil verschiedener ätherischer Öle nachweisen können. Marine Algen bilden es, wenn sie der Luft ausgesetzt oder mit Alkali behandelt werden. Letzteres trifft auch für Schachtelhalme, einige Farne und den Spargel zu. In der Alge *Polysiphonia fastigiata* bildet sich Dimethylsulfid wahrscheinlich aus Dimethyl-β-propiothetin, einer Sulfoniumverbindung (CHALLENGER und SIMPSON 1948). Ein Methylsulfoniumderivat des Methionins hat man im Kopfkohl aufgefunden. Im ätherischen Öl der Ananas ist der Methioninabkömmling Methyl-3-methylthiopropionat enthalten (HAAGEN-SMIT und Mitarbeiter 1945). Ein Sulfonium von großer physiologischer Bedeutung scheint das Adenosylmethionin zu sein, das als aktives Methionin die Methylgruppenübertragung vermittelt (CANTONI 1953). Als Produkt der Hydrolyse dieser labilen Verbindung wird der schon vor längerer Zeit in *Saccharomyces* aufgefundene Thiozucker gedeutet. Die nähere Untersuchung ergab, daß es sich um eine Adenyldesoxythiomethylpentose handelt (WEYGAND und Mitarbeiter 1950).

Aus *Allium sativum* und *A. ursinum* wurde das erste natürlich vorkommende, Alliin genannte Sulfoxyd dargestellt und als Allylcysteinsulfoxyd identifiziert (STOLL und SEEBECK 1948). Es könnte bei der enzymatischen Oxydation von S-Allyl-L-Cystein entstehen, das vielleicht aus Methionin gebildet wird. Aus Alliin geht durch enzymatische Spaltung das zuerst von CAVALLITO und Mitarbeiter (1944) isolierte antibiotisch wirkende Allicin (STOLL und SEEBECK 1949), das als Monosulfoxyd des Diallylsulfids erkannt wurde, hervor. Bei der Destillation zersetzt es sich unter Bildung von Diallylsulfid und den oben genannten höheren Sulfiden. Noch näherer Untersuchung bedarf unter anderem das in Asa foetida enthaltene Gemisch von ungesättigten, zum Teil optisch aktiven Disulfiden (Semmler 1891). Als Sulfoxyd erwies sich das in den Samen von *Raphanus sativus* vorkommende, nicht flüchtige Senföl Sulphoraphen (Schmid und Karrer 1948), dessen durch das asymmetrische S-Atom bedingte optische Aktivität bemerkenswert ist. Das Senföl des Glykosids Glucoiberin aus *Iberis amara* enthält ebenfalls eine Sulfoxydgruppe (Schultz und Gmelin 1954). Das Sulfoxyd des S-Methyl-L-Cysteins wurde unlängst in den Rüben von *Brassica rapa* und im Kopfkohl festgestellt (Synge und Wood 1955, Morris und Thompson 1955). Es handelt sich also um eine ziemlich verbreitete Gruppe von Stoffen, deren Nachweis und Strukturaufklärung durch eine geringe Stabilität erschwert sind. Wahrscheinlich bilden sie sich durch enzymatische Oxydation aus den entsprechenden Sulfiden. Experimentelle Unterlagen dafür fehlen noch, auch ist nichts über ihre Rolle im Stoffwechsel bekannt.

Oxydationsprodukte von Sulfiden und Sulfoxyden sind wahrscheinlich auch die beiden schon vor längerer Zeit in Pflanzen aufgefundenen Sulfone, die Senföle Cheirolin aus dem Samen von *Cheiranthus cheiri* L. (Schneider 1910) und Erysolin, das in den Samen von *Erysimum perofskianum* vorkommt (Schneider und Kaufmann 1912).

Von der auffälligsten Gruppe der sekundären S-Verbindungen, den Senfölen, sind zur Zeit etwa 20 verschiedene Vertreter bekannt. Mindestens in der überwiegenden Menge liegen sie in der Pflanze als Glykoside vor, und diese sind von der Myrosinase begleitet, einem aus Thioglucosidase und Sulfatase zusammengesetzten Enzymkomplex (Neuberg und Wagner 1926), der alle natürlichen Senfölglykoside in Senföl, Zucker und Schwefelsäure zerlegt. Die Untersuchungen über die Senfölglykoside wurden vor mehr als 100 Jahren mit der Isolierung des Sinalbins aus dem weißen und des Sinigrins aus dem schwarzen Senf eingeleitet (Bourton und Robiquet 1831 bzw. Bussy 1840) und um die Jahrhundertwende und ein Jahrzehnt später mit Arbeiten über das Glucotropäolin, Gluconasturtin (Gadamer 1899), Glucocheirolin und Glucoerysolin (Schneider 1910) weitergeführt. Bemerkenswerte Fortschritte wurden in den letzten Jahren mit Hilfe der Papierchromatographie erzielt, besonders von Kjaer und Mitarbeitern, die unter anderem die bisher unbekannten Methyl-, Äthyl-, Methylthiopropyl-, Methylthiobutyl- und Methylthiopentyl-Isothiocyanate gewinnen und charakterisieren konnten (Kjaer, Gmelin und Larsen 1955a; Kjaer und Larsen 1954; Kjaer, Gmelin und Larsen 1955b; Kjaer und Gmelin 1955; Kjaer, Larsen und Gmelin 1955). Die Senföle lassen sich als Ester der hypothetischen Isorhodanwasserstoffsäure auffassen. Die Alkyle können also einfache Kohlenwasserstoffe mit 1—6 C-Atomen sein, es kommen gesättigte und ungesättigte, geradlinige und verzweigte Ketten vor, deren Wasserstoffatome zum Teil durch Methyl ersetzt sind; in einem Fall ist der Substituent eine OH-Gruppe. Andere sind durch endständige Methylthiogruppen gekennzeichnet, und schließlich kann das Alkyl auch eine aromatische Verbindung sein. Es ist nicht bekannt, wann und wie die Pflanze den Schwefel einbaut, die Isothiocyanatgruppe an die Alkyle anheftet, feststeht nur, daß im ganzen gesehen die Pflanze nicht wählerisch hinsichtlich der Beschaffenheit der Kohlenwasserstoffe ist, an die die Anlagerung erfolgt, wenn auch von den einzelnen Arten bestimmte Kohlenwasserstoffe ausgewählt oder bevorzugt werden. Ein Überblick über die verschiedenen Senfölglykoside mit Angaben über ihr Vorkommen findet sich im Beitrag von Stoll und Jucker, Band VI.

Zu den sekundären S-Verbindungen muß auch eine Reihe von Schwefelsäureestern gerechnet werden. So enthalten Senfölglykoside neben Senföl auch Schwefelsäure im Molekül. Bei verschiedenen marinen Algen kommen Polysaccharidschwefelsäureester vor (vgl. den Beitrag Stacey und Foster, Band VI).

Der bekannteste Vertreter ist der Agar, ein Zellwandbestandteil von Arten der Gattung *Gelidium*. Die Ketteneinheit dieses Polysaccharids besteht wahrscheinlich aus 9 D-Galaktopyranoseeinheiten, in denen die primären Alkoholgruppen eines jeden Galaktoserestes mit Schwefelsäure verestert sind (JONES und PEAT 1942). Schwefelsäureester finden sich ferner im Schleim bestimmter Rotalgen, z. B. im Carrageenschleim. Es wird angenommen, daß es sich um einen Polygalaktosido-Schwefelsäureester handelt (DEWAR und PERCIVAL 1947). Im Mycel von *Aspergillus sydowi* ist Cholinschwefelsäureester aufgefunden worden (WOOLLEY und PETERSON 1937).

Neben den cyclischen Sulfiden Biotin und Aneurin, deren Konstitution und Funktion im Stoffwechsel der Pflanze bereits angedeutet wurde, gibt es weitere, denen mehr der Charakter sekundärer Pflanzenstoffe zukommt. Große Bedeutung haben in neuerer Zeit die von FLEMING bereits 1928 als antibiotisch wirksam erkannten Penicilline erlangt, in deren Grundgerüst, dem Penin, eine S-haltige Aminosäure (β,β'-Dimethylcystein) vorliegt. An die primären Aminogruppen des Penins sind je nach Penicillintyp verschiedene Säuren — Capronsäure, Phenylessigsäure, p-Oxyphenylessigsäure oder n-Caprylsäure — amidartig angeheftet (Näheres s. Band VIII). *Aspergillus fumigatus* und einige andere Pilze scheiden in das Kulturmedium das Antibioticum Gliotoxin, ein Pyrazinoindol mit Disulfidgruppe, aus (JOHNSON, BRUCE und DUTCHER 1943). Als zusammen-

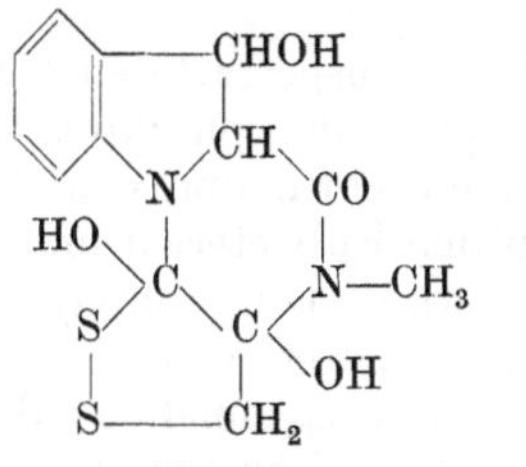

Gliotoxin 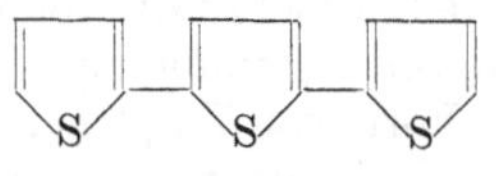   Terthienyl

gesetzter cyclischer Thioäther erwies sich das Terthienyl aus den Blütenblättern von *Tagetes erecta* L. (ZECHMEISTER und SEASE 1947). Einen Thioäther stellt vermutlich auch das im Milchsaft einiger Asclepiadaceen vorkommende Uscharin dar (HESSE, REICHENEDER und EYSENBACH 1938).

Die Schwefelspezialisten unter den Mikroorganismen (Abschnitt II E). Der S-Stoffwechsel der meisten Mikroorganismen ähnelt im Prinzip dem der höheren Pflanze. Wie diese nehmen sie das Element in Form von Sulfat auf und führen es auf wahrscheinlich ähnlichen Wegen in die physiologisch wirksamen Verbindungen über. Unter den Mikroorganismen gibt es aber auch eine Reihe von Vertretern mit Besonderheiten des S-Stoffwechsels, die sich vor allem im anorganischen Teil des Kreislaufes zeigen und Oxydationen und Reduktionen zwischen den Sauerstoffverbindungen des Schwefels, dem elementaren Schwefel und Schwefelwasserstoff betreffen. Da wo in der Natur der für die meisten Organismen giftige Schwefelwasserstoff vorkommt, entwickeln sich die Lebensgemeinschaften des Sulphuretums, dessen charakteristische Glieder die Schwefelmikroben sind. Süß- oder Meerwasser, Gewässer mit höherer Salzkonzentration und deren Ablagerungen können diese Stellen sein. Meist ist Faulschlamm die H_2S-Quelle, doch kann dieser auch vulkanischen Ursprungs sein oder durch desulfurizierende Bakterien aus Sulfaten gebildet werden.

Zu den Schwefelmikroben gehören jene Mikroorganismen, die sich Umsetzungen von anorganischem Schwefel bei der Photo- und Chemosynthese zunutze machen; der Schwefel wird dabei selbst zur elementaren Form oder zur Schwefelsäure oxydiert. Im ersteren Fall, der bei den Chlorobacteriaceen und den Thiorhodaceen (vgl. Band V, Abschnitt „Die Photosynthese der Photobakterien") gegeben ist, dient H_2S oder eine andere oxydierbare Schwefelverbindung als H-Donator für die Kohlensäurereduktion. Die Thiorhodaceen und Chlorobacteriaceen enthalten Bakteriochlorophyll, erstere außerdem Carotinoide. Beide Gruppen sind vielgestaltig und nur wenige Formen näher untersucht. Die Thiorhodaceen gedeihen besser bei niedrigem H_2S-Gehalt, während die Chlorobacteriaceen höhere Konzentrationen vorziehen. Von letzteren sind zwei auf bestimmte S-Verbindungen spezialisierte Arten bekannt, *Chlorobium limicola* (van Niel 1931), das bevorzugt H_2S verwertet, jedoch auch S und H_2 als reduzierende Agenzien benutzen kann, während die Art *Chlorobium sulfatophilum* (Larsen 1952, 1953) Thiosulfat oder Tetrathionat als S-Quelle braucht und diese immer in Sulfat überführt. Bei *Chl. limicola* entsteht neben Sulfat auch elementarer Schwefel, wenn die reduzierte Form reichlich zur Verfügung steht.

Winogradsky (1888) berichtet wohl als erster über diese pigmentierten S-Spezialisten, ohne zunächst Näheres über die Rolle des Schwefels aussagen zu können. Wesentliche Erkenntnisse über seine Funktion, die Verknüpfung der Sulfatreduktion mit der Photosynthese und die thermodynamischen Verhältnisse haben vor allem die Untersuchungen van Niels (1931, s. auch Bavendamm 1936) gebracht, nachdem es gelungen war, Reinkulturen dieser Organismen herzustellen.

Bei den mit der Fähigkeit zur Chemosynthese ausgestatteten S-Spezialisten handelt es sich um fakultativ bis streng autotrophe chlorophyllfreie Organismen, die in einer großen Mannigfaltigkeit der äußeren Form und des Stoffwechsels anzutreffen sind. Sie gehören hauptsächlich zu den Eubacterien und Beggiatoen. Das klassische Beispiel ist die von Winogradsky studierte Cyanophycee *Beggiatoa*, die unter aeroben Bedingungen aus Schwefelwasserstoff elementaren Schwefel bildet und in Form von Tröpfchen innerhalb der Zelle ansammelt. Ist der Schwefelwasserstoff aufgebraucht, wird der elementare Schwefel zu Sulfat weiteroxydiert und als solches ausgeschieden. Entsprechend verhalten sich *Thiotrix*, *Thiospirillopsis* und *Thioploca*, die ebenfalls zu den Beggiatoen gehören.

S-Spezialisten aus der Gruppe der Eubacterien sind die *Thiobacillus*-Arten. Ihr markantester Vertreter, *Thiobacillus thiooxydans*, wurde 1922 von Waksman und Joffe entdeckt. Er ist an höhere Säurekonzentrationen angepaßt und gedeiht noch in 0,5 m H_2SO_4. *Th. thiooxydans* oxydiert bevorzugt elementaren Schwefel, ist aber in der Lage auch Thiosulfat umzusetzen. Dagegen gedeiht die Art *Th. thioparus* auf elementarem Schwefel nur mangelhaft, entwickelt sich aber wie Beggiatoa gut, wenn Schwefelwasserstoff zugegen ist. Der dabei entstehende elementare Schwefel wird nicht in der Zelle abgelagert, sondern aus dieser ausgeschieden. *Thiobacillus thioparus* ist auch in der Lage, Thiosulfat zu oxydieren, desgleichen die Art *Th. novellus* (Starkey 1935). Zur Oxydation von elementarem Schwefel (Beijerinck 1904) und Thiosulfat (Lieske 1912) unter anaeroben Bedingungen ist *Thiobacillus denitrificans* befähigt. Der Sauerstoff für die Oxydation wird in diesem Fall Nitraten entnommen. Eine besondere Gruppe von S-Spezialisten sind schließlich die Mikroorganismen, die den Schwefel der in Abwässern von Gasanstalten enthaltenen Thiocyanate mit Hilfe des Luftsauerstoffes in Sulfat überführen, aber auch Sulfid und Thiosulfat oxydieren können (Happold und Key 1937; Youatt 1954). — Die bei der Schwefeloxydation frei werdende Energie wird, wie zuerst bei *Thiobacillus thiooxydans* gezeigt wurde, zur Bildung energiereicher Phosphate verwendet. Erst diese phosphatgebundene Energie benutzt die Zelle zur CO_2-Assimilation und zur Synthese der verschiedenen Zellbausteine (vgl. den Beitrag Umbreit im Abschnitt Chemosynthese von Band V). Es liegen also grundsätzlich dieselben Verhältnisse wie bei der Photosynthese vor, wo das Licht durch Vermittlung des Chlorophylls die Bildung energiereicher Phosphate bewirkt (ARONOFF 1957).

Die Hauptmenge des freien Schwefelwasserstoffs entsteht in der Natur durch die Tätigkeit desulfurizierender Mikroorganismen, deren wichtigste Vertreter zur Gattung *Desulfovibrio* zusammengefaßt sind. Die Desulfurizierer, bei denen diese Stoffwechselform obligat geworden ist, oxydieren mit Hilfe des Sauerstoffes von S-Verbindungen (Sulfat, Sulfit, Thiosulfat) die verschiedensten organischen Stoffe (s. z.B. BUTLIN, ADAMS und THOMAS 1949). Nach der Art dieser Stoffe, die Aminosäuren, Alkohole, Zucker und Säuren z. B. sein können, hat man unter Heranziehung anderer Merkmale die Gruppe zu klassifizieren versucht. Auch Schwefel selbst kann zur Oxydation (Dehydrierung) organischer Verbindungen benutzt werden. Von besonderem Interesse ist, daß manche Desulfurizierer elementaren Wasserstoff zur Reduktion oxydierten Schwefels heranziehen. Diese Reaktion befähigt sie zur Chemosynthese und damit zur Autotrophie (BUTLIN und ADAMS 1947). Sulfatreduzierende Organismen wurden schon von BEIJE-RINCK (1895) erkannt und beschrieben. Die Desulfurizierer zeichnen sich in der Regel durch eine hohe H_2S-Toleranz aus, deren Grenze bei 2,5 g H_2S je Liter noch nicht erreicht ist (MILLER 1949, 1950). Sie sind sehr verbreitet und an die verschiedensten Lebensbedingungen angepaßt. Bemerkenswert ist die Thermo-philie mancher Formen (STARKEY 1938).

Der Schwefelkreislauf (Abschnitt II F; vgl. auch den Beitrag „Bedeutung der Mikroorganismen für den Kreislauf der Stoffe" von JENSEN in Band IV, Abschnitt X). Bestimmte organische S-Verbindungen, die den Schwefel in redu-zierter Form enthalten, sind also Zellbestandteile von hoher physiologischer Aktivität. Bei der Entwicklung der Organismen werden diese Stoffe laufend benötigt, um die Stoffwechselsysteme neuer Zellen aufzubauen. Jeder Orga-nismus muß, solange er sich in der Entwicklung befindet, diese Stoffe beschaffen. Die höhere Pflanze nimmt Sulfat auf und reduziert es unter Einbau in organische Verbindungen. Ebenso verfahren die meisten Mikroorganismen; andere Vertreter können weitere Oxydationsstufen, Sulfit, Thiosulfat, elementaren Schwefel und Schwefelwasserstoff, absorbieren und diese zur Herstellung der physiologisch wichtigen S-Verbindungen benutzen. Auch die höhere Pflanze ist dazu imstande, vorausgesetzt, daß diese Stoffe in niedriger, nicht toxischer Konzentration geboten werden. Das normale Ausgangsmaterial ist für sie jedoch das Sulfat.

Nicht verwertbar ist indessen der Schwefel in anorganischer Form für die tierischen Organismen, da ihnen die Fähigkeit zur Reduktion und Überführung in die organische Form fehlt. Dasselbe ist der Fall bei vielen an heterotrophe Lebensweise angepaßten Mikroorganismen. Sie übernehmen fertige organische S-Verbindungen (Aminosäuren, S-haltige Vitamine usw.). Die Bereitstellung dieser Substanzen ist eine Leistung der höheren und niederen autotrophen Pflanzen. Letztere bilden in Gestalt der Algen die Ernährungsgrundlage für die Tierwelt der Gewässer. Eine Deponierung der physiologisch aktiven Formen des Schwefels findet in nennenswertem Ausmaß nicht statt; Ablagerungen orga-nischer S-Verbindungen, die den Schwefel in reduzierter, zuweilen auch in oxy-dierter Form enthalten, stellen die auf wenige Familien der höheren Pflanzen beschränkten Lauch- und Senföle dar.

Dem Reduktionsprozeß stehen die ebenfalls von den Organismen durch-geführten Schwefeloxydationen gegenüber. Das Tier oxydiert z. B. im Überschuß aufgenommene S-haltige Aminosäuren und scheidet Schwefel als Sulfat, frei oder mit Phenol verestert, oder als Taurocholsäure wieder aus. Bei der höheren Pflanze scheint die Rückoxydation von untergeordneter Bedeutung zu sein und tritt wahrscheinlich nur unter abnormen Bedingungen ein, z. B. wenn die Pflanze zur Eiweißhydrolyse und Oxydation der dabei anfallenden Aminosäuren gezwungen wird. Immerhin schließen sich damit die S-Umsetzungen, zu denen die höhere Pflanze befähigt ist, zu einem Kreis. Die Reaktionsfolge bei der Reoxydation ist noch ungeklärt. Wir wissen nicht, ob sie ganz oder teilweise schon vor oder erst nach Lösung des Schwefels aus der organischen Bindung vor sich geht. Es ist aber wahrscheinlich, daß dabei dieselben Zwischenstufen durchlaufen werden wie bei der Reduktion des Schwefels. Der weitaus größere Teil des im Laufe ihres Lebens von den Organismen assimilierten Schwefels bleibt jedoch von dieser Rückoxydation durch diese selbst ausgeschlossen; denn bis zum Augenblick des Absterbens muß der Stoffwechselapparat funktionsfähig sein, was nur der Fall ist, wenn auch die S-haltigen Komponenten des Mechanismus bis zu diesem Zeitpunkt vorhanden sind.

Die laufend anfallenden Pflanzen- und Tierkadaver stellen Anhäufungen von reduziertem Schwefel in organischer Bindung dar. Durch autolytische Prozesse und die Tätigkeit von Mikroorganismen werden sie zersetzt und schließlich vollständig mineralisiert. Soweit der Schwefel dabei in Sulfat übergeht, wird die höchste Oxydationsstufe, die zugleich das Ausgangsmaterial des S-Stoffwechsels für die Mehrzahl der pflanzlichen Organismen ist, wieder erreicht und damit durch das Handinhandarbeiten der großen Organismengruppen unserer Erde der Kreislauf des Schwefels geschlossen.

Schon als Winogradsky die Beggiatoen als S-Spezialisten erkannte, deren Energiestoffwechsel auf der Oxydation von reduziertem Schwefel beruht, wurde ihm klar, daß der für die Oxydation bevorzugte Schwefelwasserstoff nicht von diesen selbst, sondern von anderen mit ihnen vergesellschafteten Organismen produziert wird. Liefern die Desulfurizierer den Schwefelwasserstoff, so ergeben bereits die von beiden S-Spezialisten durchgeführten S-Umsetzungen einen Kreislauf der verschiedenen Oxydationsstufen des Schwefels. Beide Organismen finden sich im Sulphuretum neben Fäulniserregern, die H_2S aus organischen Verbindungen in Freiheit setzen, und neben anderen S-oxydierenden pigmentierten und nichtpigmentierten S-Spezialisten, die durch S-Oxydationen die Photo- bzw. Chemosynthese in Gang bringen.

Durch die Stoffwechseltätigkeit der Organismen ist also ein Teil des S-Vorrates der Erde dauernden Umsetzungen unterworfen, die insgesamt einen Kreislauf ergeben. Dieser kann in kleinerem Maßstab auch zwischen einzelnen miteinander vergesellschafteten Organismengruppen und sogar innerhalb eines Individuums ablaufen. Dabei werden die verschiedenen Oxydationsstufen zwischen Sulfat und Schwefelwasserstoff durchschritten und im Zuge dieser Reaktionen der Schwefel in organische Verbindungen eingebaut und wieder aus diesen herausgelöst. Bis auf erste Versuche, die einzelnen Glieder und Reaktionsschritte zu analysieren, sind uns nur die großen Linien dieser Umsetzungen bekannt. Für die Reduktion des Schwefels und ebenso für seine Reoxydation ist jeweils eine Reihe von Reaktionsschritten erforderlich. Wir wissen noch nicht, ob sich Reduktion und Reoxydation bei den einzelnen Organismengruppen in grundsätzlich verschiedener Weise vollziehen oder die Varianten des S-Stoffwechsels nur dadurch zustande kommen, daß entsprechend der erblichen Konstitution die Einzelreaktionen verschieden stark ausgeprägt sind oder der eine oder andere Reaktionsschritt oder auch Teile der Kette ganz fehlen können, sei es, daß sie nicht ausgebildet oder durch Mutation wieder verlorengegangen sind. Der Schwefel tritt in mehreren Wertigkeits- und Oxydationsstufen auf, und die Zahl der Wasserstoff- und Sauerstoffverbindungen ist im Vergleich zu anderen Elementen relativ groß, für die Organismen ist von den möglichen chemischen

Umsetzungen aber nur eine begrenzte Zahl durchführbar. Zu vielen dieser Reaktionen ist die höhere Pflanze mit ihrem hochentwickelten S-Stoffwechsel befähigt, der nicht nur von der verbreitetsten anorganischen Schwefelverbindung ausgeht, sondern die zahlreichen Produkte der Schwefelassimilation auch wieder in diese zurückführen kann. Weniger vollkommene Formen des S-Stoffwechsels und künstlich hergestellte Varianten (Defektmutanten) unter den Mikroorganismen, welche einzelne Reaktionsschritte häufig deutlicher erkennen lassen, wird man auch zur Aufklärung der Reaktionsfolge des Schwefelkreislaufes bei der experimentell schwerer zu handhabenden höheren Pflanze heranziehen können.

Literatur.

ARONOFF, S.: Photosynthesis. Bot. Rev. **23**, 65—107 (1957).

BARRIEN, B. S., and J. G. WOOD: Studies on sulfur metabolism of plants. II. New Phytologist **38**, 257—264 (1939). — BARRON, E. S. G.: Thiol groups of biological importance. Adv. Enzymol. **11**, 201—266 (1955). — BAVENDAMM, W.: Die Physiologie der schwefelspeichernden und schwefelfreien Purpurbakterien. In Ergebnisse der Biologie, Bd. XIII, S. 1—53. Berlin: Springer 1936. — BEIJERINCK, N. M.: Über *Spirillum desulfuricans* als Ursache von Sulfatreduktion. Zbl. Bakter. II **1**, 1, 49, 104 (1895). — Über die Bakterien, welche sich im Dunkeln mit Kohlensäure als Kohlenstoffquelle ernähren können. Zbl. Bakter. II **2**, 539 (1904). — BERSIN, TH.: Amidasen und Proteasen. In NORD-WEIDENHAGEN, Handbuch der Enzymologie, S. 574—632. Leipzig: Akademische Verlagsgesellschaft 1940. — BOUTRON et ROBIQUET: Sur la semence de moutarde. J. Pharmacie (II) **17**, 279—308 (1831). Zit. nach GILDEMEISTER-HOFFMANN, Die ätherischen Öle. II. Leipzig: Schimmel 1929. — BUSSY, A.: Untersuchungen über die Bildung des ätherischen Senföles. Liebigs Ann. **34**, 223—230 (1840). — BUTLIN, K. R., and M. E. ADAMS: Autotrophic growth of sulphate reducing bacteria. Nature (Lond.) **160**, 154 (1947). — BUTLIN, K. R., M. E. ADAMS and M. THOMAS: The isolation and cultivation of sulphate reducing bacteria. J. Gen. Microbiol. **3**, 46 (1949).

CALVIN, M.: Chemical and photochemical reactions of thioctic acid and related disulfides. I. Federat. Proc. **13**, 697—711 (1954). — Der Photosynthese-Cyclus. Angew. Chem. **68**, 253—264 (1956). — CALVIN, M., and J. A. BARLTROP: A possible primary quantum conversion act of photosynthesis. J. Amer. Chem. Soc. **74**, 6153 (1952). — CANTONI, G. L.: S-Adenosyl-methionine, a new intermediate formed enzymatically from L-methionine and adenosintriphosphate. J. of Biol. Chem. **204**, 403—416 (1953). — CAVALLITO, C. J., J. S. BUCK and C. M. SUTER: Allicin, the antibacterial principle of *Allium sativum*. II. Determination of the chemical structure. J. Amer. Chem. Soc. **66**, 1952—1954 (1944). — CHALLENGER, F.: Biological methylation. Adv. Enzymol. **12**, 429—491 (1951). — CHALLENGER, F., and M. J. SIMPSON: Studies on biological methylation. Part XII. A precursor of the dimethyl sulfide evolved by *Polysiphonia fastigiata*. Dimethyl-2-carboxyethylsulphonium hydroxide and its salts. J. Chem. Soc. Lond. **1948**, 1591—1597. — CZAPEK, FR.: Biochemie der Pflanzen, Bd. III. Jena: Gustav Fischer 1925.

DEWAR, E. T., and E. G. V. PERCIVAL: Polysaccharides of carragean. II. J. Chem. Soc. Lond. **1947**, 1622.

FRIES, N.: X-ray induced mutations in the physiology of *Ophiostoma*. Nature (Lond.) **155**, 757 (1945).

GADAMER, J.: Das ätherische Öl von *Tropaeolum majus*. Arch. Pharmaz. **237**, 111—120 (1899). — Über ätherische Kressenöle und die ihnen zu Grunde liegenden Glukoside. Arch. Pharmaz. **237**, 507—521 (1899). — GREGORY, J. D., G. D. NOVELLI and F. LIPMANN: The composition of coenzyme A. J. Amer. Chem. Soc. **74**, 854 (1952). — GRISEBACH, H.: Chemie und Biochemie der α-Liponsäure. Angew. Chem. **68**, 554—559 (1956). — GUNSALUS, J. C.: The mechanism of enzyme action. Bull. Johns Hopkins Press **1954**. Zit. nach GRISEBACH.

HAAGEN-SMIT, A. J., J. G. KIRCHNER, C. L. DEASY and A. N. PRATER: Chemical studies of pineapple (*Ananas sativus* L.). II. Isolation and identification of a sulfur-containing ester in pineapple. J. Amer. Chem. Soc. **67**, 1651—1652 (1945). — HAPPOLD, F. C., and A. KEY: The bacterial purification of gas-works liquors. Biochemic. J. **31**, 1323 (1937). — HARRISON, B. F., M. D. THOMAS and G. R. HILL: Radioautographs showing the distribution of radiosulfur in wheat. Plant Physiol. **19**, 245—257 (1944). — HEISERICH, E.: Schwefelstoffwechsel in Mais und Tabak. Z. Pflanzenernährg **37**, 55—72 (1935). — HESSE, G., F. REICHENEDER u. H. EYSENBACH: Die Herzgifte im *Calotropis*-Milchsaft. 2. Mitt. Liebigs Ann. **537**, 67—68 (1938). — HOCKENHULL, D. J. D.: Studies in the metabolism of mould fungi. I. A preliminary study of the metabolism of carbon, nitrogen, and sulphur by *Aspergillus nudilans*. J. of

Exper. Bot. 1, 194—200 (1950). — Horecker, B. L., and P. Z. Smyrniotis: The coenzyme function of thiamine pyrophosphate in pentose phosphate metabolism. J. Amer. Chem. Soc. 75, 1009 (1953). — Horecker, B. L., P. Z. Smyrniotis and H. Klenow: The formation of sedoheptulose from pentose phosphate. J. of Biol. Chem. 205, 661—682 (1953). — Horowitz, N. H.: Biochemical genetics of Neurospora. Adv. Genet. 3, 33—71 (1950).

Johnson, J. R., W. F. Bruce and J. D. Dutcher: J. Amer. Chem. Soc. 65, 2005 (1943). Zit. nach A. Stoll und E. Jucker in Paech-Tracey, Moderne Methoden der Pflanzenanalyse, Bd. IV. Berlin-Göttingen-Heidelberg: Springer 1955. — Jones, W. G. M., and S. Peat: Constitution of agar. J. Chem. Soc. Lond. 1942, 225.

Kjaer, A., and R. Gmelin: iso-Thiocyanates. XI. 4-Methylthiobutyl isothiocyanate, a new naturally occurring mustard oil. Acta chem. scand. (Copenh.) 9, 542—544 (1955). — Kjaer, A., R. Gmelin and I. Larsen: (a) iso-Thiocyanates. XIII. Methyl isothiocyanate, a new naturally occurring mustard oil, present as glucoside (glucocapparin) in Capparidaceae. Acta chem. scand. (Copenh.) 9, 857—858 (1955). — (b) iso-Thiocyanates. XII. 3-Methylthiopropyl isothiocyanate (Ibervirin), a new naturally occurring mustard oil. Acta chem. scand. (Copenh.) 9, 1143—1147 (1955). — Kjaer, A., and I. Larsen: iso-Thiocyanates. IX. The occurrence of ethyl isothiocyanate in nature. Acta chem. scand. (Copenh.) 8, 699—701 (1954). — Kjaer, A., I. Larsen and R. Gmelin: iso-Thiocaynates. XIV. 5-Methylthiopentyl isothiocyanate, a new mustard oil present in nature as a glucoside (glucoberteroin). Acta chem. scand. (Copenh.) 9, 1311—1316 (1955). — Koolhaas, D. R.: Das Vorkommen von Methylmercaptan in den Blättern der Lasianthus-Arten. Biochem. Z. 230, 446—450 (1931). — Kostytschew, S.: Lehrbuch der Pflanzenphysiologie. In Chemische Physiologie, Bd. I. Berlin: Springer 1926.

Lampen, I. O., R. R. Roepke and M. J. Jones: Studies on the sulfur metabolism of Escherichia coli. III. Mutant strains of Escherichia coli unable to utilize sulfate for their complete sulfur requirements. Arch. of Biochem. 13, 55—66 (1947). — Larsen, H.: On the culture and general physiology of the green sulfur bacteria. J. Bacter. 64, 187 (1952). — On the microbiology and biochemistry of the photosynthetic green sulfur bacteria. Kgl. norske Vidensk. Selsk., Skr. 1953, Nr. 1. — Lichstein, H. S.: Functions of biotin in enzyme systems. Vitamins a. Hormones 9, 27—74 (1951). — Lieske, R.: Untersuchungen über die Physiologie denitrifizierender Schwefelbacterien. Ber. dtsch. bot. Ges. 36, 12 (1912). — Lipmann, F.: Acetylation of sulfamide by liver homogenates and extracts. J. of Biol. Chem. 160, 173—190 (1945). — Lohmann, K.: Beitrag zur enzymatischen Umwandlung von synthetischem Methylglyoxal in Milchsäure. Biochem. Z. 254, 332—341 (1932).

Miller, L. P.: DL-Methionine as a source of sulfur by growing plants. Contrib. Boyce Thompson Inst. 14, 443—456 (1947). — Rapid formation of high concentrations of hydrogen sulfide by sulfate-reducing bacteria. Contrib. Boyce-Thompson Inst. 15, 437—465 (1949). — Tolerance of sulfate-reducing bacteria to hydrogen sulfide. Contrib. Boyce Thompson Inst. 16, 78—83 (1950). — Morris, C. J., and J. F. Thompson: Isolation of L(+)-S-methylcysteine sulphoxide from turnip roots (Brassica rapa). Chem. a. Ind. 1955, 951. — Mothes, K.: Über den Schwefelstoffwechsel der Pflanzen. II. Planta (Berl.) 29, 67—109 (1939). — Mothes, K., u. W. Specht: Über den Schwefelstoffwechsel der Pflanzen. Planta (Berl.) 22, 800—803 (1934).

Nakamura, N.: Über das Vorkommen von Methylmercaptan in frischer Raphanuswurzel. Biochem. Z. 164, 31—33 (1925). — Neuberg, C., u. J. Wagner: Über die Verschiedenheit der Sulfatase und Myrosinase. VIII. Mitt. Über Sulfatase. Biochem. Z. 174, 457—463 (1926). — Niel, C. B. van: On the morphology and physiology of the purple and green sulfur bacteria. Arch. of Microbiol. 3, 1—112 (1931). — Nightingale, G. T., L. G. Schermerhorn and W. R. Robbins: Effects of sulfur deficiency on metabolism in tomato. Plant Physiol. 7, 565—595 (1932).

Pfeffer, W.: Pflanzenphysiologie, Bd. I. Leipzig: Engelmann 1897.

Reed, L. J., and B. G. de Busk: Lipothiamide pyrophosphate: Coenzym for oxidative decarboxylation of α-ketoacids. J. Amer. Chem. Soc. 74, 3964—3965 (1952). — Reed, L. J., B. G. de Busk, J. C. Gunsalus and C. S. Hornberger: Cristalline α-lipoic acid: A catalytic agent associated with pyruvate dehydrogenase. Science (Lancaster, Pa.) 114, 93—94 (1951).

Schmid, H., u. P. Karrer: Über Inhaltsstoffe des Rettichs. I. Über Sulphoraphen, ein Senföl aus Rettichsamen (Raphanus sativus L. var. alba). Helvet. chim. Acta 31, 1017 bis 1028 (1948). — Schneider, W.: Über Cheirolin, das Senföl des Goldlacksamens. Sein Abbau und Aufbau. Liebigs Ann. 375, 207—254 (1910). — Schneider, W., u. H. Kaufmann: Untersuchungen über Senföle. II. Erysolin, ein Sulfonsenföl aus Erysimum Perowskianum. Liebigs Ann. 392, 1—15 (1912). — Schultz, O.-E., u. R. Gmelin: Das Senfölglukosid „Glukoiberin" und der Bitterstoff „Ibamarin" von Iberis amara L. (Schleifenblume). Arch. Pharmaz. 287, 404—411 (1954). — Seifter, E.: The occurrence of coenzyme A in plants.

Plant Physiol. **29**, 403—406 (1954). — SEMMLER, F. W.: Über schwefelhaltige ätherische Öle. Asa foetida Öl. Arch. Pharmaz. **229**, 1—31 (1891). — STARKEY, R. L.: Isolation of some bacteria which oxidize thiosulphate. Soil Sci. **39**, 197 (1935). — A study of spore formation and other morphological characteristics of *Vibrio desulfuricans*. Arch. of Microbiol. **9**, 268—304 (1938). — STOLL, A., u. E. SEEBECK: Über Alliin, die genuine Muttersubstanz des Knoblauchöles. Helvet. chim. Acta **31**, 189—210 (1948). — Über den enzymatischen Abbau des Alliins und die Eigenschaften der Alliinase. Helvet. chim. Acta **32**, 197—205 (1949). — SYNGE, R. L. M., and J. C. WOOD: A new free amino acid in cabbage. Biochemic. J. **60**, XV—XVI (1955).

THOMAS, M. D.: Agricultural research with radioactive sulfur and arsenic. Proc. Auburn Conference on Use of Radioactive Isotopes in Agricultural Research. Auburn, Ala. 1948. — THOMAS, M. D., R. H. HENDRICKS and G. R. HILL: Some chemical reactions of sulfur dioxide after absorption by alfalfa and sugar beets. Plant Physiol. **19**, 212—226 (1944). — TURRELL, F. M.: Physiological effects of elemental sulfur dust on citrus fruits. Plant Physiol. **25**, 13—62 (1950). — TURRELL, F. M., and M. CHERVENAK: Metabolism of radioactive elemental sulfur applied to lemons as an insecticide. Bot. Gaz. **111**, 109—122 (1949). — TURRELL, F. M., and J. R. WEBER: Elemental sulfur dust, a nutrient for lemon leaves. Science (Lancaster, Pa.) **122**, 119—120 (1955).

WAKSMAN, S. A., and J. S. JOFFE: Microorganisms concerned in the oxidation of the soil. II. *Thiobacillus thiooxydans*, a new sulphur-oxidizing organism isolated from the soil. J. Bacter. **7**, 239—256 (1922). — WEYGAND, F., O. TRAUTH u. R. LÖWENFELD: Konstitutionsaufklärung des Thiozuckers der Adenylthiomethylpentose. Chem. Ber. **83**, 563—567 (1950). — WINOGRADSKY, S.: Beiträge zur Morphologie und Physiologie der Schwefelbakterien. Leipzig: Felix 1888. — WOOLLEY, D. W., and W. H. PETERSON: J. of Biol. Chem. **122**, 213 (1937). Zit. nach A. STOLL und E. JUCKER in PAECH-TRACEY, Moderne Methoden der Pflanzenanalyse, Bd. IV. Berlin-Göttingen-Heidelberg: Springer 1955.

YOUATT, J. B.: Studies on the metabolism of *Thiobacillus thiocyano-oxidans*. J. Gen. Microbiol. **11**, 139—149 (1954).

ZECHMEISTER, L., and J. W. SEASE: Blue fluorescing compound, terthienyl, isolated from marigolds. J. Amer. Chem. Soc. **69**, 273 (1947).

Der Anteil der S-haltigen Verbindungen am Stoffwechsel*.

Von

Th. Bersin.

1. Einleitung.

Höhere Pflanzen und Mikroorganismen enthalten neben S-haltigen Aminosäuren verschiedene andere Verbindungen, in denen der Schwefel, stoffwechselchemisch gesehen, eine bedeutende Rolle spielt. Teilweise beeinflußt er durch seine Gegenwart in den Enzymproteinen Richtung und Ausmaß katalytischer Umsetzungen, teilweise dient er aber auch der Enzym- und Substrataktivierung.

Der erste Fall begegnet uns bei den sog. *Thiolenzymen* (BARRON). Es sind das die vielfach intracellulär oder in den Pflanzensäften vorkommenden Biokatalysatoren, deren Aktivität an die Gegenwart einer bestimmten Anzahl von freien HS-Gruppen in einer bestimmten Anordnung gebunden ist. Zu dieser Gruppe gehören sowohl Transpherasen als auch Hydrolasen und Redoxasen. Auf die Bedeutung der Schwefelbrücken für den Aufbau, den Zusammenhalt, die Reaktionsfähigkeit und Form vieler pflanzlicher Proteine sei hier nur nebenbei hingewiesen.

Die der Substrataktivierung dienenden Thiole werden vor allem durch zwei Verbindungen repräsentiert: das *Coenzym A* und das *Glutathion*. Die Aufgabe des Coenzyms A besteht im wesentlichen darin, die wasserunlöslichen Fettsäuren durch Bildung von reaktionsfähigeren und wasserlöslichen Acyl-CoA-Thiolestern der Wirkung von Enzymen zugänglich zu machen. Auch das Glutathion vermag durch Anlagerung an gewisse Substrate, diese dem Einfluß von Enzymen zugänglich zu machen; darüber hinaus ist es aber auch als *Enzymaktivator* wirksam, da es durch thioloprive Substanzen inaktivierte Thiolenzyme wieder reaktiviert.

Zum Schluß müssen noch die in der Pflanzenwelt vorkommenden schwefelhaltigen Antibiotica und Vitamine erwähnt werden, welche den Stoffwechsel zu beeinflussen vermögen.

Im Schwefelstoffwechsel der pflanzlichen Organismen spielen schwefelhaltige Enzyme eine nicht unbedeutende Rolle. Der vorzugsweise in anorganischer Form aufgenommene Schwefel wird in organische Bindung übergeführt, wobei zum Teil Verbindungen nichtkatalytischen Charakters gebildet werden, zum Teil jedoch Inhaltsstoffe von hoher Aktivität für den Stoffwechsel entstehen. Diese letzteren enthalten stets Schwefel entweder im Eiweißanteil oder in der prosthetischen Gruppe. Sie sollen an erster Stelle besprochen werden, da sich unter ihnen eine Reihe von wichtigen, für den pflanzlichen Stoffwechsel unentbehrlichen Enzymen befinden.

Nun gibt es aber im Schwefelstoffwechsel der Pflanzen noch eine Anzahl von Enzymen, die spezifische Veränderungen an niedrigmolekularen S-haltigen Verbindungen bewirken. Gemeint sind nicht diejenigen Enzyme, die in die

---

* Ohne S-haltige Aminosäuren.

Synthese oder den Abbau S-haltiger Aminosäuren eingreifen. Deren Besprechung erfolgt an anderer Stelle. Vielmehr soll hier das Augenmerk auf Enzyme gerichtet werden, die beim Aufbau, Abbau oder Umbau von S-haltigen Vitaminen, S-haltigen Antibiotica und sonstigen S-haltigen Verbindungen beteiligt sind, deren Bedeutung mehr in ihrer Natur als Arzneistoffe, Wuchsstoffe für tierische Organismen u. ä. liegt.

Ehe auf die Besprechung der vorliegenden Literatur über die Eigenschaften der S-haltigen Enzyme eingegangen wird, müssen zunächst einige Vorbemerkungen Platz finden.

Der intermediäre Stoffwechsel der niederen und höheren Pflanzen spielt sich in Kreisprozessen und Enzymsystemen ab. In zunehmendem Maße gewinnt die Anschauung an Boden, daß zellständige Enzyme nicht wahllos im Protoplasma zerstreut sind, sondern in wohlgeordneten Arbeitsgemeinschaften zusammenwirken. Diese Enzymteams sind, wie etwa das KEILIN-WARBURG-System und der Apparat des Citronensäurecyclus in Mitochondrien lokalisiert. Die Schule von E. D. GREEN u. a. haben gezeigt, daß manche individuellen Enzyme der Lehr- und Handbücher Kunstprodukte sind, weil man sie bei der präparativen Darstellung aus der „Molekül-Symbiose" herausgerissen hat.

Anders liegen vermutlich die Verhältnisse in den saftführenden Gefäßen, doch existieren darüber noch zu wenig exakte Unterlagen.

Immerhin sei festgestellt, daß die im nachfolgenden beschriebenen Enzyme, ähnlich wie das bei den Genen im Chromosom der Fall ist, wahrscheinlich vielfach nur im Verband mit ihresgleichen in der lebenden Pflanze zur Wirkung kommen. Diese Gebilde bringen Richtung und Form in das biochemische Geschehen. Die Vergänglichkeit und Empfindlichkeit dieser höheren Einheiten ist mit ein Grund dafür, weshalb synthetische Leistungen so spärlich im Reagensglas, doch so wirksam in vivo verlaufen. Synthesen sind an eine harmonische Folge einer Reihe ganz bestimmter Elementarreaktionen aus den zahlreichen möglichen gebunden.

II. Thiolgruppen und Enzymaktivität.

Zahlreiche Enzyme des Pflanzenreiches, die zu den verschiedensten Klassen gehören, haben ein gemeinsames Merkmal: sie enthalten in ihrem Proteinanteil freie Thiolgruppen. Diese Thiolgruppen beeinflussen die Aktivität, sie sind die Ursache für die Empfindlichkeit dieser Enzyme gegenüber verschiedenen Effektoren (Schwermetallen, Halogenderivaten, ungesättigten Verbindungen, Oxydationsmitteln).

Manche dieser Thiolgruppen lassen sich direkt durch Mercaptidbildung, Farbreaktion mit Nitroprussid oder polarographisch nachweisen. Andere wiederum sind durch Peptidkettenfaltung oder H-Bindungen maskiert und geben sich erst nach der Behandlung mit Harnstoff, Guanidin oder anderen Mitteln zur Aufhebung der H-Bindungen, wobei Auffaltung erfolgt, zu erkennen (FRENSDORFF, WATSON und KAUZMANN). Spezifische Reduktionsmittel, wie niedrigmolekulare Thiole, können durch Aufspaltung von disulfidischen Sprossenbindungen nach

$$\underset{\underline{\hspace{1cm}}}{RSSR} + R'SH \rightarrow R—SH \quad \underset{\underline{\hspace{1cm}}}{R'SS—R}$$

SS (intramolekular) + SH → SH + SS (intermolekular)

eine Umformung des Enzymmoleküls verursachen. Diese Austauschreaktion geht über die Mercaptidionen (BERSIN, BERSIN und STEUDEL, WIELAND und SCHWAHN).

Im folgenden soll eine Reihe von Enzymen besprochen werden, deren Aktivität von der Anwesenheit freier HS-Gruppen im Proteinanteil abhängt. Sie werden vielfach auch als *Thiolenzyme* bezeichnet. Zur besseren Abgrenzung von denjenigen

Enzymen, deren Thiolgruppen im Coenzym oder Cosubstrat sitzen, sollten die obengenannten besser als *enzymatisch aktive Thiolproteine* klassifiziert werden.

Die Disulfidbrücken des Cystins in pflanzlichen Proteinen mit Enzymaktivität werden von niedrigmolekularen Thiolen reduktiv aufgespalten, wobei sich auch gemischte Disulfide bilden können (ELDJARN und PIHL). Umgekehrt können Thiolgruppen des Cysteins in Eiweißstoffen von niedrigmolekularen Disulfiden oxydativ in Disulfidbrücken verwandelt werden.

Das Verhältnis SH/SS an einer Enzymoberfläche wird beeinflußt: 1. durch die Ladung des Proteins, 2. durch die Ladung des diffusiblen Thiols, 3. durch die Ionenstärke der gesamten Phasen, 4. durch die Gesamtkonzentration an Thiol und schließlich 5. durch das Verhältnis HS/SS in der Gesamtphase.

Folgende pflanzliche Thiole und Disulfide diffusibler Natur kommen für die Effektorwirkung auf Enzyme in Frage:

Thiole	Disulfide
Thioacetaldehyd HS CH$_2$·CHO	Disulfidacetaldehyd
Cystein	Cystin
Cysteamin (Mercaptoäthylamin)	Cystamin
γ-Glutaminylcystein	γ-Glutaminylcystin
γ-Glutaminylcysteinglycin = HS-Glutathion	SS-Glutathion
Viscotoxin-SH	Viscotoxin-SS
Allylthiol C$_3$H$_5$SH	Diallyldisulfid sekundäres Butylpropenyldisulfid CH$_3$·CH$_2$(CH$_3$)CH·S—S·CH=CH·CH$_3$
3,3′-Dithiolisobuttersäure HS·CH$_2$ 　　　＞CH·COOH HS·CH$_2$	oxydierte 3,3′-Dithiolisobuttersäure
reduziertes Thiolutin und 　reduziertes Aureothricin	Thiolutin (a) und Aureothricin (b) 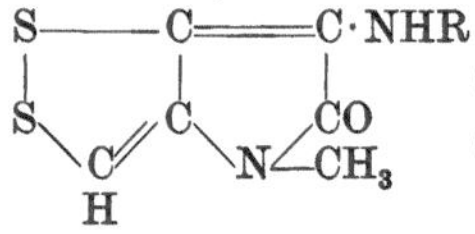 S——C——C·NHR 　　　　　　(a) R=CO·CH$_3$ S　　C　CO 　　C　N—CH$_3$　(b) R=CO·CH$_2$·CH$_3$ 　H
reduzierte Thioctsäure HOOC·CH$_2$·CH$_2$·CH$_2$·CH$_2$·CH·CH$_2$·CH$_2$ 　　　　　　　　　　SH　　SH	(+)-α-6,8-Thioctsäure (Liponsäure) HOOC·CH$_2$·CH$_2$·CH$_2$·CH$_2$·CH·CH$_2$·CH$_2$ 　　　　　　　　　　S——S
Coenzym-A-SH	Coenzym A-SS und Coenzym A-SS-Glutathion
Thiamin-SH	Thiamindisulfid und entsprechende unsymmetrische Disulfide
reduziertes Gliotoxin	Gliotoxin

Bei einigen enzymatisch aktiven Thiolproteinen wird ein Schutz der Thiolgruppen durch das Substrat gegenüber thiolopriven Reagentien beobachtet. Die Zahl der für die Aktivität maßgebenden Thiolgruppen je Enzymmolekül ist in den seltensten Fällen genau bekannt (LINDLEY). Die Untersuchung der Frage ist unter anderem dadurch erschwert, daß Schwermetalle innere Komplexe mit Nachbargruppen bilden (KLOTZ, URQUHARDT, KLOTZ und AYERS), die in langsamer Reaktion zu intramolekularen Oxydoreduktionen führen.

A. Thiolgruppen in den katalytischen Proteinen von Hydrolasen.

a) Urease.

Ein weitverbreitetes Thiolenzym aus der Klasse der Amidasen ist die Urease. Sie wurde aus Bakterien (LARSON und KALLIO), Pilzen, der Wassermelone und Leguminosen isoliert und hat die Eigenschaften eines Globulins. Der Schwefel liegt im Molekül des kristallisierbaren Enzyms zum Teil in Form von HS-Gruppen vor, die direkt und indirekt, auf Grund ihrer aktivierenden Eigenschaften, nachgewiesen werden konnten. Es scheint jedes 6. Atom Schwefel im Molekül vom Molekulargewicht 483000 als HS-Gruppe vorhanden zu sein (BERSIN 1940).

Die Urease spaltet außer Biuret (FEIGL und GENTIL) nur Harnstoff zu Ammoniak und Carbaminsäure, welch letztere sekundär in Kohlensäure und Ammoniak zerfällt. Für die Aktivität ist die Gegenwart von Thiolgruppen im Enzymmolekül unerläßlich. Oxydationsmittel und mercaptidbildende Fermentgifte, wie Organo-Quecksilberverbindungen, Schwermetallkationen (SHAW), aber auch andere thioloprive Substanzen führen die Urease in eine inaktive Form über, die teilweise oder vollständig durch Thiolverbindungen wieder reaktiviert werden kann. Das optimale Redoxpotential E_h liegt bei $+150 \, \mathrm{mV}$.

Von BERSIN und KÖSTER (1935) ist die Arbeitshypothese aufgestellt worden, daß die aktivierende HS-Gruppe eine H_2N-Wirkgruppe des Enzyms durch Dipolinduktion zur Umamidierung befähigt. Als erstes Einwirkungsprodukt von geöffneter Urease (LAIDLER und HOARE) auf Harnstoff würde eine Uraminosäure entstehen:

$$\underset{\text{Enzym}}{H_2N \cdot CO \cdot NH_2 + H_2N{-\!\!\!\overset{\overset{\displaystyle HSCH_2}{|}}{CH}}\cdot Ur} \rightarrow H_2N \cdot CO \cdot NH{-\!\!\!\overset{\overset{\displaystyle HSCH_2}{|}}{CH}}\cdot Ur + NH_3.$$

Unter der Voraussetzung der Bildung einer beim optimalen p_H der Enzymwirkung leicht spaltbaren CO—NH-Bindung wäre als nächster Schritt die Hydrolyse zur Carbaminsäure unter Rückbildung des Enzyms anzunehmen:

$$H_2N \cdot CO \cdot NH{-\!\!\!\overset{\overset{\displaystyle HSCH_2}{|}}{CH}}\cdot Ur + HOH \rightarrow H_2N{-\!\!\!\overset{\overset{\displaystyle HSCH_2}{|}}{CH}}\cdot Ur + H_2N \cdot COOH.$$

Der Spezifitätsunterschied zwischen der Urease und den analogen Thiolproteasen Papain usw. dürfte auf der Verschiedenheit der Konstitution und Form der hochmolekularen Träger der Wirkgruppen beruhen.

In den Urease führenden Pflanzen wird das Enzym durch Redoxasen in die Thiolform übergeführt; denn BERSIN und KÖSTER (2) konnten nachweisen, daß in den Schwertbohnen *(Canavalia ensiformis)* eine Succinodehydrogenase vorkommt, die Disulfid-Urease in Thiol-Urease verwandelt:

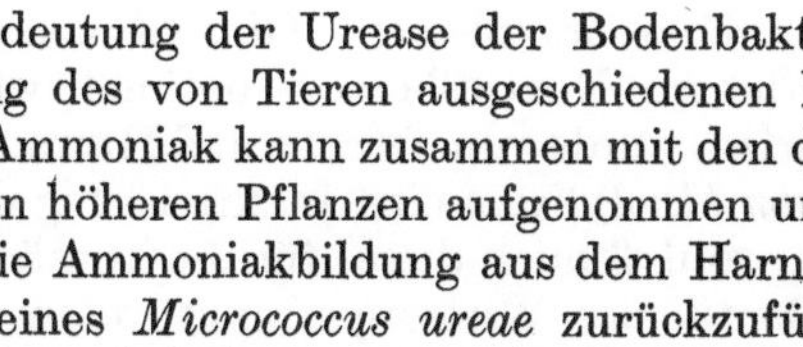

Succinat $\diagup$ Enz $\diagdown$ HS-Urease

Fumarat $\diagup$ Enz$\overset{\displaystyle H}{\underset{\displaystyle H}{\diagup}}$ $\diagdown$ SS-Urease

Die biologische Bedeutung der Urease der Bodenbakterien kommt in ihrer Fähigkeit zur Spaltung des von Tieren ausgeschiedenen Harnstoffes zum Ausdruck. Das gebildete Ammoniak kann zusammen mit den daraus mikrobiologisch gebildeten Nitraten von höheren Pflanzen aufgenommen und zum Proteinaufbau verwendet werden. Die Ammoniakbildung aus dem Harnstoff der Gülle ist auf das gleiche Ferment eines *Micrococcus ureae* zurückzuführen. Die Harnstoffzersetzer des Ackerbodens sind der sporenbildende *Bacillus probatus* und *Sarcina ureae.* Auch in den Wurzelknöllchen der Leguminosen ist Urease nachweisbar. Nach Conrad soll die an Lignoproteinverbindungen des Bodens gebundene Urease aus abgestorbenen Bakterienleibern die Ammoniakbildung aus Harnstoff verursachen.

Nach älteren Angaben nimmt der Gehalt von Samen und Früchten sowohl beim Reifen, als auch beim Keimen zu. Die Urease ist in den innersten Samenteilen, nicht in der Schale lokalisiert. Bei der *Canavalia ensiformis* wurde im Parenchym zur Zeit der Zellteilung eine starke Biosynthese des Enzyms beobachtet, welche mit dem Aufhören der Zellstreckung ihr Maximum erreicht. Mit zunehmendem Alter findet dann eine Abnahme des Gehaltes an Urease statt. Im Cambium und dem sich aus ihm ableitenden Xylem und Phloem konnte das Enzym nicht nachgewiesen werden. Samen und besonders der Embryo sind sehr reich an Urease. Welche Bedeutung die in der Rhizosphäre landwirtschaftlicher Kulturen aufgefundene Urease (Wlassjuk) besitzt, ist unbekannt. Die Rolle der Urease in höheren Pflanzen ist noch unklar. Sie hat vielleicht als Reserveprotein in manchen Früchten eine bestimmte Aufgabe im Schwefelstoffwechsel. Auffallend ist die Vergesellschaftung der Urease der Sojabohne mit Methylmethionin MMS$^+$ (Bersin, Müller und Strehler):

Methylmethioningehalt in einigen Lebensmitteln.

Ausgangsmaterial (April 1955) St. Gallen, Markt	Ansatzmenge in g	Phosphorwolframat aus	mg MMS$^+$/100 g Ausgangsmaterial	% MMS$^+$-Gehalt
Milch, roh	2	direkt	242,7	0,24
Milch, gekocht . .	30	direkt	30,9	0,03
Urease[1]	4	2-n H$_2$SO$_4$-Extrakt	2433,7	2,43
Sojamehl.	5	2-n H$_2$SO$_4$-Extrakt	218,9	0,21
Bananen, unreif .	135	2-n H$_2$SO$_4$-Extrakt	148,3	0,14
Bananen, reif. . .	130	2-n H$_2$SO$_4$-Extrakt	41,2	0,04
Wirsing	376	2-n H$_2$SO$_4$-Extrakt	603,0	0,60

			mg MMS$^+$/100 g Preßsaft	% MMS$^+$-Gehalt Preßsaft
Tomaten	93 → 24 g Preßsaft		108,3	0,10
Blaukraut	400 → 117 g Preßsaft		478,9	0,47
Weißkohl	97 → 97 g Preßsaft		597,7	0,59
Rüben	210 → 110 g Preßsaft		86,29	0,08
Wirsing	342 → 175 g Preßsaft		386,1	0,38

[1] Amerikanisches Präparat von Dr. Strehler.

b) Bakterielle Thiolproteasen.

Auf Grund der Aktivierbarkeit durch Thiolverbindungen und Inaktivierung durch thioloprive Reagentien lassen sich einige intracelluläre Proteasen von Bakterien zur Gruppe der Thiolenzyme rechnen.

Maschmann schlug für diesen Typus die Bezeichnung *Anaerobiasen* vor. Sie finden sich bei verschiedenen *Clostridium*-Arten, bei *Streptococcus Typus A* (Elliott) und anderen Mikroorganismen. Obwohl von Hoffmann-Ostenhof geäußert wurde, daß die im Neutralbereich optimal wirkenden *Kollagenasen* keines Aktivators bedürfen, wurde von Tytell und Hewson an einem gereinigten Präparat aus *Clostridium histolyticum* eine Inaktivierung durch HS-blockierende Substanzen beobachtet, so daß auch dieses für die zerstörende Wirkung auf das Muskelkollagen und Dentin des Wirtsorganismus verantwortliche Enzym HS-Gruppen enthalten dürfte. In *Escherichia coli, Bacillus mycoides, Proteus vulgaris* und *Bacillus subtilis* findet sich eine cysteinaktivierbare Protease, die die Säure-amidbindung des Antibioticums *Chloramphenicol* spaltet (SMITH, WORREL und LILLIGREN). Die Thiolprotease aus *Clostridium histolyticum* vermag augenscheinlich Bradykininogen zu Bradykinin zu spalten.

c) Papain.

Die am besten untersuchte Protease der höheren Pflanzen ist das *Papain* (Papayotin) aus dem Milchsaft der unreifen Früchte des tropischen Melonen-baumes *Carica papaya*. Aus den verletzten, Milchsaft führenden Gefäßen der Pflanze tritt eine klare, an der Luft bald milchig trübe werdende Flüssigkeit aus. Diese gerinnt, besonders in Gegenwart von Calciumionen; die Gerinnung kann durch Komplexbildner für Calcium verhindert werden. Im geronnenen und zweckmäßigerweise im Vakuum getrockneten Saft findet sich das Papain neben einer Reihe von natürlichen Aktivatoren (BERSIN 1940). Das Enzym wird durch thioloprive Reagentien inaktiviert, kann aber durch Reduktionsmittel wieder reaktiviert werden. Durch Schwermetalle inaktiviertes Papain wird durch Komplexbildner reaktiviert. Es wird daher von den meisten Untersuchern dieses Enzyms angenommen, daß das Enzymmolekül selbst im aktiven Zustand ein Thiol ist [BERSIN und LOGEMANN (1), BERSIN 1933 und 1935, BERSIN und KÖSTER (1), FINKLE und SMITH].

$$2\,\text{PaSH} \rightleftharpoons \text{PaSSPa} + 2\,\text{H}^+ + 2\,\varepsilon$$
$$\quad\text{aktiv} \qquad\quad \text{inaktiv}$$

Dem *Papain* sehr nahestehende Enzyme sind das *Bromelin* aus *Ananas sativus*, *Ficin* aus Feigenarten, *Pinguinain* aus *Bromelia pinguin*, *Asclepaein* aus *Asclepias*-Arten, *Mexicanain* aus *Pileus mexicanus*, *Tabernamontanain* aus *Tabernamontana grandiflora* und *Euphorbain* aus *Euphorbia*-Arten.

Die ältere Literatur findet sich bei HWANG und IVY sowie WEINER, die auch auf die therapeutische Bedeutung des *Papains* hinweisen. Weitere Literatur siehe BERSIN (1940).

Nach neueren Versuchen über das kristallisierte *Papain* (KIMMEL und SMITH) gelingt die vollständige Aktivierung des Enzyms am besten mittels Cystein und Äthylendiamintetraacetat. Mit diesem Gemisch aus einem Reduktionsmittel und einem Komplexbildner gelingt auch die Reaktivierung der inaktiven Verbindung Hg^{II} (Papain)$_2$. Außer seiner Fähigkeit zur Spaltung von CO—NH-Bindungen besitzt das Enzym, wie viele andere Proteasen, eine Esteraseaktivität (KIMMEL und SMITH, McDONALD und BALLS); es katalysiert die Milchgerinnung (HINKEL und ZIPPIN), die Hydroxamsäurebildung (SLAVIK), die Umamidierung (FRUTON; JONES, HEARN, FRIED und FRUTON) (vielleicht über Acylmercaptane als Zwischenprodukte) und Hydrazidbildung (SCHULLER und NIEMANN).

Das reine HS-Papain besteht aus einer einzelnen Peptidkette vom Molekular-gewicht 20700. Oxydiertes S—S-Papain existiert in wäßriger Lösung je nach der Konzentration aus einem wechselnden Gemisch von mono- und dimeren Molekülen. Der isoelektrische Punkt liegt bei p_H 8,75. Die freien Carboxyl-gruppen gehören Asparaginsäureresten an, während als basische Reste eine Isoleucingruppe am N-Ende und in der Kette 8 oder 9 Lysinaminogruppen vorkommen. Methionin fehlt. Neben 6-Cysteinresten enthält das Papain ver-mutlich noch zwei weitere HS-Gruppen eines unbekannten Proteinbestandteiles.

Neben dem eigentlichen Papain findet sich im Papaya-Milchsaft ein weiteres Thiolenzym (JAFFÉ), das *Chymopapain*, bei dem das Verhältnis von Milch-gerinnungsaktivität zu proteolytischer Aktivität etwa doppelt so groß ist wie

beim Papain. Vermutlich vermögen beide Fermente das α-Caseinogen der Kuh-
milch in einer Primärreaktion (Mattenheimer und Nitschmann) proteolytisch
geringfügig abzubauen, worauf in Gegenwart von Calciumionen die Ausfällung
von unlöslichem Calciumcaseinat, d. h. die Gerinnung erfolgt.

Welche Funktionen das Papain und ähnliche Enzyme im Milchsaft ausüben,
ist unklar. Vielleicht ist es mit seinen Thiolgruppen an der Ausbildung eines
bestimmten Redoxpotentials beteiligt (Cretin), wirkt plastifizierend auf die
Kohlenwasserstoffe (Conté) durch Anlagerung an die Doppelbindungen und
sensibilisiert das komplexe kolloidale System im Sinne einer Koagulation
(Lepetit), wie Versuche mit *Hevea*-Latex gezeigt haben.

Unter anderem sei ferner darauf hingewiesen, daß wahrscheinlich neben anderen
Thiolverbindungen in der Pflanze auch die Thiolproteasen mit Seleniten und
Telluriten unter Bildung von Mercaptiden (Bersin und Logemann, Lampson
und Klug) reagieren, wobei Enzyminaktivierung und eventuell Rot- bzw.
Schwarzfärbung wegen Zerfalls der Mercaptide zu den Elementen eintreten. Die
seit über 100 Jahren bekannte Reduktion von Selen- und Tellurverbindungen
durch lebende pflanzliche Zellen zu freiem Selen und Tellur beruht auf der
Zwischenbildung der entsprechenden Mercaptide. Unterbindet man diese durch
Thioätherbildung der Mercaptane mit Monojodacetat, so zeigen z. B. Kulturen
von *Penicillium crustaceum* auf glucosehaltigem Nährboden mit Selenitzusatz
keine Selenausscheidung; abgeschnittene Blütenstengel von *Viola tricolor* scheiden
nur in Selenitlösungen, nicht jedoch in Lösungen von Selenit und Monojodacetat
rotes Selen aus. Das Verfahren eignet sich zum histochemischen Nachweis von
Thiolverbindungen (Logemann).

d) Thiaminasen.

In Darmbakterien und Farnen läßt sich ein Thiolenzym nachweisen, das
einen Abbau des Vitamins B_1 (= Thiamin) und der Cocarboxylase bewirkt. Eine
Zusammenfassung der älteren Literatur findet sich bei Fujita. Das auch in
Fischeingeweiden nachweisbare Enzym bewirkt eine Transaminierung nach:

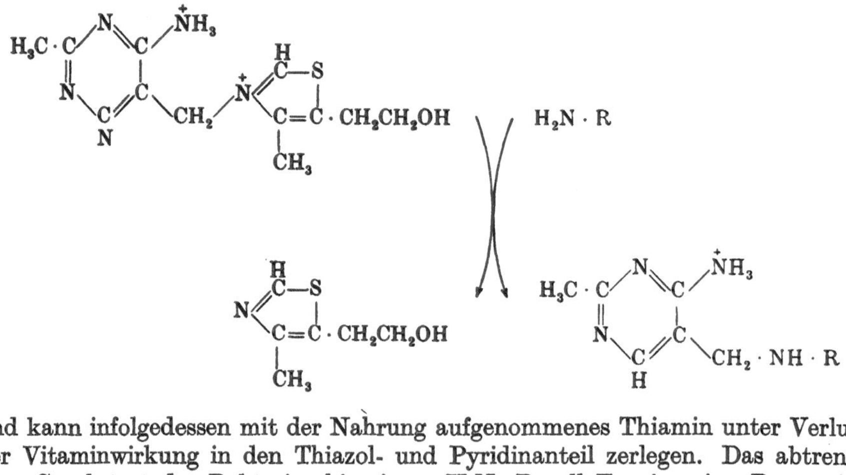

und kann infolgedessen mit der Nahrung aufgenommenes Thiamin unter Verlust
der Vitaminwirkung in den Thiazol- und Pyridinanteil zerlegen. Das abtrenn-
bare Cosubstrat der Bakterienthiaminase $H_2N \cdot R$ soll Taurin sein. Da es sich
um eine Gleichgewichtsreaktion handelt, kann mittels des Enzyms auch Thiamin
sowie eine Reihe von Homologen synthetisiert werden.

Die optimalen Reaktionsbedingungen der pflanzlichen Enzyme werden wie
folgt angegeben (Fujita) (Tabelle 1):

Tabelle 1.

Enzymquelle	pH	Temperatur °C	Inaktivierungs- zeit in Minuten (bei °C)
Bakterien:			
Bacillus thiaminolyticus Matsukawe et *Misawa* (BMM)	5,0	30	20 (100)
Bacillus aneurinolyticus Kimura et *Aoyama* (BKA)	8,0	60	10 (100)
Clostridium thiaminolyticum Kimura et *Liao* (SKL)	7,5	30	20 (100)
Pflanzen:			
Pteridium	6,0	55	15 (100)
Dicranopteris	6,0	55	15 (100)
Celosia	5,5	55	15 (100)

Das Enzym aus BKA scheint verschieden von denjenigen aus BMM und SKL zu sein. Die prozentuale Aktivierung der Enzyme durch Thiole ergibt sich aus nebenstehender Tabelle 2 (FUJITA).

Die pflanzlichen Thiaminasen sind von einem thermostabilen Faktor (SF) begleitet, der ebenfalls eine Zerstörung des Vitamins B_1 bewirkt, doch unter physiologischen Bedingungen kaum aktiv ist. Die nach Verfütterung von Farnen bei Haus-

Tabelle 2.

Aktivator	BMM	*Pteridium*	*Celosia*
L-Cysteinhydrochlorid	700	19	25
Glutathion GSH	380		
NaSH	1410		
$Na_2S_2O_3$	7	420	

tieren beobachteten Vitaminmangelerscheinungen sind im wesentlichen auf das Enzym zurückzuführen. Über die chemische Natur von SF ist wenig bekannt.

Die physiologische Bedeutung der pflanzlichen Thiaminasen ist noch unklar. Es ist möglich, daß sie intracellulär an der Vitaminsynthese beteiligt sind, die einer Energiezufuhr bedarf, da aus einem tertiären Amin eine quaternäre Ammoniumbase mit hohem Energiegehalt gebildet wird (WOOLLEY).

Außer den Thiaminasen existieren noch andere Enzyme, die am Thiamin angreifen. In Knoblauchextrakten kann ein Enzym nachgewiesen werden, das bei p_H 8 und 60° Thiamin in *Allithiamin* = 2-(2'-Methyl-4'-amino-pyrimidyl-5')-methyl-formamino-5-oxy-Δ^2-pentenyl-(3)-allyldisulfid verwandelt. Diese *Allinase* benötigt als Cosubstrat das im Knoblauch vorkommende *Allicin* $CH_2=CH \cdot CH_2 \cdot S \cdot SO \cdot CH_2 \cdot CH=CH_2$, welches sich mit der Thiolform des Thiamins

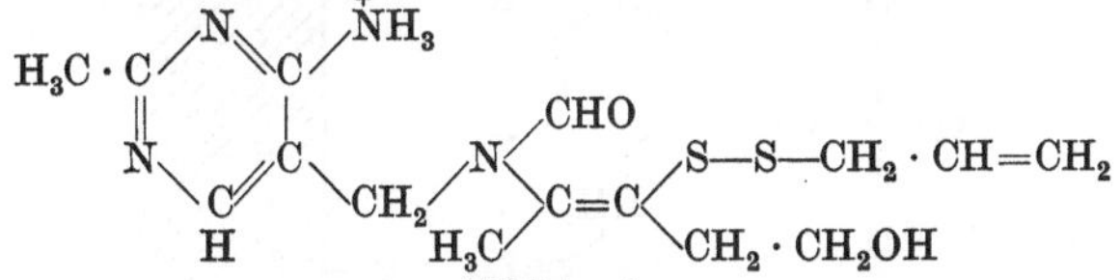

$$H_3C \cdot C \underset{\substack{|| \\ N}}{\overset{\substack{N \\ \diagdown}}{\diagdown}} \overset{N}{\underset{C}{\diagup}} \overset{\overset{+}{N}H_3}{\diagup} \quad$$

Allithiamin

umsetzt. Neben dem *Allithiamin* entsteht noch *Thiaminmethyldisulfid* und *Thiaminpropyldisulfid*. Durch Cystein kann *Allithiamin* wieder in Thiamin zurückverwandelt werden, wobei nebenbei *S-Allylmercaptocystein* entsteht.

Eine Thiamin-*Phosphorylase* wurde in *Staphylococcus aureus, Bacillus subtilis, Salmonella paratyphi, Salmonella schottmuelleri, Shigella dysenteriae, Escherichia coli, Proteus vulgaris* und *Pseudomonas aeruginosa* nachgewiesen. Das Enzym ähnelt der *Thiaminokinase* aus Rattenleber, die nach: ATP + Thiamin → Cocarboxylase + AMP das Coenzym der Carboxylase (Thiaminpyrophosphat TPP) bildet.

B. Thiolgruppen in den katalytischen Proteinen der Redoxasen.

a) Flavinenzyme.

Unter den Flavinenzymen des Pflanzenreiches, die alle das Dinucleotid (DN) Isoalloxazin-D-ribityldiphosphorsäure-D-riboseadenin als Wirkgruppe besitzen, finden sich einige, deren Proteinanteil Thiolcharakter hat.

Die *Nitratreduktase* aus Sojabohnen, deren Wirkgruppe aus 2 DN + 1 Mo besteht, vermittelt die Oxydoreduktion zwischen Nitratanionen und der reduzierten Form des Triphosphopyridinnucleotids TPNH:

$$NO_3^- + TPNH + H^+ \rightleftharpoons NO_2^- + TPN + H_2O.$$

Wie EVANS und NASON gezeigt haben, ist dieses aus Sojablättern und anderen höheren Pflanzen isolierbare Enzym durch p-Chlormercuribenzoat reversibel inaktivierbar; Cystein reaktiviert.

b) Pyridinenzyme.

Zahlreiche aktive Proteine, die Oxydoreduktionen mittels Pyridincoenzymen (I, II und III) beschleunigen, enthalten freie Thiolgruppen, die entweder für die Bindung der Coenzyme (RACKER und KRIMSKY) oder auch sonstwie für den Ablauf der enzymatischen Reaktion notwendig sind.

Nachfolgend seien einige Beispiele genannt. Nach JOYEUX und CROSON enthält die *Milchsäuredehydrogenase* aus *Bacterium anitralum* freie HS-Gruppen, da Inaktivierung durch p-Chlormercuribenzoat und Reaktivierung durch HS-Glutathion, nicht jedoch Äthylendiamintetraacetat erfolgt.

Die *Alkoholdehydrogenase* aus Hefe enthält bei einem Molekulargewicht von 150000 etwa 35—40 Cysteinreste je Molekül. Dieses Apoenzym bindet Diphosphopyridinnucleotid durch Komplexbildung zwischen dem Adeninteil des Coenzyms und dem Zink der Alkoholdehydrogenase. Das Zink wird durch mehrere Bindungen, von welchen mit großer Wahrscheinlichkeit zumindest eine über Schwefel erfolgt, im Protein besonders fest verankert. Auch bei der Bindung des Nicotinamidanteils dürfte eine HS-Gruppe beteiligt sein. Die Übertragung des Wasserstoffes erfolgt im Komplex als Hydridion entsprechend dem Vorgang bei der Oxydoreduktion nach MEERWEIN-PONNDORF-OPPENAUER. Der Komplex selbst wird von WALLENFELS und SUND wie folgt formuliert:

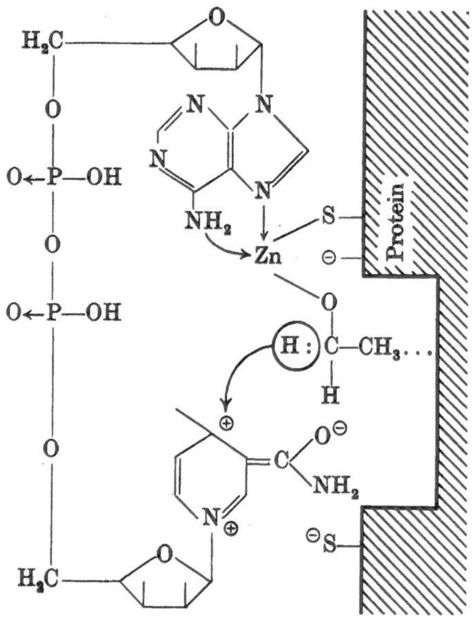

Die *Glycerinaldehyd-3-phosphat-Dehydrogenase* enthält nach KRIMSKY und RACKER fest gebundenes Glutathion im Proteinanteil. Die Oxydation des Aldehyds durch das DPN-abhängige Enzym verläuft in zwei Stufen (RACKER und KRIMSKY). Zunächst wird ein Halbmercaptal gebildet, das zum S-Acylester oxydiert wird. Dann findet eine Übertragung des Acyls unter Zwischenbildung eines Acylenzyms (RACKER) auf Phosphat statt und durch anschließende Hydrolyse entsteht Glycerinsäure-3-phosphat. Die Oxydoreduktion wird durch Monojodacetat blockiert.

c) Pyridoxalenzyme.

Die in *Escherichia coli B* nachweisbare L-Lysindecarboxylase enthält in ihrem Proteinanteil HS-Gruppen, die durch Cystein und Äthylendiamintetraacetat geschützt werden können (SHER und MALLETTE).

Cystathionase.

In verschiedenen Bakterien findet sich ein pyridoxalphosphat-abhängiges Enzym, das die Spaltung des Thioäthers Cystathionin zu Homocystein + Pyruvat + NH_3 bewirkt (BINKLEY).

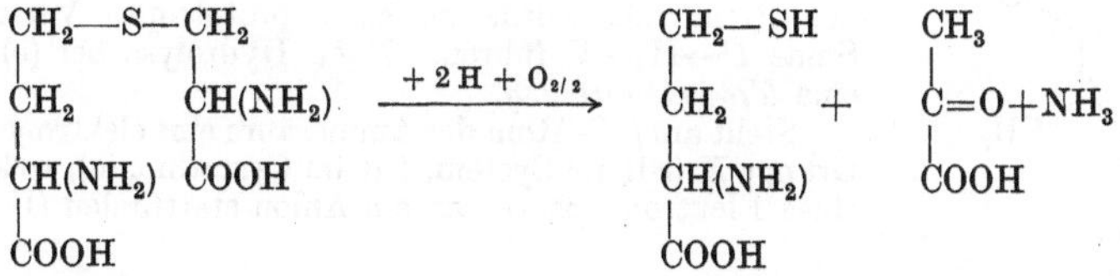

$$CH_2\text{—}S\text{—}CH_2 \qquad\qquad CH_2\text{—}SH \quad CH_3$$
$$\begin{array}{lll} CH_2 & CH(NH_2) & \xrightarrow{+\,2\,H\,+\,O_{2/2}} \quad CH_2 \quad + \quad C{=}O{+}NH_3 \\ CH(NH_2) & COOH & \qquad CH(NH_2) \qquad COOH \\ COOH & & \qquad COOH \end{array}$$

Die Wirkung dieses Enzyms konnte durch ein synthetisches Gemisch von Pyridoxal, Kupferionen sowie Äthylendiamintetraacetat, als Komplexbildner imitiert werden. Ersetzt man das Kupfer durch Zink, so läßt sich eine Spaltung von Lanthionin zu Cystein, Pyruvat und Ammoniak bewerkstelligen.

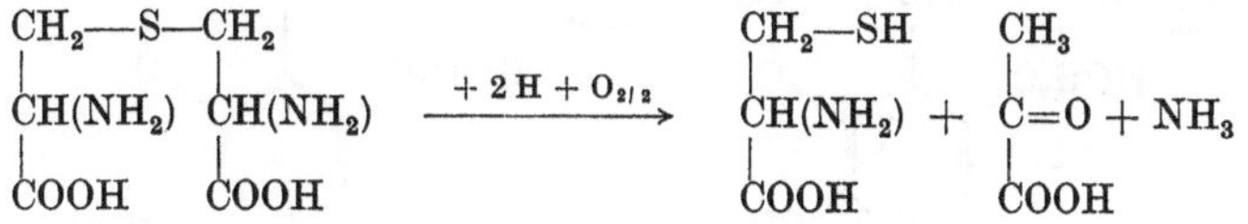

$$CH_2\text{—}S\text{—}CH_2 \qquad\qquad CH_2\text{—}SH \quad CH_3$$
$$\begin{array}{lll} CH(NH_2) & CH(NH_2) & \xrightarrow{+\,2\,H\,+\,O_{2/2}} \quad CH(NH_2) \quad + \quad C{=}O \;+\; NH_3 \\ COOH & COOH & \qquad COOH \qquad\quad COOH \end{array}$$

Augenscheinlich spielen hier innere Komplexe (chelate) beim Reaktionsmechanismus eine entscheidende Rolle. Daher sei auf diese Seite des Problems eingegangen.

Reaktionsmechanismus bei den Pyridoxalenzymen.

Einen allgemeinen Mechanismus für die durch Pyridoxalenzyme katalysierten Reaktionen haben METZLER und Mitarbeiter vorgeschlagen und zwar auf Grund von Modellreaktionen, die sie in Gegenwart von katalytischen Metallionen durchführten.

Das Pyridoxal neigt zu Elektronenverschiebungen in folgendem Sinne:

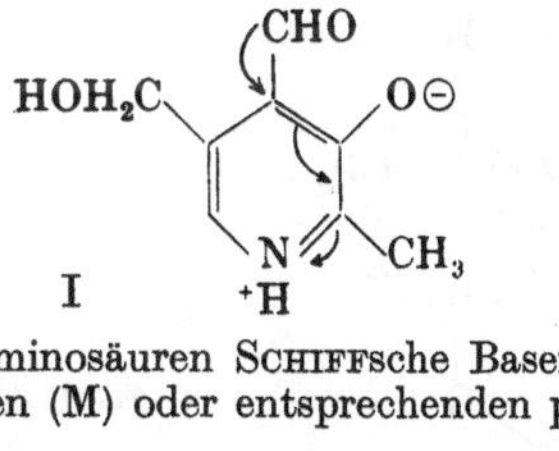

$$\text{HOH}_2\text{C}\underset{\substack{\\ \text{I}}}{\overset{\substack{\text{CHO}\\}}{\diagup}}\cdots\text{O}^{\ominus}$$
$$\text{N}\quad\text{CH}_3$$
$$\overset{+}{\text{H}}$$

I

Bildet die Aldehydgruppe mit Aminosäuren SCHIFFsche Basen, so können diese Koordinationsverbindungen mit Metallionen (M) oder entsprechenden positiv geladenen Gruppen im Apoenzym eingehen.

Folgende Fälle können eintreten:

Nach der Abgabe eines Protons bildet sich ein nucleophiles Zentrum am α-C-Atom der Aminosäure, gefolgt von der Addition des Protons (II→III→IV), wodurch sich eine *Razemase*-wirkung ergibt, wenn bei (d) Hydrolyse erfolgt:

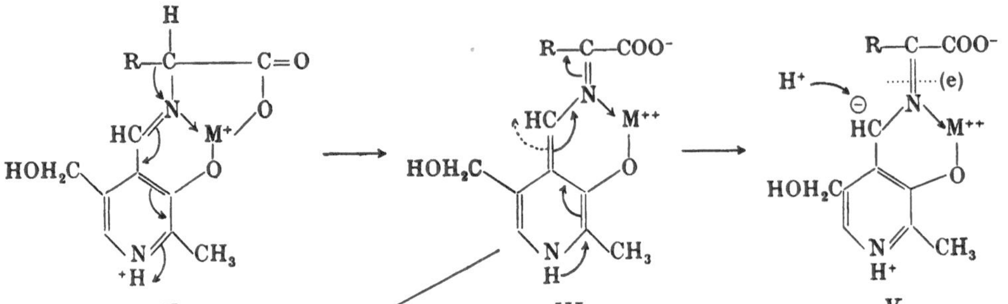

II

III

V

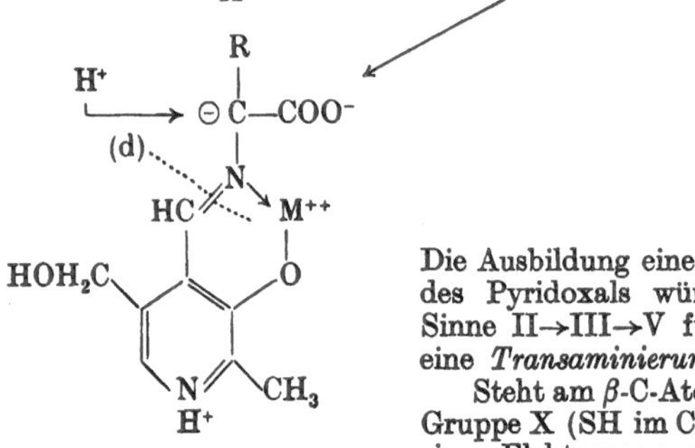

IV

Die Ausbildung eines nucleophilen Zentrums am Formyl-C-Atom des Pyridoxals würde zu einer prototropen Verschiebung im Sinne II→III→V führen. Nach Hydrolyse bei (e) ergibt sich eine *Transaminierung*.

Steht am β-C-Atom der Aminosäure eine elektronenanziehende Gruppe X (SH im Cystein, SR im Cystathionin), so kann Abgabe eines Elektronenpaares an ein Anion stattfinden (II→VI→VII):

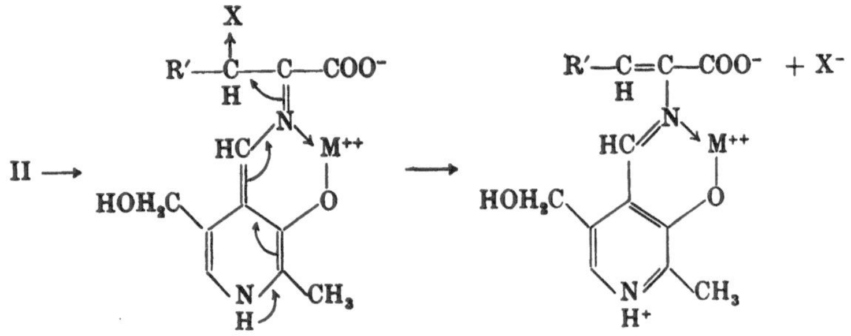

II →

VI

VII

So entsteht H_2S aus dem abgespaltenen HS^- des Cysteins und H^+ bei der *Desulfhydrase*wirkung, Allylsulfonsäure bei der *Alliinase*wirkung und Cystein bei der *Cystathioninspaltung*.

Die *Cystathioninbildung* aus Serin und Homocystein verläuft wohl auch über die Schiffsche Base VII aus Serin, an die sich Homocystein addieren kann.

Im Gegensatz zu diesen Eliminationsreaktionen am β-C-Atom existieren auch γ-Eliminierungsreaktionen, so etwa bei der Umwandlung von Homocystein zu Ketobutyrat und Ammoniak. Es wird angenommen, daß hierbei auf eine Ionisation des α-H-Atoms eine Art Wagner-Meerwein-Umlagerung nach II→VIII→IX erfolgt:

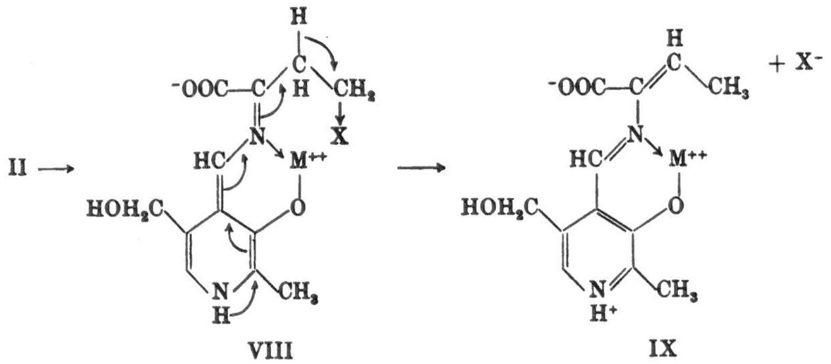

II →

VIII

IX

C. Schwefelhaltige Gruppen in Coenzymen und Cosubstraten.

a) Pyruvatoxydase und α-Ketoglutaratoxydase.

Auf die ubiquitär im Pflanzenreich verbreitete *Carboxylase*, deren physiologische Aufgabe in der Decarboxylierung des Pyruvats zu Acetaldehyd und CO_2 besteht, kann hier nur hingewiesen werden. Ihre Würdigung findet in anderem Zusammenhang statt.

Daneben vermögen Pflanzen Pyruvat auch oxydativ zu decarboxylieren, wonach das gebildete Acetat zu verschiedenen synthetischen Reaktionen herangezogen wird. Das Coenzym der *Pyruvatoxydase* ist identisch mit demjenigen der *α-Ketoglutaratoxydase* [REED und DE BUSK (1)].

Diese beiden Enzyme benötigen ein Coenzym, das von einer in *Escherichia coli* gefundenen, vermutlich aber auch sonst weitverbreiteten *Liponsäurekonjugase* [REED und DE BUSK (2)] synthetisiert wird. Aus Thiaminpyrophosphat TPP und Liponsäure LA (= Protogen) entsteht die Cooxydase, das Lipothiamidpyrophosphat LTPP:

$$
\underset{\substack{\text{H}}}{\text{N}}\text{-Struktur} \qquad NH \cdot CO \cdot CH_2 \cdot CH_2 \cdot CH_2 \cdot CH_2 \cdot CH \text{---} CH_2
$$

Nach einer Idee von REED und DE BUSK (1953) erfolgt die oxydative Decarboxylierung von Pyruvat bzw. α-Ketoglutarat in Gegenwart von LTPP, Diphosphopyridinnucleotid (DPN$^+$) und Coenzym A (CoA-SH) wie folgt:

1. Pyruvat + $\begin{smallmatrix}S\\ |\\ S\end{smallmatrix}$>LTPP ⇌ $\begin{smallmatrix}\text{Acetyl} \sim S\\ HS\end{smallmatrix}$>LTPP + CO_2

2. $\begin{smallmatrix}\text{Acetyl} \sim S\\ HS\end{smallmatrix}$>LTPP + CoA—SH ⇌ $\begin{smallmatrix}HS\\ HS\end{smallmatrix}$>LTPP + CoA—S ~ Acetyl

3. $\begin{smallmatrix}HS\\ HS\end{smallmatrix}$>LTPP + DPN$^+$ ⇌ $\begin{smallmatrix}S\\ |\\ S\end{smallmatrix}$>LTPP + DPNH + H$^+$

Es ist bemerkenswert, daß das Redoxpotential des Systems $\begin{smallmatrix}S\\ |\\ S\end{smallmatrix}$>LTPP $\big/$ $\begin{smallmatrix}HS\\ HS\end{smallmatrix}$>LTPP wesentlich negativer ist, als dasjenige des Systems DPN/DPNH.

Es existieren Hinweise darauf, daß der Schwefel des LTPP auch an der Assimilation der Kohlensäure durch grüne Pflanzen beteiligt ist. Durch Belichtung soll nach CALVIN der Eintritt des neu assimilierten Kohlenstoffs in die Verbindungen des Citronensäurecyclus verhindert werden, da durch photochemische Reduktion das cyclische Disulfid LTPP in die Dithiolform übergeht:

$$
\begin{smallmatrix}S\\ |\\ S\end{smallmatrix}>\text{LTPP} \xrightarrow[\text{+ 2 Elektronen}]{\text{Licht}} \begin{smallmatrix}\ominus S\\ \ominus S\end{smallmatrix}>\text{LTPP}.
$$

In der Thiolform vermag das Coenzym nicht die oxydative Decarboxylierung des neu aus CO_2 gebildeten Pyruvates zu katalysieren, so daß die Bildung von Acetyl-Coenzym A unterbleibt und damit dem neu eintretenden Kohlenstoff der Zugang zum Citronensäurecyclus verwehrt wird (OCHOA). Die Spaltung des Dithiolanringes durch die Überschußenergie des angeregten Chlorophylls ist jedoch aus energetischen Gründen von SUNNER bezweifelt worden.

b) Coenzym A.

Das von verschiedenen Mikroorganismen, z. B. *Lactobacillus bulgaricus*, benötigte Redoxsystem Pantethein/Pantethin (L.b.-Faktor) stellt eine Vorstufe des Coenzyms A dar. Durch Veresterung mit Adenosinmonophosphat wird aus der HS-Form das Pantetheinnucleotid (CoA-SH) gebildet:

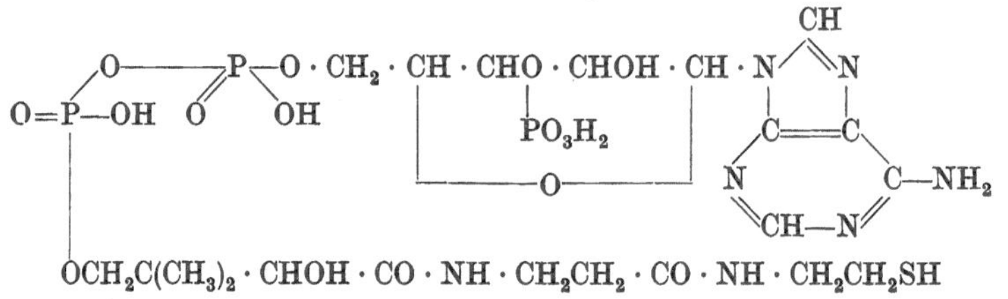

$$O=P-OH \quad O \quad P-O \cdot CH_2 \cdot CH \cdot CHO \cdot CHOH \cdot CH \cdot N \quad \overset{CH}{\underset{N}{\diagup}} N$$

$$OCH_2C(CH_3)_2 \cdot CHOH \cdot CO \cdot NH \cdot CH_2CH_2 \cdot CO \cdot NH \cdot CH_2CH_2SH$$

In der Natur kommt dieses Thiol teils frei, teils als gemischtes Disulfid (z. B. mit Glutathion als CoA-SSG) und teils in Form verschiedener Thiolester $R \cdot CO \cdot S \cdot CoA$ vor.

Das Coenzym A steht nicht nur im Knotenpunkt des Stoffwechsels der Fette und Kohlenhydrate, sondern es ist auch unerläßlich für die Biosynthese der Steroide, der Carotinoide, des Kautschuks, der Porphyrine und der Triterpene (Näheres vgl. LYNEN und DECKER, s. auch die Beiträge von HAAGEN-SMIT, HEUSNER, GOODWIN und ARREGUIN in Band X und von DECKER und LYNEN in Band XII dieses Handbuches).

Als Co-Transacetylase überträgt dieses Thiol Acylgruppen, indem zwischendurch Thiolcarbonsäureester gebildet werden. Wie alle Thiole bildet auch das Coenzym A mit seinem disulfidischen Oxydationsprodukt ein reversibles Redoxsystem $2 \, CoASH \rightleftharpoons CoASSCoA + 2 \, H^+ + 2\varepsilon$. Als Thiol bzw. Acylthiol kann es mittels der Nitroprussidreaktion bestimmt werden (LYNEN).

Am Kreislauf der Fettsäuren im sog. KNOOP-Cyclus (BERSIN 1954) ist das Coenzym A nach Arbeiten von BEINERT, LIPMANN, LYNEN u. a. maßgebend beteiligt. Darauf wird an anderer Stelle eingegangen. Die Fettbildung ist keine Veresterung, sondern zumeist eine Umesterung nach:

$$R \cdot CO \sim SCoA + Glycerin \rightarrow Fettsäureglycerinester + HSCoA.$$

Die Entstehung der Fettaldehyde (Plasmale) ist als eine enzymatische Hydrierung der Thiolcarbonsäureester aufzufassen:

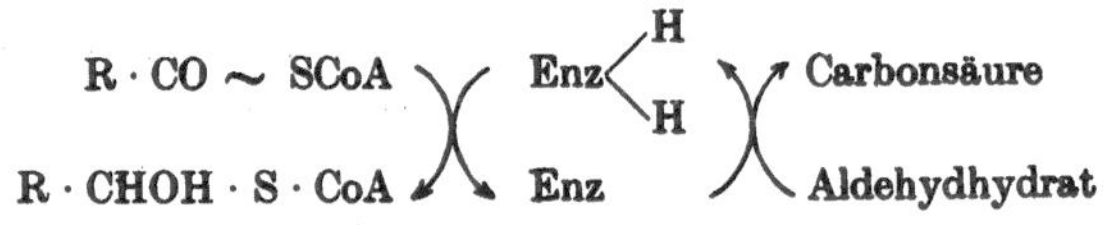

$$R \cdot CO \sim SCoA \qquad Enz \underset{H}{\overset{H}{\diagup\!\!\!\diagdown}} \qquad Carbonsäure$$
$$R \cdot CHOH \cdot S \cdot CoA \qquad Enz \qquad Aldehydhydrat$$

wobei die leicht spaltbaren Aldehydhalbmercaptale entstehen.

Die energiereiche $C \sim S$-Bindung erleichtert sehr die Reduktion von Carbonsäuren zu den Aldehyden, die ihrerseits leicht durch Pyridinredoxasen in Alkohol übergeführt werden können. Ganz augenscheinlich geht die Reduktion der Buttersäure bei der Butanol-Aceton-Gärung der *Clostridien* über Thiolcarbonsäureester.

Die Thiolcarbonsäureester können auf verschiedene Weise entstehen:
1. Durch oxydative Decarboxylierung (s. oben),

$$\alpha\text{-Ketoglutarat} \rightarrow Succinyl\text{-}CoA,$$

$$Pyruvat \rightarrow Acetyl\text{-}CoA,$$

2. durch Umesterung unter der Wirkung der Thiolase (vgl. OCHOA), z. B.

$$\text{Succinyl-CoA} + \text{Acetoacetat} \rightleftharpoons \text{Succinat} + \text{Acetoacetyl-CoA,}$$

3. durch Umsetzung mit Adenosintriphosphat ATP, z. B.

$$R \cdot COOH + CoA\text{-}SH + ATP \rightleftharpoons R \cdot CO \cdot S \cdot CoA + H_2O +$$
$$+ (\text{Phosphat})_n + \text{Adenosin-(phosphat)}_{3-n}.$$

Eine enzymatische Aktivierung von CO_2 durch Enzyme gewisser Bakterien und Hefen, die nach

a) $CO_2 + ATP \rightleftharpoons AMP - CO_2 + PP_i$

b) $AMP\text{-}CO_2 + \beta$-Hydroxyisovaleryl-Coenzym A $\rightleftharpoons$ AMP +
β-Hydroxy-β-methyl-glutaryl-Coenzym A

c) β-Hydroxy-β-methylglutaryl-Coenzym A $\rightleftharpoons$ Acetoacetat + Acetyl-Coenzym A

reagieren, haben BACHHAWAT und COON beschrieben.

Die Untersuchung dieser pflanzlichenEnzyme befindet sich erst im Anfangsstadium.

Das in kautschukführenden Pflanzen als Vorstufe gebildete β-Methylcrotonat wird offenbar aus Acetyl-Coenzym A synthetisiert (MILLERD und BONNER).

Die Bosynthese des Kautschuks, der Terpene, Carotinoide und Sterine in den Pflanzen ist nach neueren Vorstellungen ohne Coenzym A undenkbar. Es werden heute folgende Zwischenprodukte angenommen (die zugehörigen Enzyme sind noch kaum erforscht):

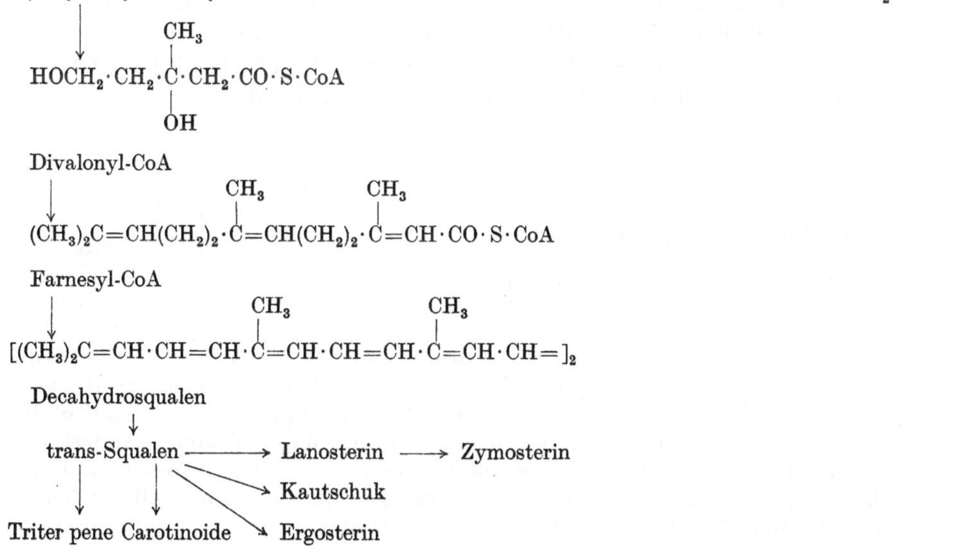

Acetat + ATP $\rightleftharpoons$ AMP-acetyl + PP_i ────────────────────→CO_2 + ATP

AMP-acetyl + HSCoA $\rightleftharpoons$ AMP + Acetyl-CoA ────→ β-Hydroxy-β-methyl-glutaryl-CoA

$$CH_3$$
$$CH_3\text{—}\overset{|}{C}\text{—}CH_2 \cdot CO \cdot S \cdot CoA$$
$$\overset{|}{O}H$$

(β-Ketothiolase)

Acetoacetyl-CoA $\longrightarrow$ Acetoacetat + HS CoA

β-Hydroxyisovaleryl-CoA $\longleftarrow$ AMP $-CO_2$

$$CH_3$$
$$HOCH_2 \cdot CH_2 \cdot \overset{|}{C} \cdot CH_2 \cdot CO \cdot S \cdot CoA$$
$$\overset{|}{O}H$$

Divalonyl-CoA

$$CH_3 \qquad CH_3$$
$$(CH_3)_2C\text{=}CH(CH_2)_2 \cdot \overset{|}{C}\text{=}CH(CH_2)_2 \cdot \overset{|}{C}\text{=}CH \cdot CO \cdot S \cdot CoA$$

Farnesyl-CoA

$$CH_3 \qquad CH_3$$
$$[(CH_3)_2C\text{=}CH \cdot CH\text{=}CH \cdot \overset{|}{C}\text{=}CH \cdot CH\text{=}CH \cdot \overset{|}{C}\text{=}CH \cdot CH\text{=}]_2$$

Decahydrosqualen

trans-Squalen ──────→ Lanosterin ──────→ Zymosterin

Kautschuk

Triter pene Carotinoide ⟶ Ergosterin

c) Biotin-Coenzyme.

Das ursprünglich als pflanzlicher Wuchsstoff entdeckte Biotin kommt in konjugierter Form in Bakterien und Schimmelpilzen vor, z. B. als Biocytin = ε-N-Biotinyl-L-lysin. Obwohl die Beteiligung des Biotins in einer Coenzym-Form bei zahlreichen Enzymsystemen wahrscheinlich ist (Lichstein), konnte

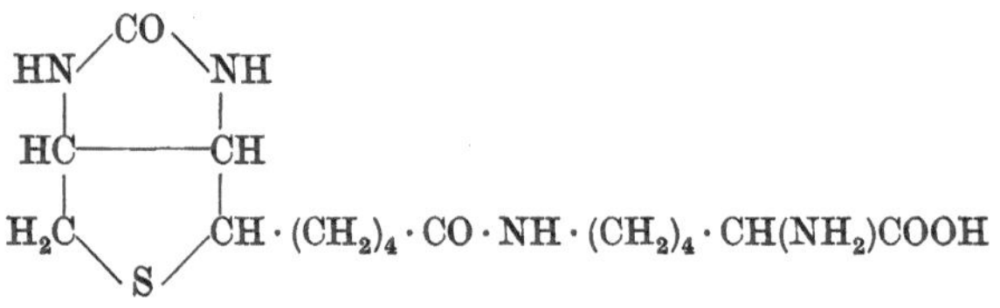

$$
\begin{array}{c}
\underset{HN}{} \overset{CO}{\diagdown} \underset{NH}{} \\
HC\text{——}CH \\
H_2C \diagdown \quad CH\cdot(CH_2)_4\cdot CO\cdot NH\cdot(CH_2)_4\cdot CH(NH_2)COOH \\
S
\end{array}
$$

nur ein Fall völlig sicher aufgeklärt werden. Es handelt sich um die *Oxalessigsäurecarboxylase*, die den Einbau des Bicarbonats in Pyruvat nach:

$$\text{Phosphopyruvat} + HCO_3^- + ADP \rightleftharpoons \text{Oxalacetat} + ATP$$

bewirkt. Dieses Enzym hat sich nach Lichstein als biotinhaltig erwiesen.

d) Ketonaldehydmutase.

Ein in der Hefe, aber auch in vielen anderen pflanzlichen Organismen vorkommendes Fermentsystem verwandelt nach Dakin bzw. Neuberg in Gegenwart von Glutathion 1,2-Ketonaldehyde in α-Oxycarbonsäuren (Nafziger). Das auch als *Glyoxalase* bezeichnete System besteht aus mehreren Komponenten, deren Natur unter anderem von Franzen aufgeklärt wurde: einer Glyoxalase I, die eine intramolekulare Oxydoreduktion, und einer Glyoxalase II, die eine Hydrolyse bewirkt. In Gegenwart von Ammoniak bilden sich α-Aminosäuren (Wieland, Franz und Pfleiderer).

Als Cosubstrat der *Ketonaldehydmutase* wurde das HS-Glutathion erkannt (Lohmann), das sich mit dem Ketonaldehyd zu einem Halbmercaptal vereinigt und nach Umwandlung in das S-Lactoylglutathion durch Hydrolyse wieder regeneriert wird.

Den Wirkungsmechanismus hat man sich (vgl. Franzen) wie folgt vorzustellen. Glyoxalase I ist ein Protein mit basischen Wirkgruppen.

Das Halbmercaptal aus Methylglyoxal und Glutathion GSH wird von Glyoxalase I, die als Anionenaustauscher wirkt, salzartig gebunden:

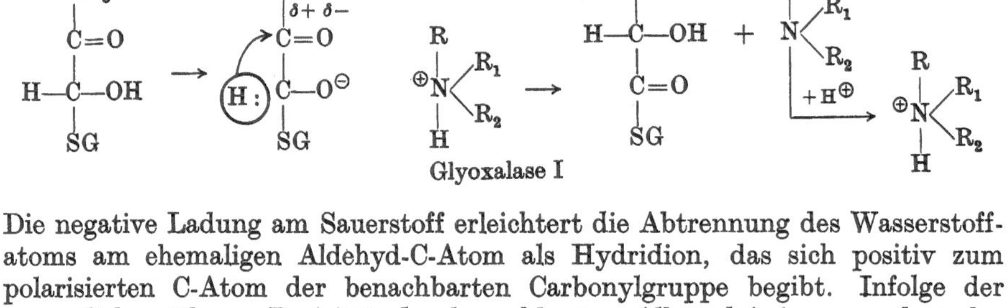

$$
\begin{array}{ccc}
CH_3 & CH_3 & CH_3 & R \\
| & |\,{\scriptstyle\delta+\ \delta-} & | & |{\diagdown}R_1 \\
C=O & C=O & H-C-OH\ + & N{-}R_2 \\
| & |\nearrow & | & R \\
H-C-OH & \to \ (H:)C-O^\ominus & C=O & |{\diagdown}R_1 \\
| & | & | & +H^\oplus\ \ ^\oplus N{-}R_2 \\
SG & SG & SG & | \\
& & & H
\end{array}
$$

Glyoxalase I

Die negative Ladung am Sauerstoff erleichtert die Abtrennung des Wasserstoffatoms am ehemaligen Aldehyd-C-Atom als Hydridion, das sich positiv zum polarisierten C-Atom der benachbarten Carbonylgruppe begibt. Infolge der wesentlich größeren Basizität des benachbarten Alkoxyl-Anions wandert das Proton von der basischen Gruppe im Ferment zum Alkoxyl-Anion, und das gebildete S-Lactoyl-glutathion löst sich vom Enzym ab. Nun greift die Thiolesterase Glyoxalase II (H-Gly II) ein, wobei Hydrolyse

$$
\begin{array}{c}
\text{CH}_3 \\
\text{H--C--OH} \\
\text{C=O} \\
\text{SG}
\end{array}
+ \text{H-Gly II} \longrightarrow
\begin{array}{c}
\text{CH}_3 \\
\text{H--C--OH} \\
\text{C} \overset{\text{OH}}{\underset{\text{Gly II}}{\big<}} \\
\text{SG}
\end{array}
\xrightarrow{+\text{H}_2\text{O}}
\begin{array}{c}
\text{CH}_3 \\
\text{H--C--OH} \\
\text{C--OH} \\
\text{O}
\end{array}
+ \text{HSG} + \text{H-Gly II}
$$

zu Milchsäure und Rückbildung von Glutathion und H-Gly II erfolgt.

Hefe, die durch plasmolytische Schädigung des Enzymsystems der alkoholischen Gärung zur Anhäufung von Methylglyoxal gebracht worden ist, bildet bei Anwesenheit von Glutathion praktisch reine D(—)Milchsäure, Hefemacerationssaft dagegen D,L-Milchsäure.

D. Weitere in den Schwefelstoffwechsel eingreifende Enzyme.

a) Glutathionreduktase und Cystinreduktase.

N. U. MELDRUM und H. L. A. TARR (1935) hatten in Hefe ein TPN-abhängiges Enzym nachgewiesen, das in Gegenwart von Hexosemonophosphat GSSG zu GSH reduziert. Diese *Glutathionreduktase* wurde später in Erbsensamen von L. W. MAPSON und D. R. GODDARD und in Weizenkeimlingen von E. E. CONN und B. VENNESLAND nachgewiesen. Das Enzym ist unwirksam gegenüber Cystin, auch vermag es nicht, DPNH auszunutzen.

Glutathionreduktase aus Weizenkeimlingen und Erbsen (CONN und VENNESLAND) bewirkt eine Oxydoreduktion zwischen Triphosphopyridinnucleotid und Glutathion:

$$\text{GSSG} + \text{TPNH} + \text{H}^+ \rightleftharpoons 2\,\text{GSH} + \text{TPN}.$$

Bei Gegenwart geeigneter H-Donatoren, die eine Affinität zu TPN-abhängigen Apofermenten zeigen, z. B. Kohlenhydraten nebst TPN-Cytochrom c-Reduktase aus Hefe, können große GSSG-Mengen zu GSH reduziert werden.

D-*Glycerinaldehydphosphat-Dehydrogenase.* Dieses DPN-abhängige Enzymprotein enthält nach RACKER und KRIMSKY (1953) Glutathion als festgebundene prosthetische Gruppe, die die Bindung des Pyridins vermittelt. Zwei von den 11 Thiolgruppen des Enzyms sind wenig reaktionsfähig (BENESCH).

ROMANO und NICKERSON wiesen nach, daß in Erbsensamen und in zwei Heferassen *(Candida albicans* und Bäckerhefe) auch eine Cystinreduktase vorkommt, die DPN-abhängig ist und bei p_H 6,2 nach:

$$
\begin{array}{c}
\text{COOH} \\
\text{H}_2\text{N--C--H} \\
\text{CH}_2 \\
\text{S}
\end{array}
\!\!-\!\!
\begin{array}{c}
\text{COOH} \\
\text{H}_2\text{N--C--H} \\
\text{CH}_2 \\
\text{S}
\end{array}
+ \text{DPNH} + \text{H}^+ \rightarrow 2\,
\begin{array}{c}
\text{COOH} \\
\text{H}_2\text{N--C--H} \\
\text{CH}_2 \\
\text{SH}
\end{array}
+ \text{DPN}^+.
$$

wirkt. L-Cystin wird spezifisch in Cystein überführt. Von L. W. MAPSON (1953) ist die DPNH-Spezifität des Erbsenenzyms bestätigt worden. Die Antibiotica Streptomycin, Penicillin und Fradicin sind ohne Einwirkung auf das Enzym.

b) Transmethylasen.

Die Übertragung von Methylgruppen mittels methylierter Oniumverbindungen erfolgt auf elementarem Schwefel durch viele Mikroorganismen (CHALLENGER).

In etiolierten Weizenkeimlingen findet sich eine Transmethylase, die Methylgruppen von Methionin auf Guanidinessigsäure überträgt:

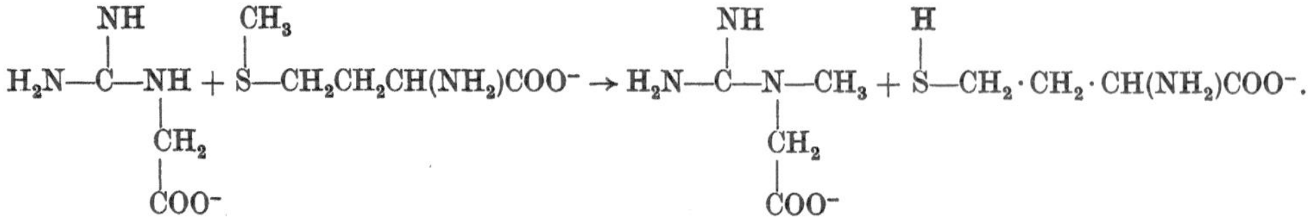

$$H_2N-\underset{\underset{\underset{COO^-}{|}}{\overset{|}{CH_2}}}{\overset{\overset{NH}{||}}{C}}-NH + \overset{\overset{CH_3}{|}}{S}-CH_2CH_2CH(NH_2)COO^- \rightarrow H_2N-\underset{\underset{\underset{COO^-}{|}}{\overset{|}{CH_2}}}{\overset{\overset{NH}{||}}{C}}-N-CH_3 + \overset{\overset{H}{|}}{S}-CH_2\cdot CH_2\cdot CH(NH_2)COO^-.$$

$$\qquad\qquad\qquad\qquad\qquad\qquad\qquad\qquad\qquad\qquad\qquad\qquad\qquad\text{Kreatin}\qquad\qquad\text{Homocystein}$$

Magnesiumionen aktivieren; Ca^{++}, Fluorid und Cyanid hemmen.

Hefeenzyme vermögen das Methyl-C des Methionins in das C (28) des Ergosterins zu verwandeln (ALEXANDER et al.).

Als Zwischenprodukt dieser Methylierungen muß ein „aktives Methionin" angenommen werden, dessen Synthese von BADDILEY und JAMIESON beschrieben wurde:

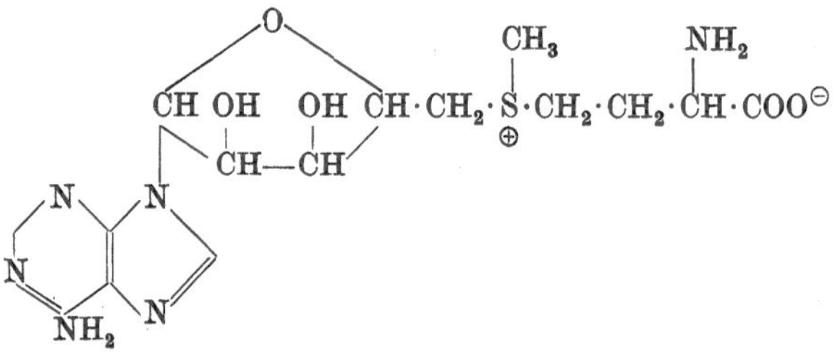

Vor allem bei Schimmelpilzen sind die Methylierungsreaktionen enzymatischer Natur eingehend untersucht worden (CHALLENGER). Nach SHAPIRO bilden *Aerobacter aerogenes* und *Torulopsis utilis* aus Homocystein und Methylmethionin, das im Pflanzenreich weit verbreitet ist, Methionin, welches zum Wachstum benötigt wird.

Es gibt aber in der Pflanze auch noch andere Methylierungsmechanismen. So bildet beispielsweise *Escherichia coli* Methionin aus Homocystein und Serin, wobei nebenher Glycin entsteht, während die CH_3-Gruppen des Lignins und Nicotins zwar hauptsächlich dem Methionin, aber zum kleinen Teil auch dem Formiat entstammen. Hier sind Folsäureenzyme beteiligt.

c) Sulfitoreduktase und Tetrathionase.

Die Reduktion von anorganischem Sulfit zu H_2S durch ein Endoenzym, die *Sulfitoreduktase* (PRÉVOT, SAISSAC und CALLAME), von *Clostridiales* erfordert keine spezifischen H-Donatoren, wie etwa die Reduktion von Nitraten zu Nitriten. Der notwendige Wasserstoff bzw. die Elektronen werden von vergärbaren Zuckern des Nährbodens, den Polypeptiden und Aminosäuren geliefert.

Die *Tetrathionase* Enz (KNOX und POLLOCK, ALLEN und VAN NIEL, VAN NIEL) kann als Elektronenüberträger in reinen Kulturen von *Pseudomonas* zusammen mit dem Redoxsystem Thiosulfat/Tetrathionat große Mengen organischer Substanz, z. B. Glycerin H_2A, oxydieren:

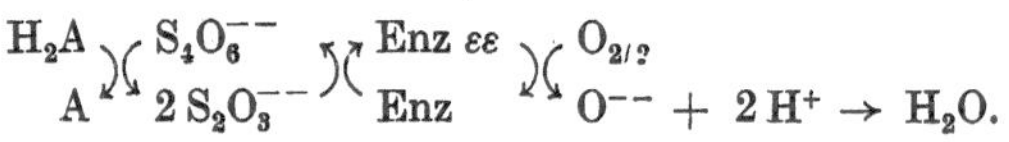

$$\begin{array}{ccccc} H_2A & \searrow & S_4O_6^{--} & \searrow & \text{Enz } \varepsilon\varepsilon & \searrow & O_{2/2} \\ A & \nearrow & 2\,S_2O_3^{--} & \nearrow & \text{Enz} & \nearrow & O^{--} + 2\,H^+ \rightarrow H_2O. \end{array}$$

Verschiedene Mikroorganismen können mit Hilfe des adaptativen Enzyms *Tetrathionase* die Einstellung des Gleichgewichtes Tetrathionat-Thiosulfat beschleunigen.

d) Sulfitoxydase.

Es war bekannt, daß verschiedene pflanzliche Organismen Sulfit in niedriger Konzentration zu Sulfat oxydieren (BERSIN 1950). Auch dürfte Sulfit, z. B. als Reaktionsprodukt der Wirkung von Rhodanese, im intermediären Stoffwechsel mancher Organismen gebildet werden (POSTGATE). Der Nachweis der Sulfitoxydase in *Escherichia coli* und *Aerobacter aerogenes* gelang HEIMBERG und Mitarbeitern. Als Coenzym der Sulfitoxydase wurde Hypoxanthin erkannt.

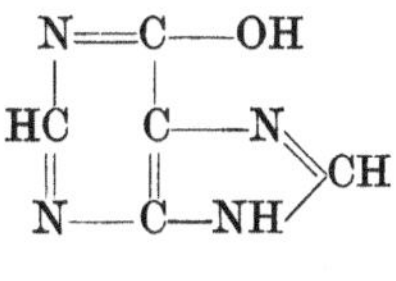

$$
\begin{array}{c}
\text{N}=\!=\!\text{C}-\text{OH} \\
\mid \qquad \mid \\
\text{HC} \qquad \text{C}-\text{N} \\
\parallel \qquad \parallel \qquad \diagdown\!\text{CH} \\
\text{N}-\text{C}-\text{NH}\diagup
\end{array}
$$

e) Rhodanese (Transsulfurase).

In *Escherichia*, *Cruciferen* und *Umbelliferen* kommt ein Enzym vor, das, ursprünglich von LANG in tierischem Material entdeckt, auch im Schwefelstoffwechsel der Pflanzen eine gewisse Rolle zu spielen scheint. Es handelt sich um ein kristallisierbares Protein, das nach SÖRBO die Übertragung des Schwefels vom Thiosulfat auf Cyanid nach:

$$
\text{S}_2\text{O}_3\text{—Enz—SS—} + \text{HCN} \rightarrow \begin{array}{c} \diagup\text{SH} \\ \text{—S—Enz—SCN} + \text{SO}_3^{2-} \\ \downarrow \\ \text{Enz—SS—} + \text{HSCN} \end{array}
$$

bewirken soll. Das Enzym scheint außer Disulfid auch Thiolgruppen zu besitzen (SAUNDERS und HIMWICH). Beide Substrate, Cyanid und Thiosulfat, sind im pflanzlichen Stoffwechsel nicht unbekannt. Auch kolloidaler Schwefel und β-Mercaptopyruvat (WOOD und FIEDLER) können an Stelle von Thiosulfat als Substrat dienen.

In gewaschenen Kulturen von *Escherichia coli*, die anaerob 5 Std bei 37⁰ vorbehandelt waren, zeigte sich eine auffallende rhythmische Periodizität von $1^1/_2$ Std bezüglich sowohl der Sauerstoffaufnahme als auch der Rhodaneseaktivität (STEARNS).

f) Kohlensäureanhydratase.

Die bisher nur im Tierreich aufgefundene Kohlensäureanhydratase (= Carbonsäureanhydrase) wurde von DAY und FRANKLIN in *Sambucus canadensis*, später von BRADFIELD sowie LUCAS und BYERRUM in zahlreichen anderen Pflanzen und von SIBLY und WOOD in Blättern des Spinats, des Hafers und von Tomaten (WOOD und SIBLY) nachgewiesen. Auch hier handelt es sich um ein Zinkproteid, das jedoch im Gegensatz zum tierischen Enzym empfindlich gegenüber thiopriven Reagentien ist und von Thiolen aktiviert wird (CHIBA, KAWAI und KONDO). Polarographisch konnten HS-Gruppen im angereicherten Enzympräparat nachgewiesen werden. Die Aktivität in dem chloroplastenfreien Saft der Blätter ist abhängig von der exogenen Zn-Zufuhr, nimmt im Laufe des Lebenscyclus der Pflanzen zu bis zu einem Maximum, um dann wieder abzufallen (WOOD und SIBLY, TOMBESI und Mitarbeiter). Die Aktivität des Enzyms in Blättern des Tabaks wurde beim Trocknungsprozeß in Abhängigkeit von der vorherigen Düngung von TOMBESI und GIOVANNOZZI untersucht. Die physiologische Bedeutung des Enzyms liegt in der Beschleunigung der Reaktion $\text{H}_2\text{CO}_3 \rightleftharpoons \text{H}_2\text{O} + \text{CO}_2$.

Literatur.

ALEXANDER, G. J., A. M. GOLD and E. SCHWENK: The methyl group of methionine as a source of C_{28} in ergosterol. J. Amer. Chem. Soc. **79**, 2967 (1957). — ALLEN, M. B., and C. B. VAN NIEL: Experiments on bacterial denitrification. J. Bacter. **64**, 397—412 (1952). BACHHAWAT, B. K., and M. J. COON: The role of adenosine triphosphate in the enzymatic activation of carbon dioxide. J. Amer. Chem. Soc. **79**, 1505—1506 (1957). — BADDILEY, J., and G. A. JAMIESON: Synthesis of "active methionine". Chem. a. Ind. **1954**, 375. — BARRON, E. S. G.: Thiol groups of biological importance. Adv. Enzymol. **11**, 201—266 (1950). — BEINERT, H., R. M. BOCK, D. S. GOLDMAN, D. E. GREEN, H. R. MAHLER, S. MII, P. C. STANSLY and S. J. WAKIL: The reconstruction of the fatty acid oxidizing system of animal tissues. J. Amer. Chem. Soc. **75**, 4111—4112 (1953). — BENESCH, R. E., H. A. LARDY and R. BENESCH: The sulfhydryl groups of crystalline proteins. I. Some albumins, enzymes, and hemoglobins. J. of Biol. Chem. **216**, 663—676 (1955). — BERSIN, TH.: Über die Einwirkung von Oxydations- und Reduktionsmitteln auf Papain. II. Z. physiol. Chem. **222**, 177—186 (1933). — Über die Thiolnatur des Papains. Biochem. Z. **278**, 340—341 (1935). — Amidasen und Proteasen. In NORD-WEIDENHAGEN, Handbuch der Enzymologie, S. 574—632. Leipzig 1940. Übersichtsreferat. — Reduktion nach MEERWEIN-PONNDORF und Oxydation nach OPPENHAUER. Angew. Chem. **53**, 266—271 1940). — Newer Methods of Preparative Organic Chemistry. Interscience, New York 1948. Übersichtsreferat. — Die Phytochemie des Schwefels. Adv. Enzymol. **10**, 223—323 (1950). — Übersichtsreferat. — Kurzes Lehrbuch der Enzymologie, 4. Aufl., S. 284. Leipzig: Akademische Verlagsgesellschaft 1954. — Exchange adsorption in man. In C. CALMON and T. R. E. KRESSMAN, Ion Exchangers in Organic and Biochemistry. New York 1957. Übersichtsreferat. — BERSIN, TH., u. H. KÖSTER: (1) Über die Einwirkung von Oxydations- und Reduktionsmitteln auf Papain. III. Z. physiol. Chem. **233**, 59—66 (1935). — (2) Die Einwirkung von Aktivatoren, Hemmungskörpern und Destruktoren auf Urease. Z. Naturwiss. **1935**, 230—242. — BERSIN, TH., u. W. LOGEMANN: (1) Über die Einwirkung von Oxydations- und Reduktionsmitteln auf Papain. I. Z. physiol. Chem. **220**, 209—216 (1933). — (2) Selen- und Tellurmercaptide. Liebigs Ann. **505**, 3—17 (1933). — BERSIN, TH., A. MÜLLER u. E. STREHLER: Methylmethioninsulfoniumsalze. Arzneimittel-Forsch. **6**, 174—176 (1956). — BERSIN, TH., u. J. STEUDEL: Polarimetrische Untersuchungen über das Thiol-Disulfid-System. Ber. dtsch. chem. Ges. **71**, 1015—1024 (1938). — BINKLEY, F.: Catalytic cleavage of thioethers. J. Amer. Chem. Soc. **77**, 501 (1955). — BRADFIELD, J. R. G.: Plant carbonic anhydrase. Nature (Lond.) **159**, 467—468 (1947).

CALVIN, M.: Der Photosynthese-Cyclus. Angew. Chem. **68**, 253—264 (1956). — CHALLENGER, F.: The biological importance of organic compounds of sulphur. Endeavour **12**, 1—9 (1953). — Biological methylation. Quart. Rev. Chem. Soc. Lond. **9**, 255—286 (1955). — CHIBA, H., F. KAWAI and K. KONDO: Plant carbonic anhydrase. V. Properties as thiol enzyme. Bull. Res. Inst. Food Sci., Kyoto Univ. Nr 13, S. 12—22, 1954; durch C. A. **48**, 9421—9422 (1954). — CONN, E. E., and B. VENNESLAND: Glutathione reductase of wheat germs. J. of Biol. Chem. **192**, 17—28 (1951). — CONRAD, J. P.: Hydrolysis of urea in the soil by thermolabile catalysis. Soil Sci. **49**, 253—255 (1940). — Nature of the catalyst effecting the hydrolysis of urea in soils. Soil Sci. **50**, 119—122 (1940). — CONTÉ, M.: The plasticization of rubber in the form of latex. Rev. gén. caoutchouc **30**, 262—264 (1953); durch C. A. **47**, 6167 (1953). — CRETIN, C. J.: Oxidation-reduction system of *Hevea brasiliensis* latex. Rubber Res. Inst. Malaya **13**. Comm. 276, 184—191 (1951); durch C. A. **46**, 9333 (1952).

DAKIN: Zit. bei NAFZIGER. — DAY, R., and J. FRANKLIN: Plant carbonic anhydrase. Science (Lancaster, Pa.) **104**, 363—365 (1946).

ELDJARN, L., and A. PIHL: On the mode of action of X-ray protective agents. II. Interaction between biologically important thiols and disulfids. J. of Biol. Chem. **225**, 499—510 (1957). — ELLIOTT, S. D.: The crystallization and serological differentiation of a streptococcal proteinase and its precursor. J. of Exper. Med. **92**, 201—218 (1950). — EVANS, H. J., and A. NASON: Pyridine nucleotide-nitrate reductase from extracts of higher plants. Plant Physiol. **28**, 233—254 (1953).

FEIGL, F., u. V. GENTIL: A new spot test for urease. Biol. Jaarbook Konink. Natuurw. Genoot. Dodonaea Gent **20**, 47—49 (1953); durch C. A. **48**, 8305 (1954). — FINKLE, B. J., and E. L. SMITH: Sulfhydryl groups of crystalline papain. Federat. Proc. **16**, 180 (1957). — FRANZEN, V.: Wirkungsmechanismus der Glyoxalase. I. Chem. Ber. **89**, 1020—1023 (1956). — FRENSDORFF, H. K., M. T. WATSON and W. KAUZMANN: The kinetics of protein denaturation. IV. J. Amer. Chem. Soc. **75**, 5157—5166 (1953). — The kinetics of protein denaturation. V. J. Amer. Chem. Soc. **75**, 5167—5172 (1953). — FRUTON, J. S.: The rôle of proteolytic enzymes in the biosynthesis of peptide bonds. Yale J. Biol. Med. **22**, 263—271 (1950). — FUJITA, A.: Thiaminase. Adv. Enzymol. **15**, 389—421 (1954).

HEIMBERG, M., J. FRIDOVICH and PH. HANDLER: The enzymatic oxidation of sulfite. J. of Biol. Chem. **204**, 913—926 (1953). — HINKEL, F. T., and C. ZIPPIN: Correlation of the results obtained by beef-digestion, gelatine-digestion, and milk-clotting methods of measuring the proteolytic activity of papain. Ann. New York Acad. Sci. **54**, 228—235 (1951).— HOFFMANN-OSTENHOF, O.: Enzymologie, S. 328. Wien: Springer 1954. — HWANG, K., and A. C. IVY: A review of the literature on the potential therapeutic significance of papain. Ann. New York Acad. Sci. **54**, 161—207 (1951).

JAFFÉ, W. G.: The activation of papain and related plant enzymes with sodium thiosulfate. Arch. of Biochem. **8**, 385—393 (1945). — JONES, M. E., W. R. HEARN, M. FRIED and J. S. FRUTON: Transamidation reactions catalyzed by cathepsin. C. J. of Biol. Chem. **195**, 645—665 (1952). — JOYEUX, Y., et M. CROSON: Lactic dehydrogenase from a strain of Bacterium anitralum. C. r. Acad. Sci. Paris **239**, 1439—1440 (1954); durch C. A. **49**, 4786 (1955).

KIMMEL, J. R., and E. L. SMITH: Crystalline papain. J. of Biol. Chem. **207**, 515—574 (1954). — KLOTZ, I. M., J. M. URQUHARDT, T. A. KLOTZ and J. AYERS: Slow intramolecular changes in copper complexes of serum albumin. J. Amer. Chem. Soc. **77**, 1919—1925 (1955). — KNOX, R., and M. R. POLLOCK: Bacterial tetrathionase: Adaptation without demonstrable cell growth. Biochemic. J. **38**, 299—304 (1944). — KRIMSKY, I., and E. RACKER: Glutathione, a prosthetic group of glyceraldehyde-3-phosphate dehydrogenase. J. of Biol. Chem. **198**, 721—729 (1952).

LAIDLER, K. J., and J. P. HOARE: The molecular kinetics of the urea-urease system. II. Heats and entropies of complex formation and reaction. J. Amer. Chem. Soc. **72**, 2489 bis 2494 (1950). — LANG, K.: Das Enzym Rhodanese. Z. Vitamin-, Hormon- u. Fermentforsch. **2**, 288—291 (1949). — LAMPSON, G. P., and H. L. KLUG: Preparation of selenium derivatives of sulfhydryl compounds. Proc. South Dakota Acad. Sci. **27**, 47—49 (1948). — LARSON, A. D., and R. E. KALLIO: Purification and properties of bacterial urease. J. Bacter. **68**, 67—73 (1954). — LEPETIT, F.: Natural, thermal, and biochemical degradation of ammoniated *Hevea* latex to give heat sensitivity. Trans. Inst. Rubber Ind. **23**, 104—117 (1947); durch C. A. **42**, 2796—2797 (1948). — LICHSTEIN, H. C.: Functions of biotin in enzyme systems. Vitamins a. Hormones **9**, 27—74 (1951). — LINDLEY, H.: The mechanism of action of hydrolytic enzymes. Adv. Enzymol. **15**, 271—299 (1954). — LIPMANN, F.: Development of the acetylation problem, a personal account. Science (Lancaster, Pa.) **120**, 855—865 (1954). — LOGEMANN, W.: Untersuchungen über die Autoxydation von Mercaptanen. Zugleich ein Beitrag zur chemischen Natur des Papains. Inaug.-Diss. Marburg a. d. Lahn 1935. — LOHMANN, K.: Beitrag zur enzymatischen Umwandlung von synthetischem Methylglyoxal in Milchsäure. Biochem. Z. **254**, 332—341 (1932). — LUCAS, E. H., and R. U. BYERRUM: Plant carbonic anhydrase. I. Occurence, distribution and properties. Papers Mich. Acad. Sci. **37**, 55—61 (1951); durch C. A. **47**, 9433 (1953). — LYNEN, F.: Quantitative Bestimmung von Acyl-mercaptanen mittels der Nitroprussid-Reaktion. Liebigs Ann. **574**, 33—37 (1951). — Der Fettsäurecyclus. Angew. Chem. **67**, 463—470 (1955). — LYNEN, F., u. K. DECKER: Das Coenzym A und seine biologischen Funktionen. Erg. Physiol., biol. Chem. exper. Pharmakol. **49**, 327—424 (1957).

MAPSON, L. W.: Estimation of oxidized glutathione. Biochemic. J. **55**, 714—717 (1953). — MAPSON, L. W., and D. R. GODDARD: Reduction of glutathione by plant tissues. Biochemic. J. **49**, 592—601 (1951). — MASCHMANN, E.: Über Bakterienproteasen. I. Biochem. Z. **294**, 1—33 (1937). II. Biochem. Z. **295**, 1—10 (1937). — MATTENHEIMER, H., u. Hs. NITSCHMANN: Das Lab und seine Wirkung auf das Casein der Milch. VIII. Die Abspaltung von Nicht-Protein-Stickstoff (NPN) aus Casein durch verschiedene proteolytische Fermente, verglichen mit der Abspaltung durch Lab. Helvet. chim. Acta **38**, 687—698 (1955). — McDONALD, C., and A. K. BALLS: Esterase action of papain. Federat. Proc. **13**, 262 (1954).— MELDRUM, N. U., and H. L. S. TARR: Reduction of glutathione by the Warburg-Christian system. Biochemic. J. **29**, 108—112 (1935). — METZLER, D. E., M. IKAWA and E. E. SNELL: A general mechanism for vitamin B_6-catalyzed reactions. J. Amer. Chem. Soc. **76**, 648—652 (1954). — MILLERD, A., and J. BONNER: Acetate activation and acetoacetate formation in plant systems. Arch. of Biochem. a. Biophysics **49**, 343—355 (1954).

NAFZIGER, H.: Experimentelle Studien zur Kenntnis der Glyoxalase. Inaug.-Diss. Marburg a. d. Lahn 1937. — NEUBERG: Zit. bei NAFZIGER. — NIEL, C. B. VAN: Introductory remarks on the comparative biochemistry of microorganismes. J. Cellul. a. Comp. Physiol. **41**, Suppl. 1 (1953).

OCHOA, S.: Enzymic mechanisms in the citric acid cycle. Adv. Enzymol. **15**, 183—270 (1954).

POSTGATE, J. R.: Reduction of sulfur compounds by *Desulfovibrio desulfuricans*. J. Gen. Microbiol. **5**, 725—738 (1951). — PRÉVOT, A. R., R. SAISSAC et B. CALLAME: Études sur l'origine de l'hydrogène pour la réduction des sulphites par les anaerobies. Ann. Inst. Pasteur **79**, 93—94 (1950).

RACKER, E.: Glutathione as a coenzyme in intermediary metabolism. Glutathione Proc. Symposium, Ridgefield, Conn. **1953**, 165—183; durch C. A. **49**, 5552 (1955). — RACKER, E., and I. KRIMSKY: The mechanism of oxidation of aldehydes by glyceraldehyde-3-phosphate dehydrogenase. J. of Biol. Chem. **198**, 731—743 (1952). — REED, L. J., and B. G. DE BUSK: Lipothiamide pyrophosphate: coenzyme for oxydative decarboxylation of α-keto acids. J. Amer. Chem. Soc. **74**, 3964—3965 (1952). — Lipoic acid conjugase. J. Amer. Chem. Soc. **74**, 4727—4728 (1952). — Mechanism of enzymatic oxidative decarboxylation of pyruvate. J. Amer. Chem. Soc. **75**, 1261—1262 (1953). — ROMANO, A. H., and W. J. NICKERSON: Cystine reductase of pea seeds and yeasts. J. of Biol. Chem. **208**, 409—416 (1954).

SAUNDERS, J. P., and W. A. HIMWICH: Properties of the transsulfurase responsible for conversion of cyanid to thiocyanate. Amer. J. Physiol. **163**, 404—409 (1950). — SCHÖBERL, A., u. A. WAGNER: Methoden zur Herstellung und Umwandlung von Thiosäuren und ihren Derivaten. In HOUBEN-WEYL, Methoden der organischen Chemie, 4. Aufl., Bd. IX, S. 745—771. Stuttgart: Georg Thieme 1955. — SCHULLER, W. H., and C. NIEMANN: The papain-catalyzed synthesis of acyl-D- and L-phenylalanylphenylhydrazides from a series of enantiomorphic pairs of acylated phenylalanines. J. Amer. Chem. Soc. **73**, 1644—1646 (1951). — SHAPIRO, S. K.: Biosynthesis of methionine from homocysteine and methylmethionine sulfonium salt. Biochim. et Biophysica Acta **18**, 134—135 (1955). — SHAW, W. H. R.: The inhibition of urease by various metal ions. J. Amer. Chem. Soc. **76**, 2160—2163 (1954). — SHER, I. H., and M. F. MALLETTE: Purification and study of L-lysine decarboxylase from *Escherichia coli B*. Arch. of Biochem. a. Biophysics **53**, 354—369 (1954). — SIBLY, P. M., and J. G. WOOD: The nature of carbonic anhydrase from plant sources. Austral. J. Sci. Res. B **4**, 500—510 (1951); durch C. A. **46**, 3101 (1952). — SLAVIK, K.: Die enzymatische Bildung von Hydroxamsäuren aus Amiden und Peptiden. Collect. czechoslov. chem. Commun. **16**, 380—390 (1951); durch C. A. **1955**, 1020—1021. — SMITH, G. N., C. S. WORREL and B. L. LILLIGREN: Enzymic hydrolysis of chloramphenicol (chloromycetin). Science (Lancaster, Pa.) **110**, 297—298 (1949). — SÖRBO, B. H.: The active group in rhodanese. Acta chem. scand. **5**, 1218—1219 (1953). — STEARNS, R. N.: Respiration, rhodanese, and growth in *Escherichia coli*. J. Cellul. a. Comp. Physiol. **41**, 163—170 (1953). — SUNNER, ST.: Thioctsäure. Chemie und Energetik. Svensk. kem. Tidskr. **67**, 513—522 (1955); durch Chem. Zbl. **1957**, 6990. — *Symposium on Chemistry and Functions of Coenzym A*. Federat. Proc. **12**, 673—715 (1953).

TOMBESI, L., A. BAROCCIO, T. CERVIGNI, S. FORTINI, M. TARANTOLA e M. E. VENEZIAN: Oxidase, catalase, carbonic anhydrase and peroxidase activity and content of reduced glutathione and of ascorbic acid during the maturation of fruits and seeds. Ann. sper. agr. (Roma) **6**, 857—874 (1952); durch C. A. **47**, 2835 (1953). — TOMBESI, L., and M. GIOVANNOZZI: Enzyme activity during the cure by indirect fire. Tobacco **56**, 53—61 (1952); durch C. A. **47**, 1791 (1953). — TYTELL, A. A., and K. HEWSON: Production, purification, and some properties of *Clostridium hystolyticum* collagenase. Proc. Soc. Exper. Biol. a. Med. **74**, 555—558 (1950).

WALLENFELS, K., u. H. SUND: Zum Wirkungsmechanismus der Alkoholdehydrogenase aus Hefe. Angew. Chem. **67**, 517 (1955). — Über den Mechanismus der Wasserstoffübertragung mit Pyridinnucleotiden. I. Freie SH-Gruppen und Aktivität bei Alkoholdehydrogenase aus Hefe. Biochem. Z. **329**, 17—30 (1957). — WEINER, S.: Papain. A review of literature. Paul-Lewis Labs. Inc. 1955; durch C. A. **49**, 7190 (1955). — WIELAND, TH., J. FRANZ u. G. PFLEIDERER: Über die Bildung von Aminosäuren aus α-Keto-aldehyden. Chem. Ber. **88**, 641—646 (1955). — WIELAND, TH., u. H. SCHWAHN: Zur Struktur und Reaktionsweise organischer Disulfide. Chem. Ber. **89**, 421—428 (1956). — WLASSJUK, P. A., K. M. DOBROTWORSKAJA u. SS. A. GORDIJENKO: Aktivität der Urease in der Rhizosphäre von landwirtschaftlichen Kulturen. Ber. Allunions landwirtsch. (Lenin-Orden) Lenin-Akad. **21**, 28—31 (1956); durch Chem. Zbl. **1957**, 6182. — WOOD, J. G., and P. M. SIBLY: Carbonic anhydrase activity in plants in relation to zinc content. Austral. J. Sci. Res. B **5**, 244—255 (1952); durch C. A. **46**, 7182 (1952). — WOOD, J. L., and H. FIEDLER: β-Mercaptopyruvate a substrate for rhodanese. J. of Biol. Chem. **205**, 231—234 (1953). — WOOLLEY, D. W.: Biosynthesis and energy transport by enzymic reduction of onium salts. Nature (Lond.) **171**, 323—328 (1953).

ZELITCH, I.: The isolation and action of crystalline glyoxylic acid reductase from tobacco leaves. J. of Biol. Chem. **216**, 553—575 (1955).

Assimilation of sulfur
and physiology of essential S-compounds.

By

Moyer D. Thomas.

With 9 figures.

Sulfur is an essential element in plants. Plants actually require, on the average, nearly the same amount of sulfur as of phosphorus, amounting to 3–20 lbs per ton of crop. Since sulfur tends to leach from the soil more readily than phosphorus, even larger amounts may be needed. As much as 40 lbs of sulfur per acre per year has been measured in percolation water. The importance of sulfur as a nutrient has often been underestimated. ALWAY (1940) referred to sulfur as a nutrient element slighted in agricultural research. More recently GILBERT (1951) reviewed the place of sulfur in plant nutrition. He discussed the responses of plants to sulfur fertilization, particularly in the United States. In colonial times, widespread deficiency developed on the Atlantic seaboard which could be corrected by gypsum. Later, with the industrial development in the East and Middle West, this deficiency disappeared, partly because of the fact that industry wasted into the air large amounts of sulfur compounds which were subsequently brought back to earth in rain and snow, or as sulfur dioxide, partly also because phosphate fertilizers inadvertently supplied a great deal of sulfate. Sulfur in precipitation ranges from less than 5 lbs to more than 250 lbs per acre per year. Many rural areas receive 15–30 lbs. Unless corrected by fertilization, sulfur deficiency is now present in many soils of Florida, California, the Pacific Northwest, northern Minnesota, Saskatchewan, Manitoba and Ontario, among other areas where meteoric sulfur values are low.

In extreme sulfur deficiency, dramatic increase in growth rate follows sulfur fertilization, for example, with alfalfa in Oregon (POWERS *et al.* 1923). The other extreme involving the uptake of too much sulfate so as to cause premature senescence of the leaves has also been observed (THOMAS *et al.* 1950). This can occur in plants growing in gypsum soils, or in an industrial atmosphere with a very low level of sulfur dioxide. Both sulfate and chloride are taken up in excess from alkali soils.

Sulfur compounds in plants. The essentiality of sulfur in plants is due primarily to the fact that it is a constituent of the amino acids, cystine and methionine, the peptide glutathione and related compounds, including coenzyme A, and the newly recognized photosynthetic enzyme lipoic acid—6-8 thioctic acid (CALVIN 1953). Cytochrome c is an iron enzyme with thioether linkages between the hematin and protein. The vitamins thiamin and biotin contain sulfur. The former is a thiazole derivative. Th e latter has a reduced thiophene ring. Cocarb - oxylase is the pyrophosphate of thiamin. Other sulfur-containing organic compounds are present in different plants but their roles in the plant are obscure and their essentiality has not been established. The glucosides sinigrin and sinalbin and their hydrolytic products, the mustard oils, vinyl and allyl isothiocyanate, together with methyl, allyl, and vinyl sulfides, disulfides and mercaptans, are important constituents of a number of plant families such as the mustards and the onions, but are unimportant or absent in other plants. Dithiol isobutyric acid has been found in asparagus.

Sulfur is normally supplied to the plant by uptake of sulfate through the roots. In the leaves, part of the sulfate is reduced and built into proteins. Absorption and reduction in measurable amount require only a few minutes, as indicated by studies using radiosulfur. Sulfate in excess of that required for organic sulfur compounds remains unchanged in the plant. Sulfur can also be supplied to the plant as sulfur dioxide through the leaves if the gas is added in sublethal concentrations. However, sulfur dioxide is not as effective as sulfate for nutritional purposes probably because it tends to be fixed in the leaves and is not so free to move to other parts of the plant. Plants growing in a sulfur-deficient medium but supplied with low concentrations of sulfur dioxide in the air grow much better than comparable plants without the sulfur dioxide but not so well as plants supplied with adequate sulfate. Cystine (BARRIEN and WOOD 1939) and methionine (MILLER 1947) have been experimented with as sulfur nutrients. They behave like sulfate.

TURRELL and WEBER (1955), noting that sulfur-dusted lemon trees gave increased yields of first grade fruit, conducted dusting experiments with radio-sulfur, which indicated that part of the elemental sulfur entered the leaves and was incorporated into leaf proteins. It was also translocated to some extent to leaves on adjacent stems. Evidently it can serve as a nutrient for the leaves.

Sulfur reactions in normal plants. It seems clear that part of the economy of sulfur in plants is due to the oxidative and reductive properties of the labile sulfide linkage, $R—S—S—R \rightleftharpoons R—S—H$, exemplified by the cystine-cysteine reaction, and also by glutathione. Another important part of its economy is due to its close association with nitrogen in the proteins. Protein synthesis may be limited by the amount of sulfur available if the latter is small, even when adequate nitrogen is present. This is explained in part by the fact that cystine and methionine are constituents of essential proteins, which cannot be produced in adequate amounts if these two amino acids are in short supply.

The mechanism of the reduction of sulfate to sulfide in plants is poorly understood. The reaction requires 118.5 kcal at 25° C. There is evidence that it is not photochemical. In bacteria the reaction goes anaerobically with the energy from respiration of carbohydrates. In higher plants, reduction of the sulfate by combination with energy-rich substances analogous to the phosphate esters has been suggested, but this mechanism remains to be studied.

WOOD and BARRIEN (1939) postulate a sequence of more or less reversible reactions starting with ammonia nitrogen and sulfate sulfur together with the glycolytic products of respiration and ending with synthesized proteins. Intermediate reversible reactions involve: (1) reduction of sulfate to sulfide and reoxidation to sulfate; (2a) amination and amidation of the glycolytic compounds, also (2b) combination of sulfide with these products to form cystine, etc., and a reversal of these reactions; and finally (3) protein synthesis and proteolysis.

The reversible nature of these reactions was illustrated by a series of experiments with the grasses *Phalaris tuberosa* and *Lolium multiflorum* in sand culture with a sulfur supply sufficient to avoid deficiency symptoms. Additional sulfate raised the sulfate level in the plants, but a significant increase in the cystine or protein sulfur or the nitrogen fractions could not be demonstrated. Addition of cystine to the nutrient solution likewise raised the sulfate level in the leaves but had no other effect. However, addition of ammonium chloride to the nutrient solution caused a significant increase in protein and amino nitrogen, and also cystine, protein, and soluble organic sulfur. Sulfate sulfur decreased correspondingly, suggesting balanced reactions between these constituents. It was also observed that these reactions could be reversed by placing the plants in the dark. Within a few days, sulfate increased, whereas soluble organic and protein sulfur decreased correspondingly. Changes of about 30 percent occurred in 15 days.

In an experiment with Sudan grass grown in sand culture having adequate sulfur in the nutrient solution to prevent deficiency symptoms, and enriched with sodium nitrate, BARRIEN and WOOD (1939) noted that the total weight of protein sulfur and protein nitrogen in the leaves increased to a maximum in about two months, then fell off, the nitrogen more rapidly than the sulfur. The curves followed the trend of the dry matter, except that the latter remained constant after reaching the maximum. Increasing the nitrate supply 3-fold, with the same initial amount of sulfate, nearly doubled the weight of dry matter and tripled the weight of the protein sulfur and protein nitrogen. When the nitrate supply was increased 6-fold, the factors for increase of dry matter, protein sulfur, and protein nitrogen in the leaves were 2.2, 4, and 4 respectively. Evidently nitrate, like ammonium salts, stimulates protein synthesis with a fixed supply of sulfate. This suggests that the nitrate caused a shift of the apparent equilibrium and an increasing percentage of the total available sulfur was incorporated into the proteins.

HEISERICH (1935) observed that maize and tobacco gave 50–75 percent more growth and much higher sulfate concentrations in the plants (4- and 8-fold respectively) but less protein-S when the nitrogen was supplied as nitrate as compared with ammonium salt. He suggested that since both sulfate and nitrate metabolisms require initial reductive processes involving considerable energy, it is necessary to take up an excess of sulfate when nitrate is used, so as to shift the energy balance between the two and permit adequate reduction of sulfate. This is not necessary when ammonium salt is used because there is no competition between sulfate and ammonium for the available energy of the leaves. While this postulate is attractive, there may well be an entirely different explanation. The postulate is hardly in accord with the previously mentioned Sudan grass experiment in which additional nitrate was attended by the utilization of an increasing percentage of a fixed sulfate supply.

Symptoms of sulfur deficiency. A study of sulfur deficient plants is useful for elucidating the role of sulfur in the nutritional processes. Sulfur deficient (—S) plants are characterized by small yellow or yellowish-green leaves; by stems of small diameter, with or without reduced length; and by extensive small roots. In the tomato (NIGHTINGALE et al. 1932), deficiency symptoms developed slowly and the plants looked as if they had been gradually but not completely deprived of nitrogen. The lower leaves became chlorotic first and later developed a yellowish-brown color. The plants were not shortened in length but the stems were of small diameter, hard and woody, with no active cambium except at the tip. The walls of the internal and external fibers were extremely thick, as were also the walls of the xylem and collenchyma tissue. The roots likewise were of small diameter, lacking cambium or secondary thickening but without deformity or injury.

In black mustard (EATON 1942), the reduced growth of the leaves of the (—S) plants was evident in two weeks after transferring the seedlings to the sulfur-deficient solution, but chlorosis did not appear for five weeks. The upper leaves became yellow first after being greener than the (+S) leaves. Later anthocyanin developed in the leaves and stems, together with hairiness, a pimply character on the younger leaves, and a stiff texture on all the leaves. The stems became thin, hard and woody, with low water content. Flowering was hastened by sulfur deficiency.

Different plants developed (—S) symptoms at different rates. Sunflower developed small chlorotic leaves early. Mustard formed small leaves early but the chlorosis was delayed. Both symptoms were delayed in soybean. Usually the entire leaf was uniformly chlorotic, but in teabush (STOREY and LEACH 1933) the leaves were mottled. The upper leaves showed the chlorosis first in soybean, mustard, tobacco and teabush. The lower leaves became yellow first in tomato, according to NIGHTINGALE et al. but EATON (1951) observed the first chlorosis in the upper leaves. There were no gradations in sunflower. The leaves of mustard and teabush became stiff-textured, but not those of soybean and sunflower. The stems of tomato, mustard, and soybean became harder but not those of sunflower. Stem elongation was slight in mustard before flowering, appreciable in soybean and sunflower, and equal to the (+S) plants in tomato (NIGHTINGALE 1932). The latter observation on tomato was not confirmed by EATON (1951), whose sulfur-deficient plants were only half as tall as the controls.

With cotton (ERGLE and EATON 1951), the intensity of the deficiency symptoms depended on the concentration of sulfate in the nutrient solution. When the concentration was increased from 0.1 to 1.0 ppm, the fresh weight increased 4-fold in 65 days and the height nearly doubled. Increase to 10 ppm caused a further 4.7-fold increase in fresh weight and another doubling in height. Additional sulfate caused relatively much smaller effects (Table 8).

It seems probable that the variability in response to (—S) noted above was caused in part by variability in the sulfur supply either unintentionally due to sulfate in the seeds or nutrient salts or contamination from the atmosphere, or intentionally due to a desire to avoid too drastic (—S) effects (EATON 1951). Species and environmental differences also played a part in the responses. The chemical reasons for these symptoms will be discussed later.

Chlorophyll and sulfur. The chlorotic condition of the leaves, which is a characteristic symptom of (—S) plants, might mean either failure to synthesize chlorophyll in sufficient amount or too-rapid destruction of the pigment. Many analyses indicate that the chlorophyll concentration is lower in chlorotic than in normal leaves. Since chlorophyll contains no sulfur, chlorosis due to lack of sulfur must be an indirect effect of the deficiency unless the sulfur-containing chloroplast proteins are directly involved in the synthesis or stability of the chlorophyll which is attached to the protein as a prosthetic group (SMITH 1940).

Analysis of isolated chloroplasts from mature leaves indicates that chloroplast protein represents a fairly constant percentage of the total protein in the cell: 35—45 percent in tobacco and tomato, GRANICK (1938); 48 percent in spinach, MENKE (1938); 30—40 percent in oat leaves, GALSTON (1943); 35—40 percent in Sudan grass, HANSON et al. (1941); and 36 percent in *Phalaris tuberosa*, HANSON (1941). WILDMAN and BONNER (1947) have separated the leaf proteins of spinach into two fractions, one of which, comprising 70—80 percent of the total protein, was electrophoretically homogeneous. The other fraction was a mixture of proteins. Spinach chloroplasts had 7.7 percent chlorophyll. The protein-chlorophyll ratio was 6:1.

Chloroplast proteins in Sudan grass were rich in sulfur. In the mature leaves they contained 70 percent of the total protein sulfur in the cell. The ratio of protein N to protein S in the chloroplasts was constant, suggesting that the chloroplast contains a single or closely related group of allied proteins. ERGLE and EATON (1951) found in cotton that the total protein in (—S) leaves was reduced to one-third the amount in (+S) leaves, but the percentage of protein S in the (—S) protein was increased by 60 percent. The chlorotic (—S) leaves therefore had a reduced amount of sulfur enriched protein. Evidently the chloroplast protein is a preferred location for the available sulfur supply in (—S) plants. Possibly the chlorosis was due to sulfur deficiency in the cytoplasmic protein rather than the chloroplast protein. However, NOACK and TIMM (1942) and CHIBNALL (1939) found only slight differences between these proteins. Further work is needed to clarify the sulfur-protein-chlorophyll relationships in the leaf.

Sulfur deficiency chlorosis in *Chlorella* was studied by MANDELS (1943). As the sulfur supply became limiting, both chlorophyll synthesis and cell division gradually decreased in rate, the former somewhat more rapidly than the latter. At the time cell division ceased, chlorophyll in the cells began to decompose. Large quantities of fat accumulated in the cells which normally store starch.

Addition of sulfate resulted in rapid recovery from chlorosis. First there was a lag period of about five hours, then chlorophyll began to be formed in the cells. The rate of formation increased until it was 2.6 times the rate during normal growth in complete nutrient solution. The greater the degree of chlorosis, the faster this rate. Finally the rate returned to normal. Evidently factors normally operating to control the rate of chlorophyll formation were not operating during the recovery process. In the nutrient solution, whose concentration ranged from 0.01 to 1.93 ppm S, each mole of sulfate produced 0.34 mole of chlorophyll in the light and 0.42 mole in the dark. This stoichiometric relationship suggests that the sulfate does not act catalytically in chlorophyll formation. Cell division was resumed about 24 hours after the addition of sulfate and was much more rapid in the light than in the dark, whereas recovery of chlorophyll formation occurred at about the same rate in the dark as in the light.

It was found that cysteine, d-l-methionine, glutathione, thioglycollic acid, and potassium thiocyanate could not substitute for sulfate in the chlorophyll-forming reaction. On the other hand, thio-, pyro-, and persulfate, sulfite and sulfide could be used. The latter group of compounds may readily be converted to sulfate, the former group only with difficulty. Presumably compounds like cysteine and methionine are not involved in the processes of cell recovery and chlorophyll formation in *Chlorella*.

Fractionation of sulfur compounds. It is necessary to fractionate the sulfur compounds in the plants in order to understand the reactions involved in metabolism. Unfortunately, the complexity of the system makes a satisfactory fractionation difficult, and many of the separations that have been made are crude at best. Different investigators have employed different procedures.

Satisfactory analytical methods are available for total sulfur, such as the Parr and oxygen bombs, sodium peroxide fusion, and combustion with magnesium nitrate. Sulfate may be determined in the water or alcohol extract of the plant material. The difference between total sulfur and sulfate sulfur is assumed to be organic sulfur. The sulfur in the insoluble residue from an alcohol extract is assumed to be protein sulfur (NIGHTINGALE *et al.* 1932, EATON 1941, and ERGLE and EATON 1951). The sulfur in the extract after precipitation of sulfate is designated soluble organic sulfur. KYLIN (1953) expressed the juice and obtained four fractions, *viz.*, sulfate, water soluble organic, coagulable, and insoluble organic sulfur. The two latter fractions were considered to be protein sulfur. WOOD and BARRIEN (1939) made a hot water digestion of the dry tissue, added tannic acid to the cooled mixture, and filtered. Sulfate was determined in the filtrate. The sulfur in the precipitate, analyzed by the oxygen bomb, was designated protein sulfur.

More definite means of characterizing the proteins would be valuable. A promising procedure consists of isolation of proteins from the chloroplasts or cytoplasm (WILDMAN and BONNER 1947). These can be studied electrophoretically to determine their homogeneity, and chromatographically for sulfur amino acids. NIGHTINGALE *et al.* (1932) tested for cysteine and other sulfhydryl groups using SULLIVAN's reagent and nitroprusside.

In some studies, the organic sulfur was further fractionated into "labile" sulfur and soluble and insoluble organic sulfur (THOMAS and HENDRICKS 1944). "Labile" sulfur was largely derived from free or combined cystine and closely related compounds. It was obtained by macerating the fresh plant material in a blendor, adding a mixture of magnesium and cadmium hydroxides, and digesting for 48 to 96 hours with gentle boiling under reflux in an atmosphere of hydrogen (or nitrogen). Hydrogen sulfide formed by dismutation and hydrolysis of cystine was effectively collected as cadmium sulfide. The mixture was acidified with hot dilute hydrochloric acid and the hydrogen sulfide swept with hydrogen into zinc acetate absorbers. The sulfide was finally titrated iodometrically or determined colorimetrically as methylene blue. It should be noted that the dismutation of cystine yields some sulfite and sulfate, and the labile sulfur value found by iodometric titration must be increased 10 percent to give the equivalent cystine value.

Table 1. *Sulfate and organic sulfur fractions in alfalfa leaves fumigated with sulfur dioxide, and comparable unfumigated leaves on eight pairs of plots.* (THOMAS, HENDRICKS and HILL 1944.)

Fraction	Sulfur (dry basis) in		$\dfrac{\text{Fum}}{\text{check}}$	P[1]
	check leaves (%)	fumigated leaves (%)		
1. Total (Parr bomb) . .	0.53	1.47	2.77	< 0.01
2. Sulfate	0.27	1.19	4.41	< 0.01
3. Organic (1–2)	0.256	0.283	1.105	< 0.01
Organic fractionation				
4. Labile (cystine) . . .	0.075	0.087	1.16	< 0.01
5. Acid soluble organic (methionine)	0.102	0.111	1.09	0.07
6. Acid insoluble organic	0.086	0.091	1.05	0.10
7. Total organic (4+5+6)	0.263	0.289	1.10	< 0.01

[1] Probability P, from FISHER's table of "t", of significance of differences between check and fumigated plots.

After the digestion, the mixture was filtered. The residue was burned in the Parr bomb for sulfur determination. The filtrate was first cleared of sulfate, and then made alkaline and evaporated to dryness in a nickel crucible. Its sulfur content was found after a sodium peroxide fusion.

Results obtained by this method of fractionation are illustrated in Table 1, which gives the average values of the different fractions of eight pairs of alfalfa plots, comparing plants fumigated in various ways with sulfur dioxide with comparable unfumigated plants. The data indicate that due to the fumigation treatments there was a 4.4-fold increase in the sulfate content of the leaves, but only a 10 percent increase in the organic sulfur. The increments of labile sulfur and total organic sulfur, though small, were statistically highly significant. The soluble and insoluble organic fractions, which were more subject to analytical errors, gave differences slightly above the 5 percent level of significance. There was close agreement between the organic sulfur determined either as the difference between total sulfur and sulfate sulfur or as the sum of the soluble, insoluble, and labile organic sulfur.

Table 2 gives "cystine" sulfur values for the leaves and roots of check and fumigated alfalfa and sugar beet plants growing in (+S) and (—S) nutrient solutions. Addition of either sulfate or sulfur dioxide raised the "cystine" sulfur level significantly, particularly when the sulfur level or the general level of nutrient solution was low. The concentrations in the sugar beet roots were very low, but the low sulfur plants showed a significant increase of labile sulfur due to fumigation.

Evidently these methods of fractionation give reproducible results. In the absence of detailed knowledge of the organic sulfur compounds present, the interpretation of the fractionation cannot be complete. Proteins are at least partly hydrolysed by the alkaline digestion, and some proteins are completely hydrolysed. The labile sulfur fraction involves cystine and glutathione primarily and any other similar sulfhydryl compounds that might be produced by protein hydrolysis. Methionine is present in the acid-soluble organic fraction, as has been found by paper chromatography (STEWARD *et al.* 1951). For brevity it will be referred to as the "methionine" fraction. The insoluble organic fraction presumably contains the more refractory unhydrolysed proteins along with other unspecified insoluble organic sulfur compounds. An alcohol extract of alfalfa leaves, subjected to paper chromatography (STEWARD *et al.* 1951), showed some cystine and probably glutathione which yielded cystine on hydrolysis. Methionine was not present. Other amino acids in the alcohol extract included aspartic and glutamic acids, alanine and serine, together with asparagine and glutamine. Valine, the leucines, tyrosine, and arginine were present in small amounts.

Table 2. *Labile sulfur in the leaves of alfalfa and the leaves and roots of sugar beets, with various nutrient solutions and with or without fumigation with sublethal concentrations of sulfur dioxide.* (THOMAS, HENDRICKS and HILL 1944.)

Year	No. crops	Nutrient solution level	Labile sulfur (dry basis)			
			high S + med. S		med. low S + low S	
			check (%)	fum. (%)	check (%)	fum. (%)
Alfalfa leaves						
1939	4	high p_H	0.128	0.132	0.088	0.121[1]
1939	4	low p_H	0.129	0.134	0.091	0.131[1]
1940–41	7	high	0.102	0.103	0.057	0.090[2]
1940–41	7	low	0.098	0.106[2]	0.075	0.093[2]
Sugar beet leaves						
1943	1	high	0.096	0.100	0.050	0.093[1]
1943	1	low	0.104	0.109	0.080	0.100[1]
Sugar beet roots						
1943	1	high	0.011	0.011	0.010	0.013[1]
1943	1	low	0.010	0.011	0.011	0.012[1]

[1] Difference between check and fumigated plots significant at or below 5 percent level.

[2] Difference between check and fumigated plots significant at 1 percent level.

Tracer studies of sulfur metabolism. Using these methods of fractionation analysis in conjunction with radiosulfur S^{35}, a study was made of the metabolism of sulfur in wheat, barley, corn, sugar beets, alfalfa, and tomatoes (THOMAS *et al.* 1944). Radiosulfur was added either to the nutrient solution as sulfate* or to the air as sulfur* dioxide. Uptake was then followed by appropriate sampling and fractionation.

Fig. 1 shows the disposition with time of radiosulfur in the leaves of (—S) vegetative winter wheat growing in large sand culture plots under glass. Uptake of sulfate* was rapid. Samples of wheat taken 5 hours after the addition of sulfate* to the nutrient solution showed about the same amount of absorption and transformation to organic forms as was found after 1 or 2 days. Thereafter

---

* The asterisk refers to sulfur compounds containing radiosulfur S^{35}.

there was a steady increase in the absorption until the concentrations were nearly doubled at 8 days. Then the concentrations began to fall as the sulfate* in the nutrient solution was depleted and dilution in the tissue by growth predominated over uptake of the sulfate*. There was 75 percent depletion of the nutrient solution in 8 days and 94 percent in 16 days. It will be noted that the sulfate* concentration in the wheat was always low. The three organic fractions were approximately equal but there was a tendency for the insoluble organic fraction to fall off more slowly with time than the others.

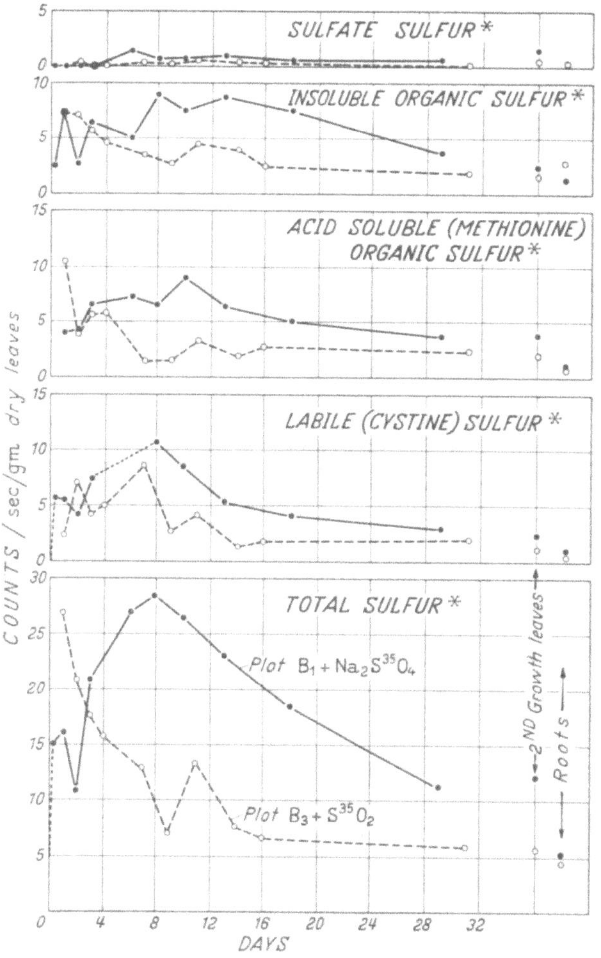

In a similar experiment with corn growing in (+S) nutrient solution, absorption of sulfate* was much slower than by the wheat because a large amount of inactive sulfate was already present in the plants and in the nutrient solution before the sulfate* was applied. In spite of this, measurable activity was found in the leaves in 1–2 hours. Uptake increased steadily for 3 weeks, then did not change appreciably during the following 3 weeks, due in part to the lateness of the season. The sulfate* fraction was approximately equal to each of the 3 organic fractions in the early samples. Later the cystine* fraction fell to about half the others. The nutrient solution retained 77 percent of its initial activity after 6 weeks.

Fig. 1. Radiosulfur in the leaves of sulfur-deficient, vegetative winter wheat at various times after treating plots B–1 and B–3 with sodium sulfate* and sulfur* dioxide, respectively. Fractionation of the sulfur* into sulfate* and 3 organic sulfur* fractions is shown. The roots are also analyzed. (THOMAS, HENDRICKS, BRYNER and HILL.)

The behavior of sulfur* dioxide differed from sulfate* in that the initial concentration of the former was maximum because the fumigation lasted only 2.5 hours. In the wheat there was immediate conversion of the sulfur* dioxide to the organic forms since practically no sulfate* could be found even in the earliest samplings. There was no appreciable difference between the 3 organic fractions except on the 7th day when a high cystine* value was obtained. This was probably without significance because a low value was found on the 9th day.

* The asterisk refers to sulfur compounds containing radiosulfur S^{35}.

In the corn, the sulfate* formed by oxidation of sulfur* dioxide represented over half the total sulfur* present in the first sampling, and was 40 percent on the third day. Later it became approximately equal to the methionine* and insoluble fractions which were somewhat larger than the cystine* fraction. At six weeks, sulfate* and cystine* were very low and the methionine* fraction predominated over the insoluble organic fraction.

Fig. 2 gives data for the leaves of large alfalfa plants that were slightly sulfur deficient. The sulfate* in the nutrient solution was taken up rapidly and changed to organic forms. Maximum leaf concentration was reached in 16 days, followed by a decrease due to growth of the plants and depletion of the sulfate* supply. The three

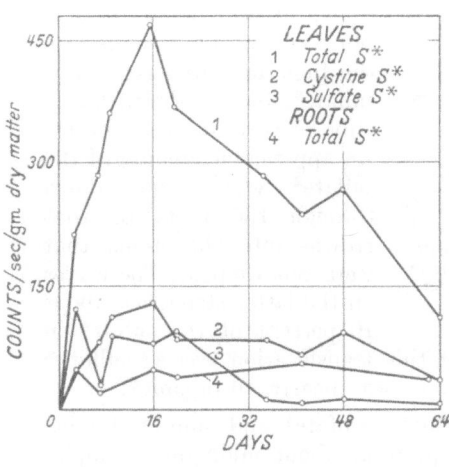

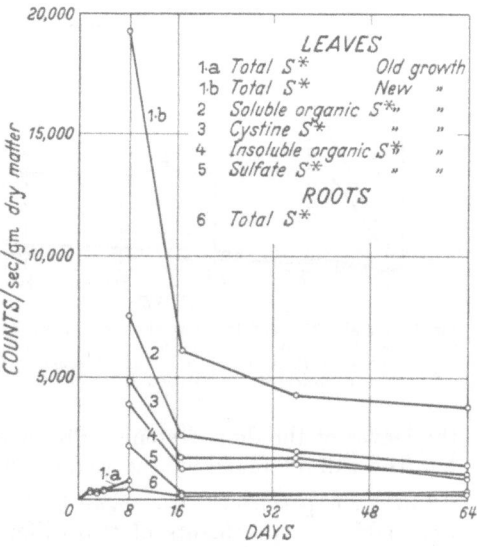

Fig. 2.   Fig. 3.

Fig. 2. Absorption of radiosulfur in old-growth leaves and in roots of sulfur-deficient alfalfa plants and transformation of sulfate into organic sulfur compounds. Curves for soluble and insoluble organic fractions similar to the cystine fraction are omitted. (THOMAS, HENDRICKS and HILL 1950.)

Fig. 3. Absorption and transformation of radiosulfate in old- and new-growth leaves and in roots of alfalfa. The new-growth, 8-day leaves were formed after the radiosulfate was added to the nutrient solution at 0 time. (THOMAS, HENDRICKS and HILL 1950.)

organic sulfur fractions were similar. The sulfate* fraction was like the organic fractions for two weeks, then fell off to a low value. The roots remained constant at a low level throughout the experiment.

In another experiment with alfalfa (Fig. 3), sulfate* was added to the nutrient solution of a (—S) plot, some plants of which had been harvested that same day. Rate of uptake of sulfate* into the leaves of the old growth was only about 4 percent of the rate into the new growth. After 8 days, growth of the latter was so rapid that the concentrations began to fall sharply. Sulfate* represented only about 10 percent of the mixture in the leaves at this time and it fell off more later. The three organic fractions were not greatly different but the methionine* fraction predominated slightly. Root samples indicated that in the plants with old growth, the root concentration was about 50 percent of the leaf concentration and sulfate* was very low. The corresponding value for the plants with new growth was 2 percent. In a third experiment in which no top growth was permitted for 6 days after adding the sulfate* to the nutrient solution, the roots took up the activity at about the same rate as before, but

---

* The asterisk refers to sulfur compounds containing radiosulfur S^{35}.

the sulfur* all remained as sulfate*. Evidently there is no chemical transformation of sulfate* in the roots without top growth.

The terminal portions of several alfalfa stems on each of three plants were dipped in a solution of sulfate*. After 1 to 5 days, the solutions were removed and the parts that had been immersed were cut off and discarded. Fig. 4 gives sulfur* taken up by the stems which had had their ends immersed, and also stems on the same plants that were not treated. Uptake of sulfur* was rapid the first day. Thereafter uptake was nearly proportional to the time of immersion. Of the sulfur* taken up the first day, 75 percent was sulfate*, but on the fifth day only 30 percent was sulfate*. An appreciable amount of the sulfate* was carried down through the roots or root crowns into the stems that were not dipped. The leaves on the latter stems had about 10 percent of the sulfur* in the leaves of the dipped stems.

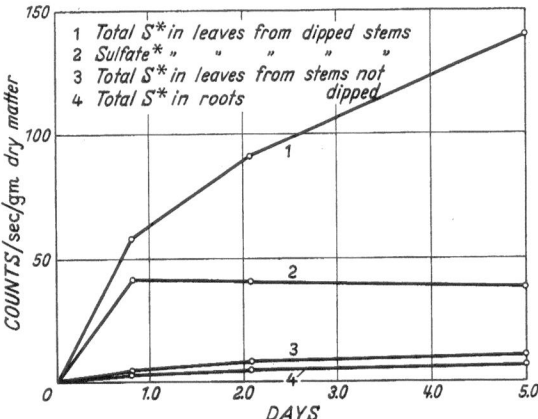

Fig. 4. Translocation and transformation of radiosulfate in alfalfa after absorption following dipping of the ends of a few stems of three plants for various times in a strong solution of radiosulfate. (Thomas, Hendricks and Hill 1950.)

Presumably this translocation was as sulfate* but the sulfur* found in the roots and leaves was nearly all organic*.

In an experiment with sugar beets, Table 3, sulfate* and sulfur* dioxide were added to the plants of two different plots at about midnight. Samples of the fumigated leaves taken before daylight showed 50 percent transformation to organic sulfur compounds, suggesting that light is not necessary for these reactions. Samples from the sulfate* treated plants taken at 8 A.M. suggested the same conclusion though less definitely because they were taken after daylight. The relative amounts of the different organic fractions from the sulfate* and sulfur* dioxide were similar. The methionine* fraction was greatest at first but it fell off with time while the insoluble organic fraction increased. The cystine* fraction did not exceed 22 percent of the total at any time.

Table 3. *Radiosulfur fraction in sugar beet leaves after addition of sulfate* or sulfur* dioxide to the plants at midnight.*
(Thomas, Hendricks and Hill 1950.)

Treat-ment	Time of sampling	Total S* in leaves C/S/G[1]	Percentage of S* as			
			cystine S*	acid sol. organic (methionine) S*	insol. organic S*	sulfate S*
A[2]	1–6 hours[3]	300	7	29	13	51
	3–8 days	330	21	37	21	21
	80 days	300	17	26	41	16
B[4]	7 hours	54	8	37	13	42
	1–8 days	400	22	47	25	6
	48 days	180	17	30	51	2

[1] Counts per second per gram (dry basis).
[2] S*O₂ fumigation at 11 P.M.
[3] Before daylight.
[4] Na₂S*O₄ added to nutrient solution at 12:30 A.M.

* The asterisk refers to sulfur compounds containing radiosulfur S³⁵.

Table 4 gives the distribution of sulfur* in various tissues of spring wheat a week before harvest, while some of the leaves and stems were still somewhat green, and also at harvest. These plants had been treated a month earlier with Na_2S*O_4 in the nutrient solution. The table also includes average sulfur* distribution data for the kernels alone of 3 plots of wheat treated with Na_2S*O_4 4–5 weeks before harvest, and 3 plots treated with $S*O_2$ 4–7 weeks before harvest. Barley growing in the same plots gave similar values, which are therefore not presented. The individual values for each of the 3 plots averaged were closely concordant except that in the plot treated with $S*O_2$ 7 weeks before harvest, the cystine value was about 10 percent lower and the "methionine" value 10 percent higher than in the plots treated 4–5 weeks before harvest. Evidently the tagged atoms were distributed somewhat differently depending on the stage of development

Table 4. *Distribution of radiosulfur fractions in wheat treated in June with $S*O_2$ or Na_2S*O_4 final harvest July 29–30.* (THOMAS, HENDRICKS, BRYNER and HILL 1944.)

S* treatment	Date of sampling	Tissue	Count C/S/G	Percentage of radiosulfur as			
				labile (cystine) S*	acid sol. organic (methionine) S*	insol. organic S*	sulfate S*
Wheat[1]							
Na_2S*O_4	July 22	leaves	13	11	31	39	19
	July 22	sheaths	5	9	36	26	29
	July 22	stems	6	6	19	14	61
	July 22	heads	12	31	20	30	17
	July 30	kernels	24	32	47	9	12
		straw	9	10	36	26	28
		roots	5	15	26	50	9
Wheat[2]							
$S*O_2$	July 29–30	kernels	11	36	46	11	7
Na_2S*O_4	July 29–30	kernels	25	37	43	11	9

[1] Detailed data on one plot.
[2] Primary and secondary kernels of wheat on six plots.

of the kernel when the sulfur* was added. However, primary and secondary kernels from the same plants had nearly the same distribution, and even the immature "tertiary" kernels were not greatly different from the mature grain.

In all treatments the cystine* and methionine* fractions predominated in the kernels. There was good agreement between corresponding grain samples that had been treated with $S*O_2$ and Na_2S*O_4, indicating that there was no significant difference in the grain whether the sulfur* was supplied from the air or through the roots. The vegetative portions of the plant were quite different from the kernels. In the leaves the methionine* and insoluble organic fractions predominated, in the stems, sulfate*; and in the roots, the insoluble organic fraction.

It is of interest to note that the relative proportions of the organic fractions appear to be related to the physiological activity of the plant tissues. The sugar beet leaves in Table 3 were initially very high in the methionine* fraction. The cystine* and insoluble organic fractions were lower and about equal in amount. As the leaves aged, however, the insoluble organic fraction increased markedly, the methionine* fraction fell off correspondingly, and the cystine* fell slightly.

In Table 4 the old leaves and the roots of the wheat also had a high level of insoluble organic sulfur*, and about the same distribution for the other

* The asterisk refers to sulfur compounds containing radiosulfur S^{35}.

fractions as the old sugar beet leaves. The kernels, on the other hand, were very rich in methionine* and cystine* fractions and poor in the insoluble organic fraction. However, the stems, sheaths, and straw carried more of the methionine* than of the insoluble organic fraction. They also were very high in sulfate* (28 to 61 percent) which might modify the relationships. In any case, the aging process in the leaves of ripening grain plants is more rapid than in sugar beet leaves. In vegetative wheat, Figure 1, the insoluble organic fraction increased at the expense of the other fractions as the leaves got older.

In alfalfa leaves (Table 1), the methionine fraction predominated somewhat over the other two which were approximately equal. This was also true of the alfalfa leaves in Fig. 2 and 3. The insoluble organic fraction, which at first was smaller than the "cystine" fraction, became somewhat larger as the leaves aged. The conclusion seems definite that in active tissue like young leaves and in the seeds, the "methionine" and "cystine" fractions represent the principal organic sulfur compounds, whereas in old leaves and in less active tissue like the roots, the insoluble organic fraction increases and may predominate.

Radiosulfur in wheat seedlings. KYLIN (1953) observed similar absorption rates and chemical changes in a study of the uptake and metabolism of sulfate by deseeded wheat plants using radiosulfur. The plants were germinated on filter paper for 3 days, then deseeded and mounted on paraffined cork holders in Pyrex beakers at 25° C. They were illuminated 16 hours per day at 5000 lux. The fresh plants, when sampled, were cytolyzed with chloroform and expressed at 1200 kg/cm². The residue was soaked with 10⁻³ magnesium sulfate and expressed again. This treatment was repeated. The expressed juice, cleared of water-soluble protein by addition of alcohol and filtration, was designated fraction I. It contained sulfate and water soluble organic sulfur. The water-soluble precipitable protein was designated fraction II. The residue from the expressed juice, containing the bulk of proteins, was designated fraction III. Fraction II contained less than 4 percent of the nitrogen and sulfur of fraction III and is included with the latter in this discussion.

In one experiment, the plants were grown in complete nutrient without sulfate* for 10 days after deseeding, then transferred to a nutrient containing 10 millicuries per liter. Chromatograms and autographs were made on samples taken after various exposure times. Within 11 minutes, S* was found in all fractions. Cystine and methionine could be identified chromatographically. After one hour, 0.13 micromol per 56 plants had been taken up and 15–25 percent of the absorbed sulfate* was converted to organic form. After 10 days, one third of the sulfate* in the roots and one half in the shoots was similarly changed. Evidently conversion of sulfate* to organic sulfur* is very rapid at first, then slows down.

In another experiment, the plants were exposed to sulfate* for 3 days after deseeding, then transferred to either (+S) or (—S) solutions without sulfate*. Table 5 gives complete data on growth and uptake of nitrogen and sulfur* after 3, 14, and 24 days. Fresh weight of the shoots increased more rapidly in the (+S) than (—S) solution. The reverse was true of the roots. The top-root ratio was about 2 in the (+S) solution but fell to 0.9 in (—S) at 24 days.

Total sulfate* uptake was 9.0 microcuries per 56 plants at 3 days from a solution containing 0.25 millicuries per liter. This fell to about 7 microcuries at 24 days, indicating that some of the S* probably went back into the nutrient solution. The sulfate* was changed to organic sulfur compounds, rapidly and

---

* The asterisk refers to sulfur compounds containing radiosulfur S³⁵.

Table 5. *Nitrogen and sulfur in the roots and shoots of wheat seedlings grown 3 days in complete nutrient solution containing sulfate*, then transferred to (+ S) and (—S) solutions without sulfate*.* (KYLIN 1953.)

Time days	Sulfate supply	Fresh weight, grams	Uptake of nitrogen by fraction (micromols per 56 plants)			Uptake of sulfur* by fraction (micromols per 56 plants)			
			I	II+III	total	I sulfate	I organic	II+III	total
	Roots
0	+S	0.90	75	190	265	—	—	—	—
3	+S	1.20	86	107	193	1.02	0.18	0.22	1.42
14	+S	2.61	126	277	403	0.07	0.12	0.46	0.65
24	+S	7.40	275	635	910	0.09	0.09	0.56	0.74
14	—S	2.71	122	237	359	0.06	0.10	0.73	0.89
24	—S	8.40	700	526	1226	0.05	0.10	1.35	1.50
	Shoots
0	+S	1.14	82	342	424	—	—	—	—
3	+S	2.09	245	440	685	5.45	0.55	1.63	7.6
14	+S	7.48	590	1424	2014	4.40	0.90	2.82	8.1
24	+S	14.44	990	3230	4220	3.04	0.54	2.26	5.8
14	—S	6.25	435	1055	1490	2.69	0.65	3.98	7.3
24	—S	9.65	1210	1410	2620	1.75	0.82	3.23	5.8

nearly completely in the roots where the level was low, less completely in the shoots where the level was higher.

Fractions II + III predominated over organic fraction I after the third day by factors of 4 to 13 in the roots and 3 to 6 in the shoots. At 24 days the (+S) shoots still had half the S* as sulfate*, while the (—S) shoots had one-third. Symptoms of sulfur deficiency were noted in the (—S) plants in spite of the presence of this sulfate* in the shoots. Presumably it was not sufficiently mobile to supply the needs of the sulfate*-free roots adequately. The observation mentioned earlier, that sulfate in the leaves from sulfur dioxide fumigation is less effective in supplying the needs of the plant than sulfate from the soil, appears to have the same import. A certain amount of sulfur appears to be required in the stems and roots as well as in the leaves in order for the plant to function normally. It is of interest to note that the 24-day values of fraction II + III for S* increased in the (—S) and (+S) roots while they decreased in both the (—S) und (+S) shoots as compared with the 14-day values. Downward translocation of sulfate* and possibly also some proteolysis was occurring at this time.

Radioautographs. Sulfur autographs of KYLIN's 24-day (+S) and (—S) plants showed strong concentrations of S* in the first leaf and slight concentrations in the second. The roots of the (—S) plants were larger and had considerably more activity than the (+S) roots. These plants had the first leaf half-formed when they were removed from the sulfate* solution on the third day. They had three or four leaves when autographed. There appeared to be a little translocation from the first leaf to the second, but the third leaf had no S*. The higher activity of the (—S) roots is readily understood from the data in Table 5, since (—S) roots took up twice as much sulfur* as the (+S) roots of somewhat smaller fresh weight.

The distribution of sulfur in plants thus can be clearly observed by means of radioautographs. HARRISON *et al.* (1944) showed sections of wheat kernels and their autographs which indicate appreciable concentrations of S* in the

---

* The asterisk refers to sulfur compounds containing radiosulfur S^{35}.

embryo, the periphery of the endosperm, and the aleurone layer. THOMAS (1948) published a number of autographs illustrating the distribution of sulfur* in sugar beets and tomato. Some of these and others are reproduced in Figures 5 through 9. Fig. 5 is a thin slice of sugar beet containing S*, mounted with a photograph as a positive. The phloem and xylem in the sugar beet are closely associated in the rings. The phloem density is somewhat greater on the outside of the ring and xylem density is greater on the inside. Careful examination of the section and its autograph indicates that the sulfur* is principally in the phloem areas, particularly in the cambium cells. The parenchymatous

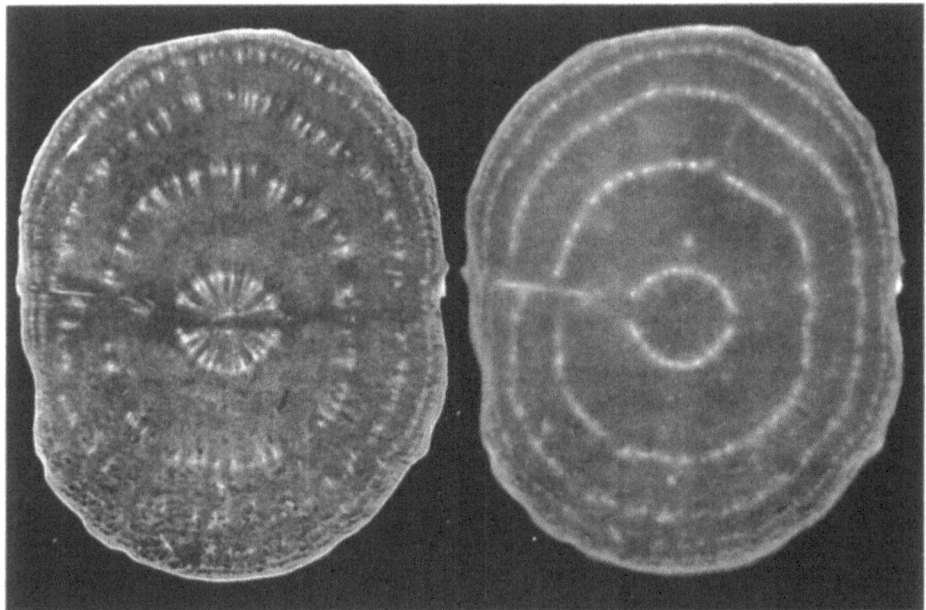

Fig. 5. Photomicrograph positive (left) and reversed autograph of a section of sugar beet root, showing accumulation of radiosulfur primarily in the phloem and cambium and in root traces. Xylem and parenchymatous tissue exhibited only slight uptake of radiosulfur.

tissue between the rings is low in sulfur* due partly to lower cell density. However, the dense center xylem cells do not show an appreciable amount of activity.

The tomato root (Fig. 6) shows accumulation in the inner and outer phloem, the rays, root traces, and especially in meristematic cells in the lacunae, while the xylem and pith are nearly free of activity. A longitudinal section of tomato stem is shown in Fig. 7 with its autograph above the section as mirror image. This plant was treated with S*O$_2$. The concentration of sulfur* in the inner and outer phloem is striking, whereas the xylem is nearly free except for slight lateral penetration into its outer layers. The detached phloem and cortical section shows how little of the sulfur* penetrated the xylem. The outer layers of xylem had considerably more activity when Na$_2$S*O$_4$ was added through the roots as shown by other autographs (THOMAS 1948).

At first, leaves show heavy accumulation in the vascular system and diffuse activity in the interveinal area when supplied with sulfate* through the roots. Later the veins empty and the activity is found in the intercostal tissue. A sulfur* dioxide fumigation may result in more uneven distribution of the sulfur*

---

* The asterisk refers to sulfur compounds containing radiosulfur S^{35}.

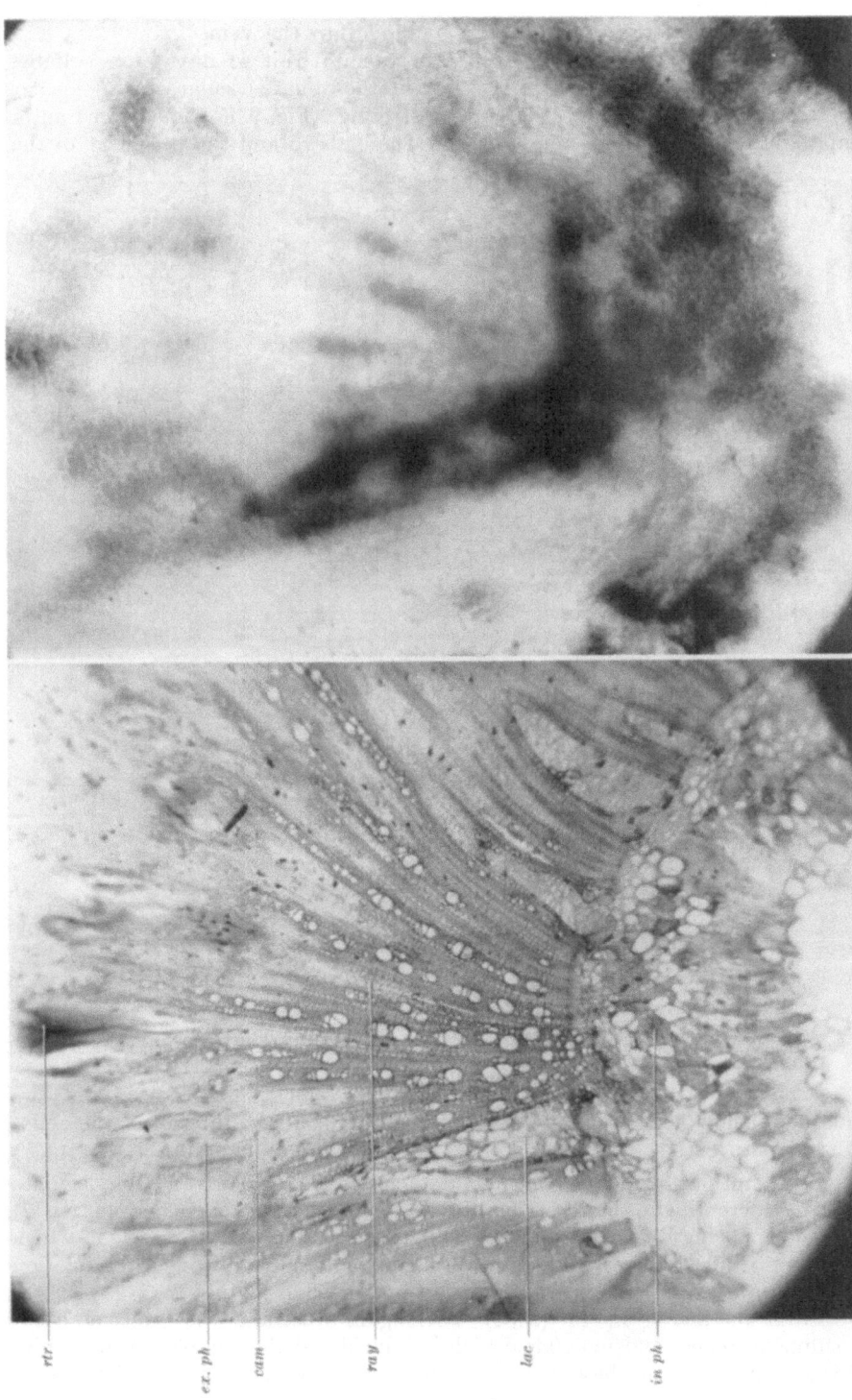

Fig. 6. Photomicrograph (left) and autograph (right) of a section of tomato root, showing the accumulation of radiosulfur in the phloem (*ph*) and cambium (*cam*), also in the meristematic tissue of the lacunae (*lac*), root traces (*rtr*), and rays.

4*

over the leaf surface. The spots disappear in a few days and the activity is spread diffusely over the leaf surface but not into the veins.

Fig. 8 is the autograph of a slice of tomato fruit 41 days after sulfate* treatment. The accumulation of sulfur* in the skin, vascular traces, locular surfaces and especially the seeds, is very striking. Fig. 9 is the enlarged autograph and micrograph of a tomato seed. The embryo and the periphery of the

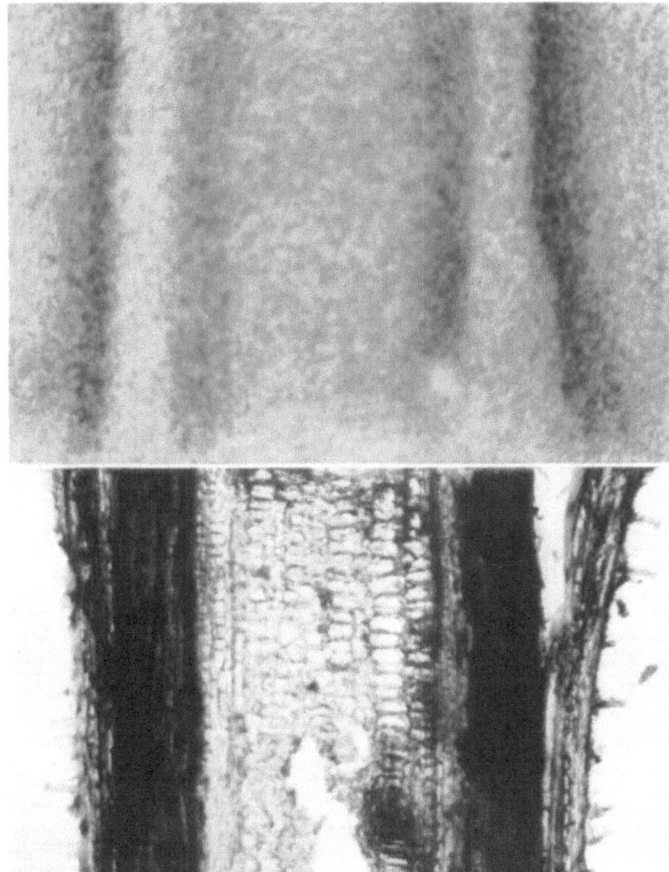

Fig. 7. Photomicrograph (below) and autograph (above as mirror image) of a section of tomato stem from a plant treated with $S*O_2$ showing the accumulation of radiosulfur in the internal and external phloem. Penetration into the xylem was slight.

endosperm show heaviest accumulation of sulfur*, but the interior of the endosperm is also active. The seed coat also has considerable sulfur*.

Sulfur, nitrogen, and molybdenum interactions. Legumes and non-legumes behave differently in their response to the supply of nitrogen and sulfur. If nitrogen is in short supply, legumes usually form nodules on their roots in which bacteria fix atmospheric nitrogen. With adequate fixed nitrogen, nodulation is slight or nonexistent. Non-legumes do not develop nodules.

ANDERSON and SPENCER (1950), working in Australia with soils deficient in sulfur, nitrogen, and molybdenum, have elucidated the complex relationships of the fertilization of these soils. Molybdenum appears to be associated with

---

* The asterisk refers to sulfur compounds containing radiosulfur S^{35}.

the activity of the symbiotic nitrogen-fixing bacteria in the nodules on legumes. It is also an essential constituent of nitrate reductase.

In (+S) media, molybdenum increased markedly the yield and nitrogen concentration of clover when no combined nitrogen was provided, but had little or no effect when nitrogen was supplied as nitric acid or ammonium sulfate. With flax, molybdenum increased the yield only when combined nitrogen as nitrate or ammonium salt was also supplied. Molybdenum deficiency decreased

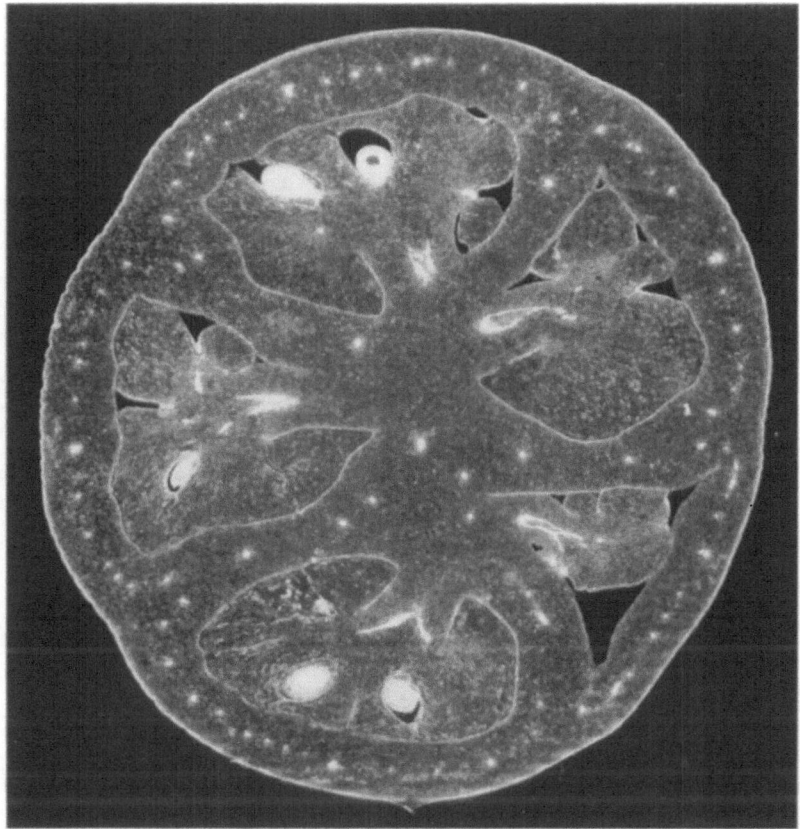

Fig. 8. Reversed autograph of a slice of tomato fruit showing accumulation of radiosulfur in the skin, vacular traces, locular surfaces and the seeds.

protein nitrogen and increased non-protein nitrogen in flax, without affecting total nitrogen. The molybdenum thus seems to be needed for utilization of nitrate in non-legumes as well as in legumes.

In (—S) media the molybdenum and nitrogen relationships are changed. Table 6 gives the yield data for subterranean clover on two soils. Soil A was more deficient in sulfur than soil B. On the former, clover responded to sulfur alone, but not on the latter. In soil A, (—S) clover plants took up only 0.06 percent sulfur, while (+S) plants took up 0.24–0.32 percent. Neither molybdenum nor nitrogen alone or in combination had an appreciable effect on growth in soil A, but in soil B each of these three treatments increased growth by a factor of 2.3. Sulfur plus nitrogen and sulfur plus molybdenum increased growth by factors of 4.4 and 6.4 in soil A and by 3.4 and 3.3 in soil B respectively. Sulfur plus nitrogen plus molybdenum gave factors of increase of 6.6 in soil A and 4.0 in soil B.

In soil A, nodulation of clover was greatly increased by every addition of sulfur but not by any of the other treatments. In soil B, clover nodulation was high in the untreated soil and in the sample treated with sodium sulfate, but rather low in all the others. The (—S) plants not provided with combined nitrogen were pale green to yellow and the nitrogen was low, indicating that

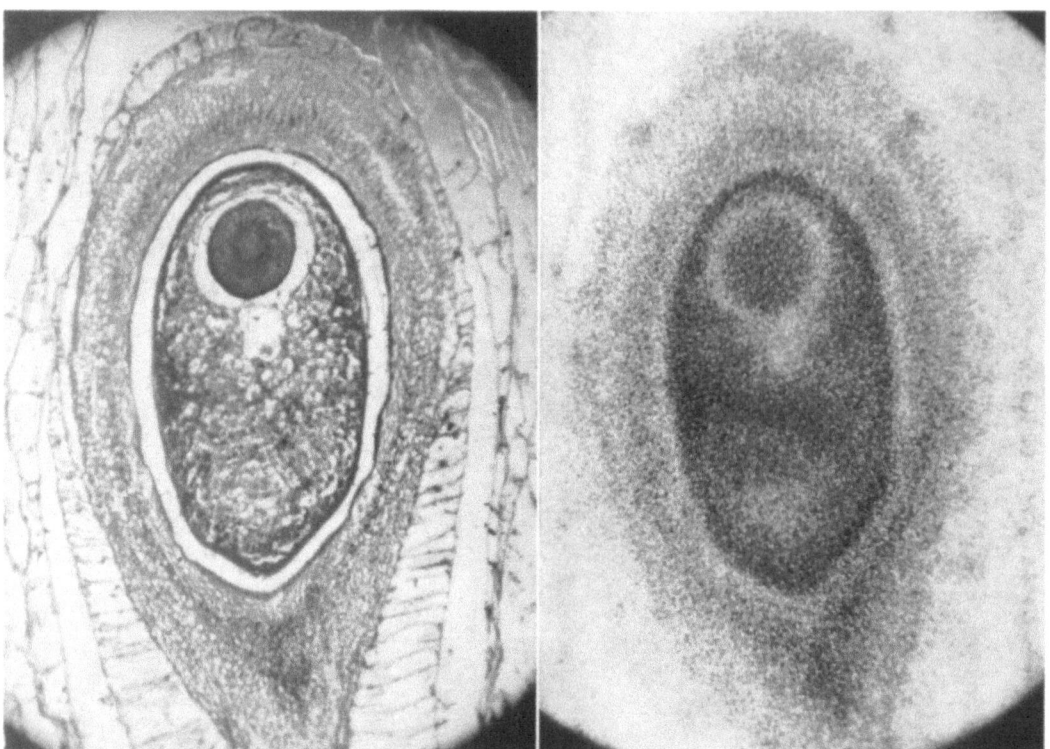

Fig. 9.   Photomicrograph (left) and autograph (right) of a section of tomato seed showing accumulation of radiosulfur in the embryo, the endosperm—especially its periphery, and the seed coat.

introgen fixation was inhibited by lack of sulfur. Protein nitrogen increased and non-protein nitrogen decreased in every sulfur treatment of soil A. Tests with diphenylamine indicated a lowering of nitrate due to sulfur. Evidently nitrate reduction as well as nitrogen fixation was inhibited by sulfur deficiency.

The data indicate that the demand for nitrogen and the intensity of nodulation are increased by deficiency of combined nitrogen or molybdenum and decreased by deficiency of sulfur. The latter decreased the response to nitrogen and produced unthrifty plants even when nitrogen was present, whereas addition of sulfur stimulated symbiotic fixation of nitrogen, as well as nitrogen utilization.

Table 7 gives data for flax in soil B similar to the clover data in Table 6. No data are available for soil A. Neither sulfur nor molybdenum added alone or together had more than a slight effect on the growth of flax, and the plants were pale green in color. Nitrogen caused a considerable increase in growth, which was improved significantly by adding either sulfur or molybdenum or both along with the nitrogen. Since ammonium nitrogen gave a better yield

Table 6. *Dry weight, nodules, protein nitrogen, and non-protein nitrogen of subterranean clover growing in soils deficient in sulfur, nitrogen, and molybdenum.* (ANDERSON and SPENCER 1950.)

Treatment	Soil B	Soil A	Soil B	Soil A	Soil A	
	dry weight		rel. estimated wt. nodules		protein N	non-protein N
	(grams)	(grams)			(%)	(%)
None	2.20	1.43	63	20	1.10	0.77
Na$_2$SO$_4$	2.13	3.87	80	100	1.61[1]	0.33[1]
NH$_4$NO$_3$	—	1.70	—	3		
HNO$_3$.	5.00	—	4	—		
Na$_2$SO$_4$ + NH$_4$NO$_3$	—	6.30	—	96		
Na$_2$SO$_4$ + HNO$_3$	7.42	—	7	—		
Na$_2$MoO$_4$	4.77	1.50	8	10	1.07	0.61
Na$_2$MoO$_4$ + Na$_2$SO$_4$	7.27	9.17	30	100	1.61[1]	0.49[1]
Na$_2$MoO$_4$ + NH$_4$NO$_3$	—	1.77	—	4	—	—
Na$_2$MoO$_4$ + HNO$_3$	5.25	—	4	—	1.19[2]	0.93[2]
Na$_2$MoO$_4$ + Na$_2$SO$_4$+NH$_4$NO$_3$.	—	9.50	—	83	—	—
Na$_2$MoO$_4$ + Na$_2$SO$_4$+HNO$_3$ . .	8.70		8	—	1.60[1]	0.63[1]

[1] Manganese sulfate used instead of sodium sulfate.
[2] Sodium nitrate used instead of nitric acid.

than nitrate nitrogen under comparable conditions, it is likely that nitrate reduction was one of the limiting factors except when both molybdenum and sulfur were present together with nitrate. This is further indicated by the high non-protein nitrogen values when nitric acid was added alone or with molybdate, and the high protein nitrogen when nitric acid and sulfate were added together or with molybdate, suggesting that interference with protein synthesis might be another limiting factor without sulfur. The role of manganese in these experiments appears to be of little significance. Manganese sulfate is regarded primarily as a carrier of sulfur. Manganese in appreciable amount is antagonistic to molybdenum.

These experiments with legumes recall an old experiment by NELLER (1926) in which sulfur fertilization of arid sulfur-deficient soils of East Central Washington increased the yield of alfalfa by factors of 2 to 4 and at the same time increased the nitrogen content of the plants 20 to 50 percent. Evidently the sulfur caused a large increase in nitrogen fixation.

Table 7. *Dry weight, protein nitrogen and non-protein nitrogen of flax growing in a soil deficient in sulfur nitrogen and molybdenum.* (ANDERSON and SPENCER 1950.)

Treatment	Flax (soil B)		
	dry weight (grams)	protein N (%)	non-protein N (%)
None	1.55	1.11	0.32
MnSO$_4$	1.50	0.92	0.40
Na$_2$MoO$_4$	1.45	0.81	0.26
MnSO$_4$ + Na$_2$MoO$_4$	1.75	0.83	0.24
HNO$_3$	3.05	0.93	2.59
Na$_2$MoO$_4$ + NO$_3$	3.40	1.19	2.51
MnSO$_4$ + HNO$_3$	4.10	1.43	1.05
(NH$_4$)$_2$SO$_4$	4.75	—	—
Na$_2$MoO$_4$ + (NH$_4$)$_2$SO$_4$ . .	4.85	—	—
MnSO$_4$ + Na$_2$MoO$_4$ + HNO$_3$	5.55	1.70	0.77

Chemical changes in nitrogen and carbohydrates due to sulfur deficiency.

The symptoms of sulfur deficiency are due to a number of chemical changes in the metabolic processes involving primarily sulfur and nitrogen. Chlorosis has already been discussed. As the supply of sulfur is reduced, reduction of nitrate is inhibited, and the plant behaves as if it were deficient in nitrogen. Protein synthesis appears to be inhibited and there is a marked increase in am-

Table 8. *Influence of sulfur deficiency on the fractions of sulfur, nitrogen and carbohydrate compounds in the stems of various plants.*

	Tomato NIGHTINGALE et al. (1934)		Tomato EATON (1951)		Soy bean EATON (1935)		Sunflower EATON (1941)		Black mustard EATON (1942)		Cotton ERGLE and EATON (1951)	
	+S (%)	−S (%)	+S (%)	−S (%)	+S (%)	−S (%)	+S (%)	−S (%)	+S (%)	−S (%)	+S (%)	−S (%)
Sulfate S	0.17	tr	[2]	[2]	0.22	0.08	[2]	[2]	[2]	[2]	0.01	0.00
70–80% alc. sol. S	tr[1]	0.11[1]	0.38	0.22			0.07	0.12	0.64	0.17	0.01	tr
Alc. insol. S.	0.17[1]	0.01[1]	0.14	0.14			0.23	0.10	0.04	0.00	0.08	0.04
Total organic S					0.05	0.05						
Volatile S									0.05	0.01		
Total S	0.34	0.12	0.52	0.36	0.27	0.13	0.30	0.22	0.74	0.18	0.10	0.04
Protein N[3]	0.77[1]	0.69[1]	0.93	0.73	0.92	1.42	0.82	0.65	1.75	1.59	0.41	0.79
Sol. organic N	0.24[1]	0.69[1]	1.29	2.44	0.14	1.25	0.38	1.64	0.53	1.40	0.28	3.11
Amino N	0.07	0.28			0.17	0.54	0.13	0.55	0.11	0.45	0.02	1.01
Amide N	0.04	0.16	0.07	0.18	0.03	0.04	tr	0.14	0.11	0.22		
Ammonia N	0.03	0.11	0.05	0.36	0.01	0.02	0.01	0.10	0.08	0.24		
Total organic N	1.01	1.38	2.22	3.17	1.06	2.65	1.20	2.29	2.28	2.99	0.69	3.90
Nitrate N	0.07	0.30	0.72	0.95	0.11	0.23	1.67	1.92	3.86	3.12	0.07	0.97
Total N	1.08	1.68	2.94	4.12	1.17	2.88	2.87	4.21	6.14	6.11	0.76	4.87
Reducing sugars	4.2	6.3	11.3	3.7	1.79	0.40	15.66	3.64	2.30	0.32	2.63	0.45
Sucrose	4.4	7.0	1.9	2.5	0.49	0.44	2.23	0.67	0.23	0.55	2.43	0.29
Starch	12.2[4]	14.8[4]	1.0	2.6	1.90	2.32	0.45	0.88	0.61	2.40	11.68	9.90
Acid-hydrolysable carbohydrate					12.07	11.61	0.64	0.93	9.66	10.50	14.58	14.18
Dry weight/plant (grams)			20.9	8.4	2.7	2.0	4.2	1.2	0.6	0.4	45.7	9.0

[1] Extraction made with water instead of alcohol. Extract boiled with dilute acetic acid to coagulate soluble protein.
[2] Sulfate not separated from organic fractions.
[3] Determined in same fraction as insoluble organic sulfur.
[4] Starch plus dextrin.

monia and the amino acids, amides, and other water-soluble organic nitrogen compounds. At the same time, there is usually an increase in the starch and hemicellulose, and a large decrease in the reducing sugars. Sucrose often decreases also.

Table 8 summarizes analyses of the sulfur, nitrogen, and carbohydrate fractions in the stems of (+S) and (—S) tomato, soybean, sunflower, black mustard, and cotton, made by NIGHTINGALE et al. (1932), EATON (1935, 1941, 1942, 1951), and ERGLE and EATON (1951). The sulfur analyses of the stems indicate a considerable degree of sulfur deficiency in all plants except possibly EATON's tomato, though the latter showed marked decrease in growth rate in spite of the fact that the (—S) plants were grown in (+S) solutions for two weeks at

Table 9. *Effect of sulfate concentration in the nutrient solution on percentage dry weight of carbohydrates, nitrogen and sulfur compounds in the leaves and stems of cotton. Growth period, 65 days.* (ERGLE and EATON 1951.)

SO₄ in nutrient (ppm)	Fresh weight plants (grams)	Sulfate S %	Organic S %	Total sugar %	Nitrate N %	Soluble organic N %	Protein N %	S in protein %
				Leaves				
0.1	13	0.003	0.11	0.0	1.39	2.23	0.96	1.57
1.0	50	0.003	0.12	0.0	1.37	2.21	1.28	1.26
10	237	0.009	0.17	1.5	0.06	1.19	2.56	0.92
50	350	0.10	0.26	3.1	0.00	0.51	3.25	1.05
200	345	0.36	0.25	3.4	0.10	0.45	3.20	0.94
Height (cm)				Stems				
0.1	24	0.000	0.04	0.0	0.97	3.11	0.79	0.73
1.0	41	0.000	0.04	0.0	0.60	2.98	0.38	1.42
10	86	0.001	0.04	5.9	0.26	0.39	0.38	1.68
50	100	0.008	0.06	7.8	0.08	0.30	0.41	1.91
200	103	0.011	0.09	7.9	0.07	0.28	0.41	3.00

first, and for one week about a month later. Unfortunately these fractionations are not all comparable since sulfate was not determined in three of the plants. However, in the other three, organic sulfur in the (—S) plants did not exceed 0.12 percent.

The nitrogen analyses are of great interest. Protein nitrogen showed variable changes due to sulfur deficiency. Cotton stems doubled their protein concentration in the (—S) plants, and soybean gave an increase of over 50 percent but the other plants decreased 10 to 20 percent. Nitrate nitrogen increased in the (—S) stems in every case except black mustard, in which the nitrate concentration was already very high. A large increase was observed without exception in the total soluble organic nitrogen. The (—S) plants had 2 to 11 times as much as the (+S) plants. Amide, amino, and ammonia nitrogen also increased by factors of 2 to 7 in all plants.

Reducing sugars were lowered by factors of 3 to 7 in all cases except NIGHTINGALE's tomatoes. The sucrose values were variable. There was a marked decrease in two cases, little change in one case, and appreciable increase in three cases. Starch and hemicellulose values, when low, increased by factors of 1.2 to 4, but when high the changes were both up and down, and of doubtful significance.

ERGLE and EATON (Table 9) used a range of sulfate concentrations in the nutrient solution and observed the effects on protein, soluble nitrogen, and sugar in the leaves as well as the stems of cotton. There were large changes in these constituents with change from extreme sulfur deficiency to abundant sulfate. The changes in the leaves and the stems were sometimes in opposite directions.

Protein concentration in the leaves increased over 3-fold with increase in the sulfur supply. Protein in the stems on the other hand decreased to half with increase in sulfur from extreme deficiency to moderate or adequate supply. The sulfur content of the proteins decreased to 60 percent in the leaves and increased 4-fold in the stems with increasing sulfur supply. The large increase in water-soluble nitrogen and the large decrease in reducing sugars due to sulfur deficiency, noted in Table 8 for the stems, were also found in the leaves. Thomas et al. (1950) observed increases of 15, 50, and 70 percent in the water soluble nitrogen of (—S) alfalfa leaves, stems, and roots respectively as compared with (+S) plants in (+N) nutrient solution. In nitrogen-deficient nutrient, (+S) and (—S) alfalfa plants had identical water-soluble nitrogen concentrations. In view of the low concentrations of simple sugars, and high concentrations of soluble organic nitrogen compounds in the (—S) leaves and stems, Ergle and Eaton suggest that the principal organic substances translocated in (—S) plants are carbon-nitrogen compounds rather than sugars.

Table 10. *Amino acid content of alfalfa as influenced by sulfur supply.*
(Mertz, Singleton, and Garey 1952.)

	+S (%)	—S (%)		+S (%)	—S (%)
			Leaves + Stems — 3rd Cutting		
Alanine	1.10	0.56	Methionine.	0.29	0.15
Arginine	0.95	1.04	Phenylalamine . . .	0.96	0.52
Aspartic acid . . .	1.90	6.22	Proline	0.90	0.49
Cystine	0.46	0.18	Serine	0.88	0.60
Glutamic acid . . .	1.41	0.55	Threonine	1.16	0.57
Glycine	0.87	0.50	Tryptophane	0.37	0.17
Histidine	0.36	0.24	Tyrosine	0.43	0.24
Isoleucine	0.86	0.54	Valine	1.09	0.62
Leucine	1.82	0.95	Total N	2.85	3.01
Lysine	0.89	0.53	Amide N	0.21	0.67
			4th Cutting		
	Leaves			Stems	
Arginine	1.48	3.31		0.47	0.97
Aspartic acid . . .	2.86	7.85		1.24	12.91
Glutamic acid . . .	2.11	1.57		1.45	0.22
Methionine.	0.40	0.22		0.14	0.09
			Nitrogen content		
Total N	4.55	5.06		1.78	4.49

Amino acids in alfalfa as affected by sulfur supply. Mertz et al. (1952), Table 10, analyzed (+S) and (—S) alfalfa for nearly all the amino acids in the leaves and stems. This very interesting study showed that in the third crop leaves plus stems, 16 amino acids were decreased in the (—S) plants by 35 to 65 percent of the (+S) values. The average decrease was 47 percent. Aspartic acid increased by a factor greater than 3 and arginine increased 10 percent. Amide nitrogen increased by a factor of 3, but total nitrogen increased only 5 percent. Evidently the large increase in the (—S) aspartic acid offset the decrease in the other amino acids.

In the fourth cutting, the aspartic acid increased by factors of 2.7 in the (—S) leaves and 10 in the (—S) stems. Arginine was doubled in both leaves and stems. Seventy-eight percent of the aspartic acid, 55 percent of the arginine,

17 percent of the glutamic acid, and 10 percent of the methionine in the (—S) plants were dialysable, and thus not a part of the alfalfa proteins. Aspartic acid calculated as asparagine accounted for 95 percent of the amide nitrogen, whereas glutamic acid accounted for only 8 percent as glutamine. Most of the free amide in the (—S) plants was therefore asparagine. The increase of arginine in the (—S) plants was due to the fact that half of this material was dialysable and was not a part of the protein.

Nitrate reductase. The data in Tables 6, 7, 8, and 9 suggest that nitrate reduction is inhibited by sulfur deficiency. ECKERSON (1932) found that the activity of the enzyme reductase was lowered by deficiencies of a number of essential elements such as phosphorus, potassium, calcium, and sulfur. If any of these elements were in short supply, nitrates accumulated without being reduced, and nitrogen starvation resulted even in the presence of abundant nitrate. Drastic changes occurred with phosphorus deficiency (ECKERSON 1931): first, loss of reductase activity as phosphate and water-soluble phosphorus compounds were used up; then, accumulation of nitrate, sugars and starch, with thickening of the cell walls and reduced rate of cell division on the meristematic tissues; finally, when phosphorus starvation became acute, the complex phosphorus compounds were broken down, the carbohydrates disappeared, and the plants soon died. Similar drastic consequences followed potassium deficiency. In sulfur deficiency, reductase activity decreased to about 15–20 percent, then remained constant for several weeks, indicating that synthesis of reductase was retarded but not completely inhibited.

More recently it has been found by EVANS and NASON (1953) that the reductase enzyme in soybean leaves and other plants is a flavoprotein of which flavin adenine dinucleotide is the prosthetic group. NICHOLAS and NASON (1955) showed that molybdenum is an integral part of the enzyme. The enzyme catalysed the reduction of nitrate to nitrite by oxidation of the reduced form of tri- or di-phosphopyridine nucleotide (TPNH and DPNH). The enzyme was sensitive to heavy metal inhibitors which can remove or bind the molybdate. It could be inactivated by dialysis against cyanide and reactivated by the addition of sodium molybdate. It could also be inactivated by p-chloromercuribenzoate, and reactivated by cysteine hydrochloride, suggesting that it has a sulfhydryl group on the active enzyme surface. The coenzymes TPNH and DPNH and the nitrate react mole for mole. The energy for the production of TPNH and DPNH is presumably supplied by the intermediates of carbohydrate metabolism, and this might explain the disappearance of carbohydrates. The enzyme is also present in the nodules of legumes (EVANS 1954). This might be inferred from the experiments of ANDERSON and SPENCER (1950). If nitrate reduction is involved in the process of nitrogen fixation, the initial step in fixation may be oxidative rather than reductive.

NIGHTINGALE (1932) noted that when (—S) plants were placed in (+S) nutrient solutions, sulfites appeared in considerable quantity, especially in the phloem of the tops and to a lesser degree in the phloem of the roots. Shortly following the appearance of sulfite, strong sulfhydryl reactions were observed. Cysteine was found and probably glutathione. These reactions were largely restricted to the phloem, cambium, and meristematic tips of the stems and roots. These observations support the work of HAMMETT (1929), who found that cell division at the root tips of peas and corn was regulated by sulfhydryl compounds. NIGHTINGALE was unable to find more than slight traces of sulfhydryl compounds in these active areas of (—S) plants, though appreciable amounts of soluble organic sulfur compounds were present.

Proteolysis and translocation. Translocation of sulfur occurs as sulfate. There is no definite evidence that other sulfur compounds are normally translocated. However, experiments in which cystine and methionine were taken up from the nutrient solution and utilized suggest that such translocations could occur. If there was an adequate sulfate supply to the roots, there would presumably be no need to reutilize the sulfur by translocating it from old to new growth or to the fruit. In sulfur deficiency, such translocation with or without proteolysis might be necessary. The experiments of Kylin (1953) indicate that these processes occur to some extent even in (+S) wheat.

The development of the seed in plants is accompanied by extensive translocation of organic compounds and the various nutrient elements from the vegetative parts of the plant to the seed. In wheat, as illustrated in Table 4, the kernels at final harvest had 65 percent of the total sulfur* in the plant: 2.5 times as much as the straw and 5 times as much as the roots. The kernels also had 40 percent of total dry weight of the plant. The high sulfate* content of the stems and sheaths as compared with the leaves and kernels suggests proteolysis in the leaves, translocation as sulfate to the heads, and resynthesis in the kernels of a different type of protein rich in cystine and methionine. It was noted that the straw and chaff associated with the primary kernels which were fully hardened at harvest had only about half the total sulfur that was present in the grain, whereas the straw and chaff of the less mature secondary stems that developed later had about the same total sulfur content as the kernels. Presumably about half the sulfur in the older straw had been translocated. There can be little doubt about the occurrence of proteolysis in ripening grain.

Amide, amino, and ammonia nitrogen accumulation in (—S) plants, together with reduced protein levels, occur, according to Nightingale et al. (1932) and Eaton (1941), because nitrate reduction can proceed slowly, resulting in simple nitrogen compounds, but protein synthesis is limited by the lack of available sulfur amino acids. Reducing sugars are also low. Stem elongation may be rapid. This is said to be the result of proteolysis in the older tissues and resynthesis in the meristematic growing points. Since the polysaccharides are usually increased in (—S) plants, in contradistinction to the general reduction of carbohydrates when proteolysis and stem elongation occur in the dark, Eaton considered proteolysis due to (—S) a special type. It is not known that the carbohydrate concentration is a controlling factor in proteolysis. Carbohydrate starvation in prolonged darkness would be expected to result in the utilization of all the readily available sources of energy. In sulfur deficiency, photosynthesis still goes on and carbohydrate production would therefore continue at a reduced rate. The reducing sugars would tend to be used up more rapidly than the polysaccharides, but the latter should accumulate to some extent as is shown in Tables 8 and 9. It is uncertain to what extent the accumulation of the soluble nitrogen compounds is due to proteolysis on the one hand or blocking of protein synthesis by an insufficient supply of sulfur amino acids, on the other.

Eaton (1941) sampled the upper, middle, and lower portions of sunflower stems to determine the concentration gradient of the various nitrogen, sulfur, and carbohydrate fractions. He found that all the constituents listed in Table 8, except sucrose and acid-hydrolysable carbohydrate, had concentrations in the upper sections of the (—S) stems 2 to 4 times greater than in the lower sections. The (+S) stems showed similar differences. Nightingale found small gradients in the same direction for sulfate and organic sulfur in tomato stems. These data do not support the hypothesis of proteolysis in the older, lower portion

---

* The asterisk refers to sulfur compounds containing radiosulfur S^{35}.

of the stem, and reutilization of the fragments, particularly the sulfur compounds, at the growing point. The hypothesis is also not supported by the observation that soybeans and mustard develop chlorotic leaves at the top of the stem while the leaves below remain green, though NIGHTINGALE observed the reverse in tomato. From a detailed consideration of the data and the literature, EATON (1941) concluded that proteolysis of this type is not precluded, but that it probably proceeds at or near the place where the soluble compounds are found, with transfer only through short distances. However, conclusive proof of this mechanism might be difficult to obtain, as suggested by ERGLE and EATON.

Summary of sulfur reactions in plants. There are still many gaps in our knowledge of the role of sulfur in plant nutrition. The pathway and mechanism of the reduction of sulfate is one of the more obvious areas for investigation. While there is a formal resemblance between nitrate reduction and sulfate reduction, little information is available on the latter. Enzyme systems must be involved. Sulfite is probably an intermediate step. It has been observed that shortly after (—S) plants are transferred to a (+S) nutrient, positive sulfite tests may be obtained, while sulfhydryl tests are still negative. Using *Neurospora*, two pathways have been suggested for the reduction of sulfate as follows (HOROWITZ et al. 1949):

Sulfate → sulfite → thiosulphate + 3—C compound → cysteine.

Sulfate + 3—C compound → cysteic acid → cysteine sulfinic acid → cysteine.

The two pathways are not independent of each other but probably in equilibrium. *Neurospora* cannot utilize sulfide. The reactions for higher plants are obscure.

A simple *in vitro* reaction in the reverse direction is well established (THOMAS and HENDRICKS 1944). This is the hydrolysis of cystine which produces pyruvic acid by deamination and equivalent amounts of sulfide, sulfite and sulfate by dismutation. These reactions go readily in an alkaline medium at boiling temperature in the presence of a sulfide collector. Hydrolysis and deamination occur simultaneously. Enzymatic deamination of cysteine and conversion to sulfide and pyruvic acid was accomplished by BINKLEY (1943) using a preparation from brewers' yeast. The enzyme could not break the disulfide bond of cystine. No sulfate or sulfite would be expected from the reaction with cysteine.

Sulfur is needed for the functioning of nitrate reductase. Beyond the suggestion that reactive sulfhydryl groups may be present on the enzyme surface, there is no information as to how the sulfur participates in the reduction process. Glutathione or related peptides are required as co-enzymes for the action of glyoxalase in converting methylglyoxal to lactic acid (BEHRENS 1941). The important enzyme cytochrome c contains 1.48 percent sulfur (THEORELL and ÅKESSON 1941). The compound has a hematin molecule linked to protein through thio-ether bonds and histadine-imidazol groups. The prosthetic porphyrin c group, when removed from cytochrome c, appears to be a di-cysteine porphyrin. In the enzyme, it is the iron rather than the sulfur that is primarily responsible for the oxidation-reduction properties. Cocarboxylase is the pyrophosphate of thiamin. It may function in the Krebs cycle. Lipoic acid (6–8 dithiooctanoic acid or 6–8 thioctic acid) functions in several enzyme systems, in conjunction with DPN. The acetylating co-enzyme A depends on an active sulfur group, which can exist in four forms (BASFORD and HUENNEKENS 1955) including sulfhydryl, two disulfides, and possibly thiazoline.

Until recently, the enzymes associated with photosynthesis were too obscure to be discussed in any satisfactory detail. CALVIN and co-workers (1953–1954) now postulate that the energy of the light-excited state of chlorophyll, about

40 kcal, is transferred to the disulfide linkage of 6–8 thioctic acid which in turn reacts with water (30–40 kcal required) or perhaps other hydrogen donors to form a thiol sulfenic acid. Dismutation of two molecules of this compound gives a disulfenic acid and a dithiol. The latter is a powerful reducing agent which is capable of implementing the necessary reductive steps in carbon dioxide assimilation. The former yields a peroxide compound from which the elemental oxygen of photosynthesis can be derived.

HENDLEY and CONN (1953) discuss photochemical reactions by which di- or triphospho pyridine nucleotides are reduced by water with the evolution of oxygen in the presence of illuminated spinach chloroplasts. The reactions require oxidized glutathione and glutathione reductase. Optimum yields in the Warburg apparatus were: reduced glutathione, 95 percent of theoretical, and oxygen, 80 percent.

Many other enzymes involving sulfur compounds, directly or indirectly, exist. The presence of sulfhydryl groups in meristematic tissue should be emphasized. Cytoplasmic and chloroplast proteins have sulfur amino acids as essential constituents. Evidently the synthesis of these proteins does not proceed in the absence of an adequate supply of sulfur. It can be inferred that cysteine and methionine groups in the proteins may contribute oxidation-reduction or methylating properties to the protoplasm, but little is definitely known about these functions. In certain enzymes, the seat of the reaction appears to be an —SH group attached to the surface of the protein moiety. This may be the case with nitrate reductase. In glyceraldehyde dehydrogenase, KRIMSKY and RACKER (1955) suggest that the enzyme has DPN attached to the protein through a thiol linkage. On reaction with an aldehyde substrate it yields DPNH and an acyl enzyme with the acyl group attached to the protein through sulfur. This thiol ester is split by inorganic phosphate to yield acyl phosphate and regenerated SH-enzyme.

Evidently sulfur compounds play a major chemical role in plant physiology.

Literature.

ALWAY, F. J.: A nutrient element slighted in agricultural research. J. Amer. Soc. Agron. **32**, 913–921 (1940). — ANDERSON, A. J., and D. SPENCER: Molybdenum in nitrogen metabolism of legumes and non-legumes. Austral. J. Sci. Res. B **3**, 414–430 (1950). — Sulfur in nitrogen metabolism of legumes and non-legumes. Austral. J. Sci. Res. B **3**, 431–449 (1950).

BARRIEN, B. S., and J. G. WOOD: Studies on sulfur metabolism of plants. II. New Phytologist **38**, 257–264 (1939). — BASFORD, R. E., and F. M. HUENNEKENS: Studies on thiols. II. J. Amer. Chem. Soc. **77**, 3878–3882 (1955). — BEHRENS, O. K.: Coenzymes for glyoxalase. J. of Biol. Chem. **141**, 503–508 (1941). — BINKLEY, F.: On the nature of serine dehydrase and cysteine desulfurase. J. of Biol. Chem. **150**, 261–262 (1943).

CALVIN, M.: Chemical and photochemical reactions of thioctic acid and related disulfides. I. Federat. Proc. **13**, 697–711 (1954). — CHIBNALL, A. C.: Protein metabolism in the plant. New Haven: Yale University Press 1939.

EATON, S. V.: Influence of sulfur deficiency on the metabolism of the soybean. Bot. Gaz. **97**, 68–100 (1935). — Influence of sulfur deficiency on the metabolism of the sunflower. Bot. Gaz. **102**, 536–556 (1941). — Influence of sulfur deficiency on the metabolism of black mustard. Bot. Gaz. **104**, 306–315 (1942). — Effects of sulfur deficiency on the growth and metabolism of the tomato. Bot. Gaz. **112**, 300–307 (1951). — ECKERSON, S. H.: Influence of phosphorus deficiency on metabolism of the tomato *(Lycopersicum esculentum* Mill.*)*. Contrib. Boyce Thompson Inst. **3**, 197–217 (1931). — Conditions affecting nitrate reduction by plants. Contrib. Boyce Thompson Inst. **4**, 119–130 (1932). — ERGLE, D. R., and F. M. EATON: Sulfur nutrition of cotton. Plant Physiol. **26**, 639–654 (1951). — EVANS, H. J.: Diphosphopyridine nucleotide-nitrate reductase from soybean nodules. Plant Physiol. **29**, 298–301 (1954). — EVANS, H. J., and A. NASON: Pyridine nucleotide-nitrate reductase from extracts of higher plants. Plant Physiol. **28**, 233–254 (1953).

GALSTON, A. W.: Isolation, agglutination and nitrogen analysis of intact oat chloroplasts. Amer. J. Bot. **30**, 331–334 (1943). — GILBERT, F. A.: The place of sulfur in plant nutrition.

Bot. Review **17**, 671–691 (1951). — GRANICK, S.: Quantitative isolation of chloroplasts from higher plants. Amer. J. Bot. **25**, 558–561 (1938). — Chloroplast nitrogen of some higher plants. Amer. J. Bot. **25**, 561–567 (1938).

HAMMETT, F. S.: The chemical stimulus essential for growth by increase in cell number. Protoplasma **7**, 297–322 (1929). — HANSON, E. A.: A note on the metabolism of chloroplast protein. Austral. J. Exper. Biol. a. Med. Sci. **19**, 157–159 (1941). — HANSON, E. A., B. S. BARRIEN and J. G. WOOD: Relations between protein nitrogen, protein sulfur and chlorophyll in the leaves of Sudan grass. Austral. J. Exper. Biol. a. Med. Sci. **19**, 231–234 (1941). — HARRISON, B. F., M. D. THOMAS and G. R. HILL: Radioautographs showing the distribution of radiosulfur in wheat. Plant Physiol. **19**, 245–257 (1944). — HEISERICH, E.: Sulfur metabolism in maize and tobacco. Z. Pflanzenernährg **37**, 55–72 (1935). — HENDLEY, D. D., and E. E. CONN: Enzymatic reduction and oxidation of glutathione by illuminated chloroplasts. Arch. of Biochem. a. Biophysics **46**, 454–464 (1953). — HOROWITZ, N. H., M. FLING, B. O. PHINNEY and S. C. SHEN: Recent experiments on the methionine requiring mutants of *Neurospora.* Abstracts of Papers, Am. Chem. Soc. Div. Biol. Chem., San Francisco, p. 450, 1949.

KRIMSKY, I., and E. RACKER: Acyl derivatives of glyceraldehyde-3-phosphate dehydrogenase. Science (Lancaster, Pa.) **122**, 319–321 (1955). — KYLIN, Å.: The uptake and metabolism of sulfate by deseeded wheat plants. Physiol. Plantarum **6**, 775–795 (1953).

MANDELS, G. R.: A quantitative study of chlorosis in *Chlorella* under conditions of sulfur deficiency. Plant Physiol. **18**, 449–462 (1943). — MENKE, W.: Protoplasm of green plant cells. I. Z. physiol. Chem. **257**, 43–48 (1938). — MERTZ, E. T., V. L. SINGLETON and C. L. GAREY: The effect of sulfu rdeficiency on the amino acids of alfalfa. Arch. of Biochem. a. Biophysics **38**, 139–145 (1952). — MILLER, H. G.: Further studies on relation of sulfates to plant growth and composition. J. Agricult. Res. **22**, 101–110 (1921). — MILLER, L. P.: DL-Methionine as a source of sulfur by growing plants. Contrib. Boyce Thompson Inst. **14**, 443–456 (1947).

NELLER, J. R.: Effect of sulfur upon nitrogen content of legumes. Industr. Engin. Chem. **18**, 72–73 (1926). — NICHOLAS, D. J. D., and A. NASON: Role of molybdenum as a constituent of nitrate reductase from soybean leaves. Plant Physiol. **30**, 135–138 (1955). — NIGHTINGALE, G. T., L. G. SCHERMERHORN and W. R. ROBBINS: Effects of sulfur deficiency on metabolism in tomato. Plant Physiol. **7**, 565–595 (1932). — NOACK, K., and E. TIMM: Proteins in chloroplasts and cytoplasm of spinach leaves. Naturwiss. **30**, 453 (1942).

POWERS, W. L.: Sulfur in relation to soil fertility. Oregon Agric. Exper. Stat. Bull. **199**, 5–45 (1923).

SMITH, E. L.: Chlorophyll as the prosthetic group in the green leaf. Science (Lancaster, Pa.) **91**, 199–200 (1940). — STEWARD, F. C., J. F. THOMPSON, F. K. MILLAR, M. D. THOMAS and R. H. HENDRICKS: The amino acids of alfalfa as revealed by paper chromatography with special reference to compounds labeled with sulfur. Plant Physiol. **26**, 123–135 (1951). — STOREY, H. H., and R. LEACH: A sulfur-deficiency disease of the tea bush. Ann. Appl. Biol. **20**, 23–56 (1933).

THEORELL, H., and Å. ÅKESSON: Studies on cytochrome c. I., II., III., IV. J. Amer. Chem. Soc. **63**, 1804–1827 (1941). — THOMAS, M. D.: Agricultural research with radioactive sulfur and arsenic. Proc. Auburn Conference on Use of Radioactive Isotopes in Agricultural Research. p. 103–117. Alabama Polytechnic Institute, Auburn, Ala., 1948. — THOMAS, M. D., and R. H. HENDRICKS: The hydrolysis of cystine and the fractionation of sulfur in plant tissues. J. of Biol. Chem. **153**, 313–325 (1944). — THOMAS, M. D., R. H. HENDRICKS, L. C. BRYNER and G. R. HILL: A study of the sulfur metabolism of wheat, barley and corn using radioactive sulfur. Plant Physiol. **19**, 227–244 (1944). — THOMAS, M. D., R. H. HENDRICKS and G. R. HILL: Some chemical reactions of sulfur dioxide after absorption by alfalfa and sugar beets. Plant Physiol. **19**, 212–226 (1944). — Sulfur metabolism of plants. Effect of sulfur dioxide on vegetation. Industr. Engin. Chem. **42**, 2231–2235 (1950). — Sulfur metabolism in alfalfa. Soil Sci. **70**, 19–26 (1950). — TURRELL, F. M., and J. R. WEBER: Elemental sulfur dust, a nutrient for lemon leaves. Science (Lancaster, Pa.) **122**, 119–120 (1955).

WILDMAN, S. G., and J. BONNER: The proteins of green leaves. I. Arch. of Biochem. **14**, 381–413 (1947). — WOOD, J. G.: Metabolism of sulfur in plants. Chronica bot. **7**, 1–4 (1942). — WOOD, J. G., and B. S. BARRIEN: Studies on the sulfur metabolism of plants. I. New Phytologist **38**, 125–149 (1939). — Studies on the sulfur metabolism of plants. III. New Phytologist **38**, 265–272 (1939).

Secondary organic sulfur-compounds of plants.
(Thiols, sulfides, sulfonium derivatives, sulfoxides, sulfones and *iso*thiocyanates).

By

Anders Kjær.

With 1 figure.

The present chapter covers selected groups of sulfur-containing products of plant origin. The subject will be discussed from a chemical rather than a physiological viewpoint, primarily because our present knowledge of the secondary sulfur-compounds has hardly passed beyond the stage of structure determination and analysis. In the discussion emphasis will be placed on subjects of actual interest.

I. Thiols (mercaptanes).

Chemical properties. Alkanethiols, RSH, of low molecular weight are volatile compounds, possessing a most offensive odour. They are soluble in alkali and form insoluble mercaptides with salts of heavy metals. A characteristic property of the thiols is their facile oxidation to disulfides, RSSR, which takes place under the influence of various oxidizing agents or atmospheric oxygen.

a) Scope of discussion.

Several organic compounds, possessing one or more thiol-groups, are of the utmost biological significance, inasmuch as they contribute to the maintenance of the steady state of living cells and play an essential rôle in the process of cell division and growth. Various soluble thiols (*e.g.* cysteine, glutathione and coenzyme A) as well as a great number of enzymically active thiol-proteins belong to this group of normal cell constituents, which has been treated previously in this volume (BERSIN, p. 16) and elsewhere (BARRON). The present discussion will be limited to various thiols of unknown biological significance occasionally encountered in higher plants and microorganisms.

b) Occurrence and biosynthesis.

The presence of the volatile *methanethiol* has been demonstrated in fresh root material of *Raphanus sativus* L. (NAKAMURA) as well as in the leaves of various *Lasianthus* species (KOOLHAAS).

A likely origin of the compound is suggested through the extensive studies by CHALLENGER *et al.* on biological methylation in moulds (CHALLENGER 1951). During this work it was observed that the wood-destroying fungus *Scopulariopsis brevicaulis* was capable of transforming DL-methionine and S-methyl-L-cysteine into methanethiol. Similar fissions of the CH₃S—C linkage were also noticed in cultures of the fungus *Microsporum gypseum* (STAHL *et. al.*) and in a gram-negative bacterium strain isolated from soil (MITSUHASHI). The presence in higher plants of a similar enzyme, capable of splitting S-alkyl-L-cysteine derivatives, but not methionine, into alkanethiols, ammonia and pyruvic acid, has recently been established by GMELIN and coworkers. During attempts to unveil the reactions responsible for the onion-like odour which appears on steeping ground seed of *Albizzia lophantha* Benth. *(Mimosaceae)* in water, the authors found djenkolic acid to function, in this particular case, as substrate for an enzyme of a rather wide and diverse range of activity. This cleavage has its counterpart in mammalian tissues where enzymic fissions of cystathionine (DU VIGNEAUD) and various S-alkylcysteines (BINKLEY) represent reactions of the same type.

An alternative route to thiols may proceed from naturally occurring sulfoxides undergoing enzymic disproportionation, according to the following scheme:

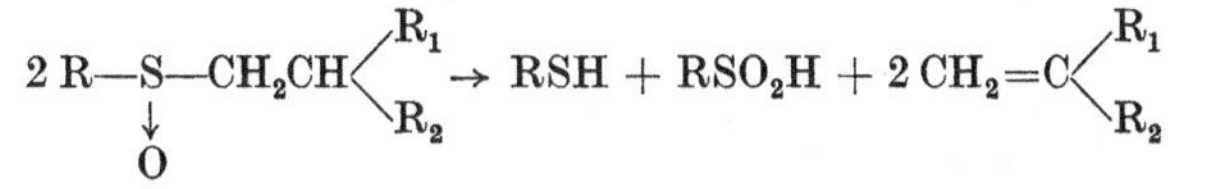

$$2\ R\!-\!S\!-\!CH_2CH{\overset{R_1}{\underset{R_2}{\diagdown}}} \rightarrow RSH + RSO_2H + 2\ CH_2{=}C{\overset{R_1}{\underset{R_2}{\diagdown}}}$$
$$\underset{O}{\overset{\downarrow}{}}$$

This type of reaction has long been recognized in classical organic chemistry. Various sulfoxides have recently been encountered in higher plants and additional ones may still appear (see IV. a).

Freshly chopped bulbs of onion (*Allium Cepa* L.) contain 1-propanethiol (CHALLENGER and GREENWOOD 1949) the origin of which is uncertain. Again, S-propylcysteine, which has only recently been found in Nature, or a related S-propyl-compound may conceivably act as precursors.

A unique natural thiol, 3,3'-dimercapto*iso*butyric acid $(HSCH_2)_2CHCOOH$, has been isolated in its disulfide form from asparagus (JANSEN). The dithiol may function as a hydrogen transfer substance in the metabolism, similar to the structurally analogous reduced form of thioctic acid and glutathione, and at the same time be of importance as a regulator of enzyme processes by controlling heavy metal inhibition. The following sequence of reactions has been suggested as a possible biosynthetic route to dimercapto*iso*butyric acid (CHALLENGER and LIU 1950):

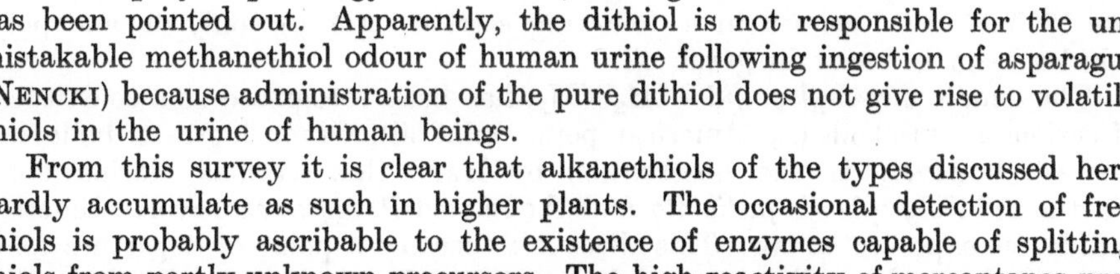

$$HSCH_2CH(NH_2)COOH \xrightarrow{[O]+H_2O} HSCH_2COCOOH \xrightarrow{CH_2COOH}$$

$$\underset{HOOCCH_2}{\overset{HSCH_2}{\diagup}}C{\overset{OH}{\underset{COOH}{\diagdown}}} \xrightarrow{-H_2O} \underset{HOOCCH}{\overset{HSCH_2}{\diagup}}C{-}COOH \xrightarrow{-CO_2} \underset{CH_2}{\overset{HSCH_2}{\diagup}}C{-}COOH$$

$$\xrightarrow{H_2S} \underset{HSCH_2}{\overset{HSCH_2}{\diagup}}CH{-}COOH$$

The step-by-step analogy to reactions, leading to citric acid in mould-cultures, has been pointed out. Apparently, the dithiol is not responsible for the unmistakable methanethiol odour of human urine following ingestion of asparagus (NENCKI) because administration of the pure dithiol does not give rise to volatile thiols in the urine of human beings.

From this survey it is clear that alkanethiols of the types discussed here hardly accumulate as such in higher plants. The occasional detection of free thiols is probably ascribable to the existence of enzymes capable of splitting thiols from partly unknown precursors. The high reactivity of mercaptanes precludes their accumulation and may account for the natural occurrence of certain disulfides and similar products, discussed in the following.

Recent results, however, point to methanethiol as an intermediate of possible metabolic significance. Thus, BLACK and WRIGHT demonstrated the presence in yeast of an enzyme catalyzing the formation of 3-phosphoglyceryl methylthiol ester (PGMTE) from 3-phosphoglyceric acid and methanethiol in the presence of adenosinetriphosphate. The further existence in yeast of a specific phosphatase, transforming PGMTE into phosphate and glyceryl methylthiol ester, supports the hypothesis that methanethiol may play a decisive rôle in sulfur metabolism. This view is further strengthened through the observation by WOLFF *et al.*

that a purified yeast enzyme can accomplish the synthesis from L-serine and methanethiol of S-methyl-L-cysteine, which may, in turn, be on the regular pathway of microbial cysteine biosynthesis.

It is interesting in this connexion that the latter amino acid, which THOMPSON *et al.* newly isolated from *Phaseolus vulgaris*, is present in and will support growth of several strains of *Neurospora crassa* (RAGLAND and LIVERMAN).

II. Sulfides and sulfonium compounds.

Chemical properties. Organic sulfides (thioethers), RSR_1, can be derived from hydrogen sulfide by substituting both hydrogen atoms with organic radicals. The simple aliphatic sulfides are volatile compounds possessing a disagreeable odour. They usually form solid addition compounds with salts of heavy metals and can be oxidized stepwise to sulfoxides, R—SO—R_1, and sulfones, R—SO_2—R_1. Upon treatment with alkyl halides, R_2X, the sulfides are easily transformed into sulfonium salts, $(RR_1R_2)S^+X^-$, containing a positively charged sulfur atom and resembling the quaternary ammonium compounds in many respects. Thus, the strongly basic sulfonium hydroxides decompose readily to give a sulfide and an unsaturated molecular fragment.

a) Scope of discussion.

From higher plants and microorganisms several sulfides have been isolated and identified. Some of these belong to the class of α-amino acids (*e.g.* S-methyl-L-cysteine, methionine, djenkolic acid, lanthionine), whereas others are cyclic sulfides occurring in antibiotics (*e.g.* penicillin, bacitracin) and vitamins (thiamine, biotin). None of these will be considered in the present discussion which will deal mainly with volatile alkyl sulfides and their natural precursors, a field of considerable recent progress.

b) Occurrence of sulfides.

For many years the term "Lauchöle" has been used as an unspecific designation for the complex mixtures of sulfides, disulfides and polysulfides which constitute the essential oils of various plant species, substantially belonging to the genus *Allium*. SEMMLER (1887) fractionated the crude oil obtained from *A. ursinum* L. and demonstrated that *divinyl sulfide*, $CH_2=CH$—S—$CH=CH_2$, was the major constituent. The same author later proved older reports of the existence of diallyl sulfide in oil of garlic (*A. sativum* L.) to be erroneous (SEMMLER 1892a). As discussed in a following section, (III. b), only di- and polysulfides seem to occur in garlic oil.

The simple *dimethyl sulfide*, CH_3SCH_3, has been recognized as a constituent of various essential oils (*e.g.* American peppermint oil). Of much greater biological interest, however, is the observation by HAAS that various marine algæ, *viz.* the red *Polysiphonia fastigiata* and the related *P. nigrescens*, evolve dimethyl sulfide on exposure to air. This discovery became a point of departure for a series of thorough studies in the laboratories of CHALLENGER at Leeds on the distribution and formation of sulfur compounds in plants [reviewed by CHALLENGER (1951) and (1953)]. Thus, the green seaweeds *Enteromorpha intestinalis* (BYWOOD *et al.* 1951) and *Spongomorpha arcta* (BYWOOD 1953), as well as various species of *Equisetaceae* (BYWOOD *et al.* 1951) were found to evolve dimethyl sulfide on exposure to alkali. The same was true for the common bracken (*Pteridium aquilinum*) and the fern *Athyrium filix-femina* (BYWOOD *et al.* 1951). In addition, asparagus has recently been shown to produce dimethyl sulfide on treatment with hot alkali (CHALLENGER and HAYWARD 1954).

Two somewhat special sulfides have lately been recognized as the acid moieties of the spasmolytically active esters, petasolester B and C, isolated by

STOLL *et al.* (1956) from rhizomes of *Petasites officinalis* Moench. Degradation and synthesis have strongly suggested their formulations as *cis-* and *trans-3-methylthioacrylic acid*.

c) Sulfonium compounds.

Several observations suggest that dimethyl sulfide is not stored as such, but rather appears as a result of enzymic or hydrolytic cleavage of some parent substance. From the marine alga *Polysiphonia fastigiata* CHALLENGER and SIMPSON (1948) isolated the likely precursor of dimethyl sulfide and proved it to be *dimethyl-β-propiothetin*, $(CH_3)_2\overset{+}{S}CH_2CH_2COOH$, the first sulfonium-compound to be found in plants. The widely varying sources of dimethyl sulfide listed above suggest a wide-spread occurrence of methyl sulfonium-compounds in Nature. The enzymic nature of the dimethyl sulfide-producing reaction has recently been verified by CANTONI and ANDERSON. From *Polysiphonia fastigiata* they isolated an enzyme which, in the presence of SH-containing activators, will bring about the fission of dimethyl-β-propiothetin to dimethyl sulfide and acrylic acid.

The sulfide, liberated on heating the urine of dogs with alkali, has been identified by LEAVER and CHALLENGER as a mixture of methyl *n*-propyl sulfide and probably methyl *n*-butyl sulfide, a finding which suggests the occurrence in the urine of additional thetines, very probably of dietary provenance. In this connexion, the recent demonstration by McRORIE *et al.* (1954) of the presence in cabbage juice of the methylsulfonium-derivative of methionine, $(CH_3)_2\overset{+}{S}CH_2CH_2CH(NH_2)COO^-$, is very interesting, because it represents an alternative carrier molecule for the extra methyl group. Its biological function is still obscure, but it appears as if the sulfonium-derivative is involved in the incorporation of one-carbon units in essential constituents of bacterial cells. S-Methylmethionine has recently been identified as a constituent of asparagus also (CHALLENGER and HAYWARD 1954). It may be identical with the so-called vitamin U of cabbage juice which appears to be of some promise in the treatment of peptic ulcers in human beings (see *e.g.* BERSIN *et al.* 1956).

Two sulfides, obviously related to methionine, have formerly been recognized as plant constituents, *viz.* *3-methylthiopropanol* (methionol), $CH_3SCH_2CH_2CH_2OH$, in soya-sauce (AKABORI and KANEKO) and *methyl 3-methylthiopropionate*, $CH_3SCH_2CH_2COOCH_3$, in the essential oil of pineapple (HAAGEN-SMIT *et al.*). Both may have common origin in 2-keto-4-methylthiobutyric acid, $CH_3SCH_2CH_2COCOOH$, formed from methionine by transamination. Decarboxylation of the keto acid, followed by enzymic disproportionation of the hypothetical aldehyde[1] to methionol and 3-methylthiopropionic acid, may represent a likely sequence of reactions. Biological methylation of the latter would result in the seaweed thetin which, in turn, under special circumstances, might undergo intramolecular transmethylation to the methyl ester of pineapple oil. The latter reaction has reasonably good chemical analogies.

The detection of naturally occurring sulfonium-compounds directed much attention to their possible metabolic rôle, especially in biological methylation. Working with liver tissue CANTONI (1953) showed that the long known enzymic transfer of methyl groups from methionine proceeded *via* an active intermediate sulfonium-compound (S-adenosylmethionine) of the following structure:

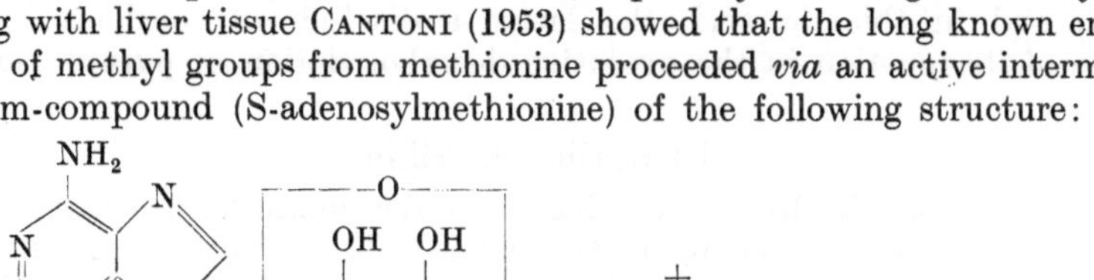

---

[1] It may be of interest that this aldehyde, $CH_3SCH_2CH_2CHO$, has recently been identified in milk after exposure to light. Although it almost certainly arises from methionine further details concerning its formation are still unknown [Anonymous author: Der Lichtgeschmack in Milch. Milchwiss. **10**, 74 (1955)].

There is good evidence that the "active methionine", obviously formed from methionine and adenosine-triphosphate, operates in plant metabolism also as a "bridge" in the transfer of methyl groups between various cell constituents. For example, this discovery offers an explanation for the long recognized existence of an adenyl-thiomethylpentose in yeast *(Sacch. cerevisiae)*. The thiosugar was finally proved by WEYGAND *et al.* (1950) and independently by SATOH and MAKINO to be *9-adenyl-(5'-desoxy-5'methylthio)-β-D-ribofuranoside:*

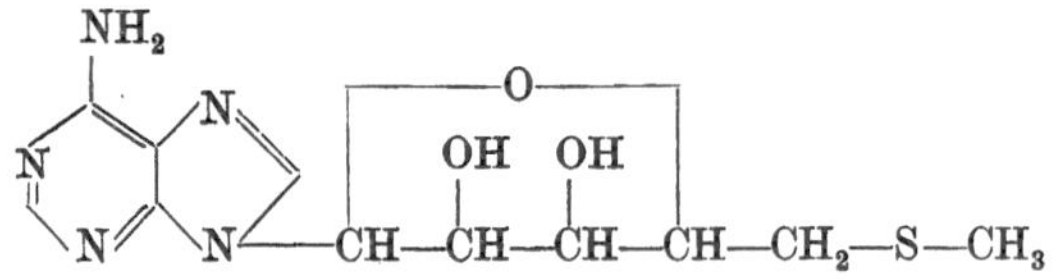

A suggestion that the presence of this thiosugar in biological materials might be the result of chemical hydrolysis of the labile "active methionine"(WEYGAND *et al.* 1952) has received support from recent chemical studies of the stability of the latter (BADDILEY *et al.*).

d) Special sulfides.

Sulfides of a rather special type (thioglucosides) are present in Nature as the mustard oil glucosides which will be discussed later (V. A). Recent results from this laboratory have disclosed the formation of *3-methylthiopropyl* (KJÆR *et al.* 1955a), *4-methylthiobutyl* (KJÆR and GMELIN 1955b) and *5-methylthiopentyl* (KJÆR *et al.* 1955c) iso*thiocyanate* by enzymic fission of glucosides present in various higher plants, a subject likewise to be covered below (V. B. b).

III. Disulfides and polysulfides.

Chemical properties. Organic disulfides, RS—SR, are well-defined compounds, easily formed by intermolecular oxidation of thiols. They are rather stable substances, susceptible however to reductive cleavage by suitable agents such as sulfides, hydrides or alkali metals. The polysulfides, containing several sulfur atoms in linear arrangement, behave chemically as mixtures of di-, tri-, tetra- and penta-sulfides. On distillation, polysulfides tend to lose sulfur with simultaneous formation of tri- and di-sulfides.

a) Scope of discussion.

The reversible reaction between thiols and disulfides, $2 RSH \rightleftharpoons RS—SR +$ [2H], is of considerable biological importance. Redox systems such as glutathione and its disulfide and cysteine-cystine represent familiar examples, which will not be treated in this chapter. Nor will any attention be given to cyclic disulfides such as occur in antibiotics (*e.g.* gliotoxin, thiolutin, aureothricin) and in thioctic acid, the latter a disulfide of considerable interest because of its function in the oxidative decarboxylation of pyruvic acid.

b) Garlic oil. Alliin.

Various simple disulfides were long ago recognized in distillates of higher plants, mainly those belonging to the genus *Allium*. Thus, SEMMLER (1892a) submitted garlic oil to fractional distillation and identified the individual constituents: *allyl propyl disulfide, diallyl disulfide, diallyl trisulfide* and, less certainly, *diallyl tetrasulfide*. According to present-day knowledge, however, all of these should probably be regarded as artifacts, secondarily formed from a common precursor by disproportionation. By careful distillation of a garlic extract CAVALLITO and BAILEY (1944a) isolated a water-soluble, antibacterial compound *allicin* for which two alternative structures were advanced (Cavallito *et al.* 1944 b). The choice between these was made by Stoll and Seebeck (1947) who demonstrated that allicin was a monosulfoxide of diallyl disulfide, $CH_2=CH—CH_2—SO—S—CH_2—CH=CH_2$. Allicin itself, however, was shown to be a product of enzymic degradation of a parent garlic constituent (Cavallito *et al.* 1945). The latter, *alliin*, was finally isolated from *A. sativum* and *A. ursinum* by Stoll and Seebeck (1948) and identified as (+)-S-allyl-L-cysteine sulfoxide, $CH_2=CH—CH_2—SO—CH_2—CH(NH_2)COOH$, representing the first sulfoxide of natural provenance. The biosynthesis of alliin is unknown. It may be formed by enzymic oxidation of S-allyl-L-cysteine which may, in turn, result as a product of reactions fundamentally analogous to those operating in the transformation of methionine *via* cystathionine to cystine in the animal cell. The enzymic cleavage of alliin to allicin and further to diallyl disulfide has been depicted as follows (Stoll and Seebeck 1949):

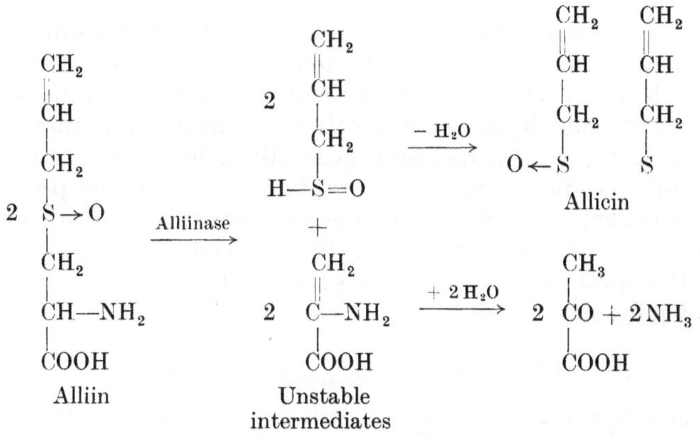

When submitted to distillation, allicin readily decomposes to yield diallyl disulfide, accompanied by higher sulfides as discussed above. The chemistry and biochemistry of alliin have been thoroughly reviewed elsewhere (Stoll and Seebeck 1951).

A sensitive test for the detection and identification in plant material of allicin, or analogously constituted compounds, has been introduced by Matsukawa *et al.* subsequent to the establishment of the reaction between thiamine and ingredients of *Allium sativum*. Ring-cleavage of the vitamin, followed by reaction of the open thiol-form with allicin, originating in enzymic fission of alliin, results in the formation of the crystalline allithiamine and analogous compounds. By this approach evidence was obtained that also S-methyl- and S-propyl-cysteine sulfoxides, structural analogues to alliin, are species of natural provenance. In fact, the methyl derivative has since been isolated from various natural sources (*cf.* IVa).

Though obviously different from alliinase, the *Albizzia*-enzyme discussed above (I b) will accomplish the same cleavage of alliin. In general, however, the *Albizzia*-enzym seems to be far less specific than alliinase in its substrate requirements.

c) Partly unidentified di- and poly-sulfides.

Various disulfides reported in the older literature may on renewed investigation prove to be secondary cleavage products of unknown precursors. Distillates of onions (*Allium Cepa* L.) were found (Semmler 1892 b) to contain an unidentified disulfide, $C_6H_{12}S_2$, in admixture with higher sulfides possessing the same carbon skeleton. Again, the essential oil of asafetida was reported as a mixture of various unsaturated disulfides (Semmler 1891) with $C_7H_{14}S_2$ and $C_{11}H_{20}S_2$ as major components, accompanied by small amounts of the disulfides $C_8H_{16}S_2$ and $C_{10}H_{18}S_2$. The latter were all found to be optically active compounds.

In 1936, MANNICH and FRESENIUS proved the major component of asafetida oil to be optically active sec-*butyl-1-propenyl disulfide*,

$$CH_3—CH(C_2H_5)—S—S—CH=CH—CH_3.$$

The biosynthetic origin of such asymmetric disulfides is difficult to visualize. They may arise, however, from a type of precursor fundamentally different from alliin, although no experimental evidence is on hand. The actual presence of volatile disulfides in plants is therefore rather questionable with the possible exception, however, of small amounts of symmetrical derivatives resulting from spontaneous oxidation of traces of alkanethiols.

A disulfide of a rather special type is tetraethylthiuram disulfide:

$$\begin{matrix} C_2H_5 \diagdown \\ \diagup \\ C_2H_5 \end{matrix} N—CS—S—S—CS—N \begin{matrix} \diagup C_2H_5 \\ \diagdown \\ C_2H_5 \end{matrix}$$

recently identified by SIMANDL and FRANC as a constituent of the smooth ink-cap *(Coprinus atramentarius)*. The compound is identical with "Antabus", a synthetic drug widely used in the treatment of alcoholism during recent years. The present case is one of the rare examples of a long known synthetic substance of established therapeutic significance being subsequently recognized as a natural product.

For some thirty years it has been generally believed that various *Brassica* species, notably varieties of *B. oleracea*, contain an antithyroid principle eliciting goiter in animals kept on a diet including large quantities of raw cabbage. The recent papers by JIROUSEK review the controversial literature on this sulfur-containing *Brassica*-factor and present evidence that it behaves chemically as a mixture of polysulfides.

IV. Sulfoxides and sulfones.

Chemical properties. Organic sulfides can be oxidized stepwise to sulfoxides, $R—SO—R_1$, and sulfones, $R—SO_2—R_1$. The sulfoxides are rather unstable, low-melting solids of slightly basic character. They possess a tetrahedral sulfur atom as evident from the resolution of unsymmetrical sulfoxides into optical antipodes. The sulfones are neutral, higher-melting solids of great stability, containing two coordinate links to the oxygen atoms.

a) Natural sulfoxides.

During recent years several sulfoxides have been identified as natural products. Allicin and alliin, discussed above, were the first representatives of this class of compounds to be found in Nature. In 1948 SCHMID and KARRER established the structure for *sulforaphene*, a non-volatile *iso*thiocyanate occurring

$$CH_3—\overset{\overset{\displaystyle O}{\uparrow}}{S}—CH=CH—CH_2—CH_2—N=C=S$$
Sulforaphene

in seeds of *Raphanus sativus* L., accompanied by the corresponding nitrile. It is remarkable by being the first natural product owing its optical activity solely to an asymmetric sulfur atom. In 1954, SCHULTZ and GMELIN (a) isolated from *Iberis amara* L. a crystalline mustard oil glucoside, *glucoiberin*, again containing a sulfoxide-group in the side chain[1].

$$CH_3\overset{\overset{\displaystyle O}{\uparrow}}{S}CH_2CH_2CH_2C \begin{matrix} \diagup S\text{-Glucose} \\ \diagdown \\ N—O—SO_2—O^- \ \ K^+ \end{matrix}$$
Glucoiberin

---

[1] The reproduced formula has been corrected so as to be in conformity with the revised general glucoside structure established by ETTLINGER and LUNDEEN (1956 b, 1957) (*cf.* V.A.).

In this laboratory evidence has been obtained for the presence in higher plants of additional glucosides of similar type. These compounds will be further discussed in a following section (V. B. b). It may be mentioned here, however, that recently one of the stereoisomeric sulfoxides of S-methyl-L-cysteine, $CH_3SOCH_2CH(NH_2)COOH$, has been detected in cabbage juice (SYNGE and WOOD) and, independently, in turnip roots (MORRIS and THOMPSON). The absolute configuration around the sulfur atom in the natural amino acid has newly been determined by X-ray crystallography (HINE and ROGERS 1956). *Biotin-l-sulfoxide* represents another recent addition to the list of naturally occurring sulfoxides. It appears to be formed in small amounts by *Aspergillus niger*; available evidence points to it as having metabolic significance although its function is poorly understood at the present (WRIGHT *et al.*).

These findings indicate that sulfoxides are rather widely distributed in Nature. It appears likely that they arise from the corresponding sulfides by stereospecific, enzymic oxidation. Their possible rôle in metabolism is unknown. As mentioned above, the occasionally observed alkanethiols may arise from disproportionation of sulfoxides.

b) Natural sulfones.

Only a few sulfones are known from plant sources. Seeds of *Cheiranthus cheiri* L. contain the glucoside *glucocheirolin*, from which enzymic hydrolysis liberates the *isothiocyanate cheirolin*, $CH_3SO_2CH_2CH_2CH_2NCS$ (SCHNEIDER 1910). Likewise, the higher homologue, *erysolin*, $CH_3SO_2CH_2CH_2CH_2CH_2NCS$, was claimed to be present as part of a glucoside contained in seeds of *Erysimum Perofskianum* Fisch. et M. (SCHNEIDER and KAUFMANN 1912). These compounds will be further discussed in a following section (V. B. b.). The sulfones are almost certainly to be regarded as secondary products, formed upon oxidation of the corresponding sulfoxides or sulfides.

V. *iso*Thiocyanates (mustard oils) and their parent glucosides.

A. *iso*Thiocyanate glucosides. Occurrence and chemical structure.

Among the sulfur-containing plant products the *iso*thiocyanate glucosides constitute a unique and interesting group, at present comprising well above thirty individual compounds. They occur preferentially in members of *Cruciferae* but are encountered regularly also in the families *Resedaceae*, *Tropaeolaceae* and *Capparidaceae*. Sporadic occurrence of *iso*thiocyanate glucosides in various other families, such as *Moringaceae*, *Salvadoraceae*, *Caricaceae*, *Limnanthaceae* and *Euphorbiaceae*, has been recorded, whereas older statements as to their presence in certain species of *Phytolaccaceae* and *Aquifoliaceae* deserve corroboration. Chemical methods for the detection and quantitative determination of *iso*thiocyanate glucosides have recently been reviewed elsewhere (STOLL and JUCKER 1955).

The glucosides have been detected in all parts of plant species belonging to the families mentioned above. No systematic studies have been made of the distribution of *iso*thiocyanate glucosides within the different parts of a given plant. In general, however, the glucosides appear to be present in vegetative tissues (especially the parenchym) as well as in the seeds. Here, microchemical methods have shown them to be located in the embryos.

The *iso*thiocyanates are liberated from the glucosides by enzymic hydrolysis with simultaneous formation of glucose and sulfuric acid, according to the following scheme:

$$\text{Glucoside} + H_2O \xrightarrow{\text{Myrosinase}} \textit{iso}\text{Thiocyanate} + \text{Glucose} + HSO_4^-$$

The enzyme, *myrosinase*, has been only partly purified and has been regarded as a mixture of a thioglucosidase and a sulfatase (NEUBERG and WAGNER, SANDBERG and HOLLY). It occurs in all plant materials containing *iso*thiocyanate glucosides and is capable of splitting all natural *iso*thiocyanate glucosides, regardless of their origin and chemical structure. On the other hand, the enzymic action is of high specificity inasmuch as synthetic substrates, structurally similar to the natural compounds, are not attacked by the enzyme (SCHNEIDER *et al.* 1914a). In seeds, the enzyme is located in special cells and exerts its action only after destruction of the cellular pattern. More work is needed to establish the details of this remarkable enzymic reaction.

Glucose seems to be a constant constituent of all glycosides, justifying their designation as *glucosides*. Sulfuric acid also is part of all known *iso*thiocyanate glucosides which are usually isolated as potassium salts. Most likely, they are of similar structure, the individual variations residing in the *iso*thiocyanate fragments only. Much biochemical interest is therefore attached to a closer study of the individual *iso*thiocyanates. As salts and sugar derivatives, the glucosides are solids, easily soluble in water and slightly soluble in most organic solvents. With a few exceptions they crystallize very poorly; up to the present only seven glucosides have been described in their crystalline state (Table 1). During the last years crystalline acetyl-derivatives have been prepared of several glucosides, thereby facilitating the purification and isolation of the latter considerably (*cf.* Table 1). A long series of names has been introduced to characterize the individual compounds, most of them derived from the Latin name of the species from which they were first isolated, preceded by the prefix *gluco* (Table 1).

Recent studies of *iso*thiocyanate glucosides have been greatly facilitated by the ingenious tool of paper chromatography. This technique permits, in a minimum of time and with exceedingly small quantities of material, the determination of the number of glucosides present in a given plant specimen and often affords, in addition, a helpful means of identification of the individual compounds (SCHULTZ and GMELIN 1952, 1953). A list of paperchromatographic data for the mustard oil glucosides of numerous plant species has recently been published, yet without assignment of the observed spots to definite glucosides (SCHULTZ and WAGNER 1956a).

*iso*Thiocyanate glucosides have been recognized as plant constituents for more than 100 years. *Sinalbin* was isolated in crystalline form from *Sinapis alba* by BOUTRON and ROBIQUET in 1831. This achievement was soon followed by the isolation of *sinigrin* from black mustard *(Brassica nigra)* (BUSSY 1840). Early suggestions as to the chemical structure of the two glucosides were revised and completed through the work of GADAMER (1897) who proposed the following formula, until recently accepted as a correct general expression for all glucosides.

$$R-N=C\left\langle \begin{array}{c} S-CH-\overset{HO}{\underset{H}{C}}-\overset{H}{\underset{OH}{C}}-\overset{OH}{\underset{H}{C}}-\overset{}{\underset{H}{C}}-CH_2OH \\ O-SO_2-O^-\ X^+ \end{array}\right.$$

R = *iso*Thiocyanate side chain; X = Potassium or sinapine.

This formula immediately accounts for the formation of *iso*thiocyanates (R—N=C=S), glucose and sulfuric acid on enzymic hydrolysis. The attachment of the sugar moiety to the sulfur atom receives support from the observation by SCHNEIDER and WREDE (1914b) that treatment of sinigrin with potassium methoxide in methanol affords 1-thio-D-glucose. Upon enzymic hydrolysis of the glucosides varying amounts of nitriles, R—C≡N, are often formed besides the *iso*thiocyanates. This fact is not easily reconcilable with the GADAMER-structure above and has, among other facts, contributed to a reformulation of the general glucoside structure.

Table 1. *Naturally occurring* iso*thiocyanates of established structure and their glucosidic progenitors.*

Name of glucoside	Formula of *iso*thiocyanate	References to the structure determination of the mustard oil
Glucocapparin[a] . . .	CH_3NCS[n]	KJÆR *et al.* (1955 d)
Glucolepidiin[b]	CH_3CH_2NCS[n]	KJÆR and LARSEN (1954 a)
Glucoputranjivin[c] . .	$CH_3CH(CH_3)NCS$[n]	PUNTAMBEKAR (1950), KJÆR and CONTI 1953 b)
Glucocochlearin[e] . . .	$(+)\text{-}CH_3CH_2CH(CH_3)NCS$[n]	HOFMANN (1874), GADAMER (1899 a)
Sinigrin[d, e, f]	$CH_2{=}CHCH_2NCS$[n]	WILL (1844)
Gluconapin	$CH_2{=}CHCH_2CH_2NCS$[n]	KJÆR *et al.* (1953 d), ETTLINGER and HODGKINS (1955)
Glucobrassicanapin .	$CH_2{=}CHCH_2CH_2CH_2NCS$[n]	KJÆR and BOE JENSEN (1956 e)
Glucoibervirin	$CH_3SCH_2CH_2CH_2NCS$[n]	KJÆR *et al.* (1955 a)
Glucoiberin[c, e, g] . . .	$CH_3SOCH_2CH_2CH_2NCS$	SCHULTZ and GMELIN (1954 a), KJÆR and GMELIN (1956 d)
Glucocheirolin[c, d] . . .	$CH_3SO_2CH_2CH_2CH_2NCS$	SCHNEIDER (1910)
Glucoerucin	$CH_3SCH_2CH_2CH_2CH_2NCS$[n]	KJÆR and GMELIN (1955 b)
Unnamed[e]	$CH_3SOCH{=}CHCH_2CH_2NCS$	SCHMID and KARRER (1948)
Glucoerysolin[h] . . .	$CH_3SO_2CH_2CH_2CH_2CH_2NCS$	SCHNEIDER and KAUFMANN (1912)
Glucoberteroin . . .	$CH_3SCH_2CH_2CH_2CH_2CH_2NCS$[n]	KJÆR *et al.* (1955 c)
Glucoalyssin[i]	$CH_3SOCH_2CH_2CH_2CH_2CH_2NCS$	KJÆR and GMELIN (1956 d) SCHULTZ and WAGNER (1956 b)
Glucoarabin	$CH_3SO(CH_2)_9NCS$	KJÆR and GMELIN (1956 h)
Glucocamelinin . . .	$CH_3SO(CH_2)_{10}NCS$	KJÆR *et al.* (1956 f)
Glucotropaeolin[c, c, j] . .	$C_6H_5CH_2NCS$[n]	GADAMER (1899 b, c)
Sinalbin[d]	$p\text{-}HOC_6H_4CH_2NCS$	SALKOWSKI (1899). KJÆR and RUBINSTEIN (1954 b)
Glucoaubrietin . . .	$p\text{-}CH_3OC_6H_4CH_2NCS$[n]	KJÆR *et al.* (1956 a)
Unnamed	$m\text{-}CH_3OC_6H_4CH_2NCS$[n]	ETTLINGER and LUNDEEN (1956 a)
Gluconasturtiin . . .	$C_6H_5CH_2CH_2NCS$[n]	GADAMER (1899 c)
Glucoerypestrin[k] . .	$CH_3OOC(CH_2)_3NCS$[o]	KJÆR and GMELIN (1957 a)
Glucomalcolmiin . . .	$C_6H_5COO(CH_2)_3NCS$[o]	KJÆR and GMELIN (1956 g)
Progoïtrin[l] (Gluco-rapiferin[m])	$CH_2{=}CHCHOHCH_2NCS$[p]	GREER (1956), KJÆR *et al.* (1956 b), SCHULTZ and WAGNER (1956 c)
Glucoconringiin[f, m] . .	$(CH_3)_2C(OH)CH_2NCS$[q]	KJÆR *et al.* (1956 b), SCHULTZ and WAGNER (1956 c)
Glucobarbarin	$C_6H_5CHOHCH_2NCS$[r]	KJÆR and GMELIN (1957 b)

[a] Isolated as a crystalline potassium salt and characterized as a crystalline tetraacetate (KJÆR and GMELIN 1956 c). [b] This name was not proposed in the original paper. [c] Characterized as a crystalline tetraacetate (KJÆR, unpublished). [d] Known for many years, as a crystalline glucoside. [e] Characterized as its crystalline tetraacetate (SCHULTZ and WAGNER 1955). [f] Characterized as its crystalline tetraacetate (KJÆR *et al.* 1956 b). [g] Isolated as the crystalline potassium salt (SCHULTZ and GMELIN 1954 a). [h] The existence in Nature of glucoerysolin has not been confirmed by work in this laboratory. [i] Characterized as its tetraacetate (SCHULTZ and WAGNER 1956 b). [j] Isolated as a crystalline tetramethylammonium salt, and characterized as the crystalline tetraacetate (ETTLINGER and LUNDEEN 1957). [k] Characterized as a crystalline tetraacetate (KJÆR and GMELIN 1957 a). [l] Isolated as the crystalline sodium salt (GREER 1956). [m] Characterized as a crystalline acetate (SCHULTZ and WAGNER 1956 c). [n] Volatile with steam. [o] Volatile in principle, but may decompose on attempted distillation with steam. [p] Cyclizes spontaneously to $(-)$-5-vinyl-2-oxazolidinethione. [q] Cyclizes spontaneously to 5,5-dimethyl-2-oxazolidinethione. [r] Cyclizes spontaneously to $(-)$-5-phenyl-2-oxazolidinethione.

Thus, ETTLINGER and LUNDEEN (1956 b) recently demonstrated by degradation that sinigrin and sinalbin, and hence probably all mustard oil glucosides, possess the following general structure:

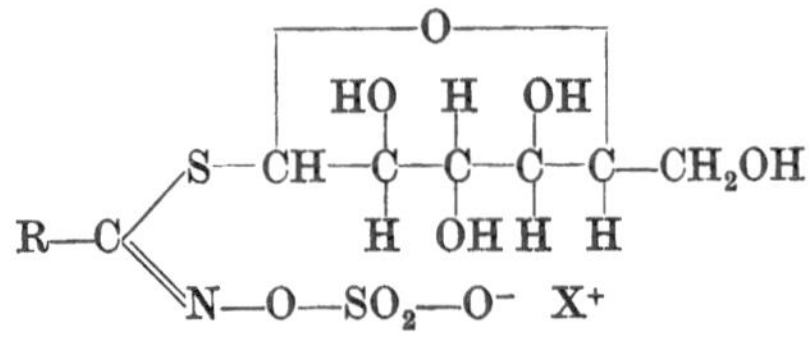

$$
\begin{array}{c}
\underset{\displaystyle R-C}{\overset{\displaystyle}{}}
\end{array}
$$

$$
R-C
\begin{cases}
S-CH-\overset{HO}{\underset{H}{C}}-\overset{H}{\underset{OH}{C}}-\overset{OH}{\underset{H}{C}}-\overset{}{\underset{H}{C}}-CH_2OH \\[2mm]
N-O-SO_2-O^- \ X^+
\end{cases}
$$

This expression deviates from the Gadamer structure above by its content of a hydroxylamine-grouping. In accord herewith, the natural glucosides split off hydroxylamine upon treatment with strong acids. The glucosidic link is of the ordinary β-type, whereas the geometric isomerism around the double bond remains to be established. As can be inferred from the revised structure, the enzymic fission of the glucosides is accompanied by an intramolecular rearrangement, comparable to the Lossen rearrangement of hydroxamic acids, a reaction which has been familiar to the organic chemist for more than half a century.

It is conceivable that the enzymic action on mustard oil glucosides is confined to a single step, effecting hydrolysis of the thioglucoside link to glucose and an unstable intermediate. This, in turn, may undergo a non-enzymic rearrangement of the Lossen type to yield the *iso*thiocyanate under simultaneous displacement of the sulfate grouping. This interpretation may necessitate a revision of the general conception of myrosinase as an enzyme mixture with dual function.

The revised formula above has been fully substantiated through total synthesis of glucotropaeolin ($R = C_6H_5CH_2$), obtained in crystalline form as the tetramethylammonium salt (ETTLINGER and LUNDEEN 1957). This achievement represents the first reported synthesis of a mustard oil glucoside.

For many years sinigrin and sinalbin represented the only crystalline mustard oil glucosides. In 1913, SCHNEIDER and SCHÜTZ succeeded in crystallizing *gluco-cheirolin* (Table 1) from seeds of wallflower (*Cheiranthus cheiri* L.) but it was not until 1954 that a new crystalline glucoside, *glucoiberin*, was added to this group (SCHULTZ and GMELIN, a). The presence of a few other natural glucosides was indicated through the identification of additional *iso*thiocyanates by the pioneers in this field (HOFMANN, GADAMER and SCHNEIDER, *cf.* Table 1). In recent years, renewed studies of this class of compounds have disclosed several new naturally occurring *iso*thiocyanates, partly derivable from uncharacterized glucosides (Table 1).

Nothing is known as to the biosynthesis and possible metabolic rôle of the mustard oil glucosides. For example, it remains to be clarified at which stage in the sequence of biosynthetic reactions glucose and sulfuric acid are incorporated.

B. Natural *iso*thiocyanates (mustard oils).

Chemical properties. *iso*Thiocyanates (mustard oils), $R-N=C=S$, can be considered as esters of the hypothetical *iso*thiocyanic acid, $H-N=C=S$, and are accessible in the laboratory by a number of standard synthetic procedures. Most of the simple mustard oils are distillable liquids with a characteristic sharp odour and biting taste, which are of great diagnostic value for their localisation in plant material. They are compounds of high chemical reactivity, easily hydrolyzed in acid or alkaline solution to the corresponding primary amines, $R-NH_2$. Reaction with alcohols (R_1OH) affords alkyl thiocarbamates, $R-HN-CS-O-R_1$. For characterization purposes, the *iso*thiocyanates are often transformed into solid thiourea-derivatives upon reaction with ammonia or suitable amines.

a) Occurrence, isolation and identification.

The *iso*thiocyanates, enzymically liberated from natural glucosides, are secondary plant products of considerable interest. Knowledge of their chemical structures usually provides the clue to the identity of the parent glucosides. As distinct from the latter, the *iso*thiocyanates can readily be purified and characterized and their structures confirmed by synthesis.

Since ancient times various *Cruciferae* have been utilized as spices (white and black mustard, radish, horseradish, garden- and water-cress *etc.*). The sharp and burning taste of the *iso*thiocyanates contained in these species is responsible for this application. A notable property of many *iso*thiocyanates is their volatility with steam (*cf.* Table 1), which considerably facilitates their isolation. In cases where the mustard oils are not steam-volatile, extraction with organic solvents must be used. For identification purposes it is often advantageous to transform the essential oils into solid thioureas by reaction with ammonia or other suitable amines. A paper-chromatographic technique has been worked out for the separation and identification of thioureas, derivable from mixtures of *iso*thiocyanates (KJÆR and RUBINSTEIN 1953 a) (Fig. 1). This method, in conjunction with the glucoside chromatography mentioned above, has been systematically employed in this laboratory for the purpose

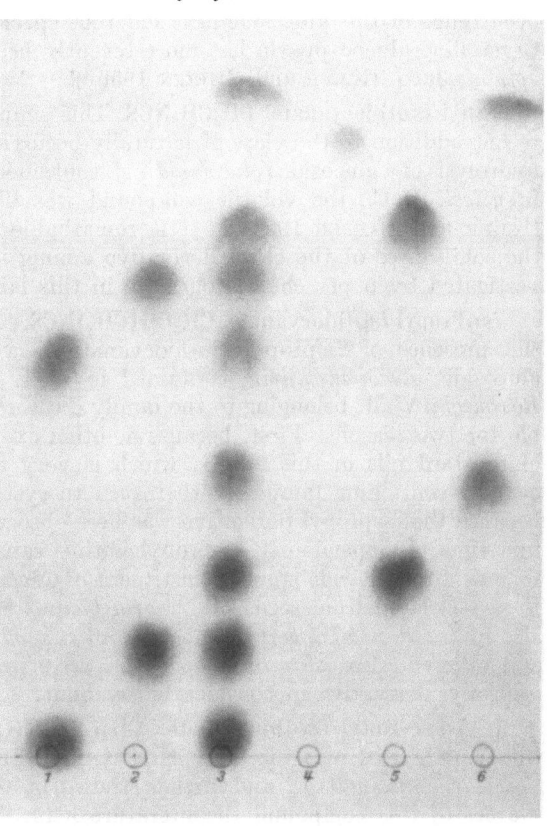

Fig. 1. Ascending paper chromatogram of thioureas (KJÆR and RUBINSTEIN 1953 a) on No. 1 Whatman paper. Solvent system: chloroform/water. Developed with GROTE'S reagent. 1. Methyl- and β-methallyl-; 2. ethyl and α-methallyl-; 3. mixture of methyl-, ethyl-, allyl-, *iso*propyl-, β-methallyl-, α-methallyl-, *n*-butyl- and β-phenylethyl-; 4. phenyl-; 5. allyl- and *n*-butyl-; 6. *iso*propyl- and β-phenylethyl-thiourea.

of augmenting and revising our knowledge of the types of *iso*thiocyanates derivable from natural glucosides. So far, several new *iso*thiocyanates of established formulae have emerged from this approach (Table 1), while structure determinations of still others are at present in progress.

In Table 1 are listed all *iso*thiocyanates which can at the moment be considered as rigorously proved constituents of natural glucosides. References are given to papers in which full documentation of the respective *iso*thiocyanates was first presented. In the following, the individual compounds will be briefly discussed with due reference to their natural sources. An extensive, critical survey of all species hitherto investigated is outside the scope of the present discussion. For the sake of convenience, the *iso*thiocyanates will be treated groupwise under separate headings, selected mainly according to their chemical nature.

b) Volatile natural *isothiocyanates*.
Simple alkyl *isothiocyanates*.

Methyl *isothiocyanate*, CH₃NCS. This compound has only recently been added to the series of naturally occurring *isothiocyanates* (KJÆR *et al.* 1955 d). It derives from the glucoside *glucocapparin* which appears as a constant seed constituent of several members of the family *Capparidaceae*, whereas no certain occurrence of this glucoside in cruciferous species has been established thus far. Crystalline glucocapparin has more recently been isolated from seeds of *Cleome spinosa* Jacq. (KJÆR and GMELIN 1956 c).

Ethyl *isothiocyanate*, CH₃CH₂NCS. This simple derivative represents another recent addition to the class of naturally occurring *isothiocyanates*. By enzymic hydrolysis of a glucoside *(glucolepidiin)*, contained in seeds of the crucifer *Lepidium Menziesii* D. C., the volatile compound was liberated and fully characterized (KJÆR and LARSEN 1954 a). It is remarkable that this seed has proved to be the sole source of the ethyl derivative among more than 250 species so far investigated by paper chromatography in this laboratory.

***iso*Propyl *isothiocyanate*, CH₃CH(CH₃)NCS.** In 1950 PUNTAMBEKAR reported the presence of *iso*propyl *isothiocyanate* as a constituent of the hypothetic glucoside, *glucoputranjivin*, contained in seeds of the Indian plant *Putranjiva Roxburghii* Wall., belonging to the family *Euphorbiaceae*. This finding is remarkable for two reasons. First, because no other examples are known of the presence of mustard oils in this family, which is very remote from traditional *isothiocyanate*-containing families with regard to systematic classification. Secondly, because the *iso*propyl derivative was here observed as a natural product for the first time. Independently, *iso*propyl *isothiocyanate* was identified in this laboratory as a rather wide-spread constituent of *Cruciferae* (KJÆR and CONTI 1953 b). It was isolated from seeds of *Lunaria biennis* Mnch., and shown to be present also in *L. rediviva* L., certain members of *Sisymbrium*, various *Cochlearia*-species and others. Also seeds of *Tropaeolum peregrinum (Tropaeolaceae)* furnish the *iso*propyl derivative in considerable amount.

(+)-*sec*-Butyl *isothiocyanate*, CH₃CH₂CH(CH₃)NCS. Secondary butyl *iso*thiocyanate was recognized by HOFMANN in 1874 as a volatile constituent of *Cochlearia officinalis* L. and further studied by GADAMER (1899 a). As additional sources of this compound the literature lists: *Cochl. danica* L., *Cardamine pratensis* L. and *Card. hirsuta* L. (BLANKSMA), *Card. amara* L. (FEIST; KUNTZE, a), *Putranjiva Roxburghii* Wall. (PUNTAMBEKAR) and *Codonocarpus cotinifolius* Desf. (BOTTOMLEY and WHITE). In this laboratory, the presence of the *sec*-butyl derivative has been established also in seeds of *Cochl. anglica* (L.) Asch. & Grb., *Draba borealis* D. C., *Lunaria annua* L., *Sisymbrium strictissimum* L., *Tropaeolum peregrinum* (KJÆR *et al.* 1953 c) and in *Lun. rediviva* L. (unpublished).

The parent glucoside, *glucocochlearin* (TER MEULEN), has not yet been described in the pure state. Its presence in various plants, *viz. Alyssum Bornmuelleri* (doubtful), *Isatis tinctoria* L. and members of *Arabis* and *Draba*, in addition to some of the species mentioned above, has been claimed from paperchromatographic evidence (SCHULTZ and GMELIN 1953). It is an interesting fact that *sec*-butyl *isothiocyanate* is regularly accompanied by varying amounts of *iso*propyl *iso*thiocyanate in plant distillates, suggesting a common pathway in their biosynthesis.

Allyl *isothiocyanate*, CH₂=CHCH₂NCS. This is the "classical" mustard oil, usually isolated from seeds of black mustard *(Brassica nigra* Koch) after enzymic hydrolysis of the parent glucoside *sinigrin*. The unsaturated, volatile compound has been reported as present in numerous other crucifers. Old statements, however, should be interpreted with caution, due to often inadequate chemical characterization. In Table 2 is presented a survey of the more important data concerning the occurrence of sinigrin or allyl *isothiocyanate.* Nothing is known about the phytochemical origin of the allyl group. It may arise by elimination of a small molecular fragment (*e.g.* H_2O, NH_3 or CH_3SH) from a saturated chain of a precursor-molecule containing such functional groups in a suitable position.

Table 2. *Sources of sinigrin or allyl* isothiocyanate.

Alliaria officinalis Andrz. . . .	S	3 (a)		*Arabis bellidifolia* Jacq.. . .	S	5 (b)
Armoracia rusticana G., M. & Sch.	S	2 (a)		*Brassica napus rapifera* . . .	S	4 (b)
Brassica juncea Czern. et Coss. .	S	1 (a)		*Capsella bursa pastoris* Med. .	P, S	6 (b)
Brassica nigra Koch	S	1 (a)		*Cochlearia officinalis* L. . . .	S	5 (b)
Brassica oleracea L..	S	1 (a)		*Isatis tinctoria* L..	S	4, 5 (b)
Br. oleracea species	S	1 (a)		*Sisymbrium officinale* (L.)		
Cakile maritima Scop..	S	2 (a)		Scop.	P, S	4, 5 (b)
Crambe maritima L..	S	2 (a)		*Sisymbrium strictissimum* L. .	S	5 (b)
Diplotaxis erucoides (L.) D.C. . .	S	3 (a)				
Diplotaxis muralis (L.) D.C. . .	S	2 (a)		*Alyssum calycinum*	?	5 (c)
Draba incana L..	S	2 (a)		*Draba aizoides*	S	5 (c)
Draba rupestris R. Br.. . . .	S	3 (a)		*Draba pyrenaica*	S	5 (c)
Erucastrum gallicum O. E. Schulz	S	2 (a)		*Erysimum alpinum*	S	5 (c)
Iberis contracta Pers..	S	3 (a)		*Erysimum helveticum*	S	5 (c)
Iberis intermedia Guersent . . .	S	3 (a)		*Kernera saxatile*	S	5 (c)
Iberis umbellata L.	S	3 (a)		*Sisymbrium loeselii*	P	4 (c)
Sisymbrium sophia L..	S	2 (a)				
Thlaspi alpestre L.	S	3 (a)				
Thlaspi arvense L..	S	2 (a)				

(a): Allyl *isothiocyanate* established by distillation and paper chromatography in this laboratory. (b): Proved *untenable* by distillation and paper chromatography in this laboratory. (c): Doubtful assignment, not checked by distillation of *isothiocyanate.* S: Seeds. P: Fresh plants. 1: Jensen, Conti and Kjær (1953); 2: Kjær, Conti and Larsen (1953 c); 3: Kjær *et al.* (unpublished); 4: Schultz and Gmelin (1952); 5: Schultz and Gmelin (1953); 6: Bodinus (1920).

In plants, the allyl derivative is often accompanied by other *isothiocyanates,* primarily thiooxazolidone precursors or compounds belonging to the methyl-thioalkyl-group. Both types will be discussed below.

3-Butenyl *isothiocyanate*, $CH_2=CHCH_2CH_2NCS$. The presence in seed of rape (*Brassica napus* L.) of an unsaturated five-carbon *isothiocyanate,* apparently a higher homologue of the allyl compound, has been known for many years. Abnormally high contents of this volatile mustard oil have been suggested as responsible for the toxic effects occasionally observed after feeding animals rape seed cake. As usually, the *isothiocyanate* originates from a still uncharacterized glucoside, *gluconapin* (Ter Meulen). It was not until recently, however, that its chemical identity as 3-butenyl *isothiocyanate* was rigorously proved by synthesis (Kjær *et al.* 1953 d). In 1955, Ettlinger and Hodgkins independently arrived at the same conclusion. In the two papers references are given to the older and rather contradictory literature on the mustard oils of rape.

Several additional sources of 3-butenyl *isothiocyanate* have been disclosed by paper chromatography in this laboratory. Thus, seeds of *Brassica rapa,* *Cardamine graeca* L., *Isatis tinctoria* L. and several species of *Alyssum* all afford 3-butenyl *isothiocyanate* after enzymic cleavage and distillation. Like the allyl compound, the present unsaturated derivative often occurs in admixture with thiooxazolidone precursors or methylthio-compounds, a fact suggesting a possible biosynthetic relationship with these substances.

4-Pentenyl *isothiocyanate*, $CH_2=CHCH_2CH_2CH_2NCS$. A minor constituent of the volatile fraction of the rape seed mustard oils has recently been identified in this laboratory as 4-pentenyl *isothiocyanate*, deriving from a hypothetic glucoside for which the name *glucobrassicanapin* has been proposed (KJÆR and BOE JENSEN 1956e). Its distribution may be wider than originally believed. Thus, glucobrassicanapin seems to occur also in certain species of the genus *Alyssum*.

The distribution of this and the two foregoing unsaturated *isothiocyanates* within seeds of numerous species and varieties of the genus *Brassica* has recently been studied by DELAVEAU, using a paperchromatographic technique.

Methylthioalkyl *isothiocyanates*, including the corresponding sulfoxides and sulfones.

In recent years a new type of volatile, natural *isothiocyanates* has been discovered, characterized by containing a terminal methylthio-group in a straight alkyl chain, $CH_3S(CH_2)_nNCS$. Such compounds, briefly mentioned in the discussion of sulfides above, have proved to be present as glucosides in several crucifers. They have in common a characteristic odour and taste, indistinguishable from those of fresh radish. So far, three representatives of the new type have been definitely identified in this laboratory. Various sulfoxides and sulfones structurally related to this type of *isothiocyanates* will be treated here also.

3-Methylthiopropyl *isothiocyanate*, $CH_3SCH_2CH_2CH_2NCS$. On paper chromatography, seeds of *Iberis sempervirens* L. proved to contain two volatile *isothiocyanates*, both different from previously known, natural compounds. One of them was identified as 3-methylthiopropyl *isothiocyanate* and the name *glucoibervirin* was proposed for the parent glucoside (KJÆR *et al.* 1955a). The other volatile compound was found to be 4-methylthiobutyl *isothiocyanate*, also a new compound, which will be discussed below.

The new mustard oil is closely related to that derived from *glucoiberin* (SCHULTZ and GMELIN 1954a), a crystalline glucoside isolated from seeds of *Iberis amara* L. On enzymic hydrolysis it affords an *isothiocyanate* (iberin) which is the sulfoxide corresponding to 3-methylthiopropyl *isothiocyanate* (*cf.* KJÆR and GMELIN 1956d). It is reasonable to assume that glucoiberin is formed by enzymic oxidation of glucoibervirin within the living cells. This suggestion receives support from the observation of a similar oxidation *in vitro*, taking place when an aqueous solution of glucoibervirin is kept at room temperature for a few days. Besides seeds of *I. sempervirens*, fresh root material of *I. amara* contains 3-methylthiopropyl *isothiocyanate* (unpublished), again pointing to the glucoside of this compound as the phytochemical precursor of glucoiberin, the sole glucoside of the seeds.

Seeds of *Cheiranthus cheiri* L. contain the *isothiocyanate* *cheirolin* in glucoside form (SCHNEIDER 1910). It represents the sulfone corresponding to 3-methylthiopropyl *isothiocyanate* and its parent glucoside is most likely formed by further enzymic oxidation of glucoiberin. Again, this view is supported by the recent observation in this laboratory that fresh leaves of *C. cheiri* contain 3-methylthiopropyl *isothiocyanate* (unpublished).

4-Methylthiobutyl *isothiocyanate*, $CH_3SCH_2CH_2CH_2CH_2NCS$. This steam-volatile compound was first isolated from seeds of *Eruca sativa* Mill. where it occurs as a glucoside for which the name *glucoerucin* has been introduced (KJÆR and GMELIN 1955b). Its presence has been established also in seeds of *Iberis sempervirens* L., as mentioned above, *Diplotaxis tenuifolia* (L.) D. C. and possibly some *Vesicaria*-species.

The natural occurrence of the sulfoxide corresponding to glucoerucin has not yet been rigorously proved although there is evidence of its existence in Nature. A natural sulfoxide, closely related to 4-methylthiobutyl *iso*thiocyanate, is the mustard oil *sulforaphene* of radish seed (*Raphanus sativus* L.), $CH_3SOCH=CHCH_2CH_2NCS$ (Schmid and Karrer). An antibacterial substance, *raphanin*, isolated from seed of radish (Ivánovics and Horváth 1947), was later proved to be identical with sulforaphene (Koczka and Ivánovics 1949). Various observations in this laboratory are consistent with the existence of the corresponding unsaturated sulfide, $CH_3SCH=CHCH_2CH_2NCS$, in *e.g.* radish roots, although the compound has not yet been fully characterized. The possibility that sulforaphene should represent an artifact of the sulfide, formed by secondary oxidation *in vitro*, is ruled out because of the optical activity of the sulfoxide-group in sulforaphene.

The sulfone *iso*thiocyanate *erysolin*, $CH_3SO_2CH_2CH_2CH_2CH_2NCS$, has been described as part of a glucoside present in seeds of *Erysimum Perofskianum* Fisch. et M. (Schneider and Kaufmann 1912). Although this finding is well in line with the natural occurrence of the above sulfide it has not been possible in this laboratory to verify the presence of glucoerysolin in *E. Perofsk.* seeds or elsewhere.

5-Methylthiopentyl *iso*thiocyanate, $CH_3SCH_2CH_2CH_2CH_2CH_2NCS$. This mustard oil represents another recent addition to the methylthio-series (Kjær *et al.* 1955c). It was isolated from seeds of *Berteroa incana* (L.) D. C. by distillation and its chemical structure was proved upon comparison with an authentic, synthetic specimen. The name *glucoberteroin* has been proposed for the parent glucoside which is present also in seeds of *Lunaria annua* L. and as a rather constant and characteristic constituent of the genus *Alyssum*.

Again, the corresponding (—)-5-methylsulfinylpentyl *iso*thiocyanate, CH_3SO $(CH_2)_5NCS$, has been unambiguously identified in this laboratory as a mustard oil of natural provenance. It was isolated from enzymically hydrolyzed seed extracts of *Alyssum argenteum* Vitm., but the corresponding glucoside, for which the name *glucoalyssin* has been introduced, appears to be a constant constituent throughout the genus *Alyssum* (Kjær and Gmelin 1956d). Independently, but on somewhat slender evidence, Schultz and Wagner (1956b) arrived at the same conclusion.

(—)-9-Methylsulfinylnonyl *iso*thiocyanate, $CH_3SO(CH_2)_9NCS$. The extensive occurrence in Nature of glucosides furnishing straight-chain alkyl *iso*thiocyanates, possessing a terminally located methylthio-grouping as such or in oxidized form, became evident when the mustard oil, liberated from a new glucoside *(glucoarabin)* in seeds of *Arabis alpina* L., was identified as (—)-9-methylsulfinylnonyl *iso*thiocyanate (Kjær and Gmelin 1956h). The distribution of glucoarabin in Nature, as well as its phytochemical formation, are at present unknown.

(—)-10-Methylsulfinyldecyl *iso*thiocyanate, $CH_3SO(CH_2)_{10}NCS$. This higher homologue of the foregoing mustard oil has been recently recognized also as an enzymic hydrolysis product of a glucoside, named *glucocamelinin*, encountered so far in seeds of *Camelina sativa* Crantz and other *Camelina* species (Kjær *et al.* 1956f). The glucoside may be of more wide-spread occurrence, however, and its remarkable structure renders a further study of its biosynthesis a matter of considerable interest.

Comparable rotatory data for the series of naturally occurring sulfoxide *iso*thiocyanates, as well as for several series of analogous derivatives, provide reasonable good evidence that they belong to the same configurational series. Unpublished results from this laboratory, however, seem to indicate that the natural sulfoxide-amino acids, such as alliin and $(+)$-S-methyl-L-cysteine sulfoxide, possess the opposite steric configuration around the sulfur atom.

Aromatic *isothiocyanates*.

Benzyl *isothiocyanate*, $C_6H_5CH_2NCS$. The presence in plants of the glucoside *glucotropaeolin*, affording benzyl *isothiocyanate* on enzymic cleavage, was first demonstrated by GADAMER (1899 b, c). Fresh parts and seeds of *Tropaeolum majus* L. (b), as well as seeds of *Lepidium sativum* L. (c), served as sources of the mustard oil. Since then, other species have been shown to contain glucotropaeolin, *e.g. Coronopus didymus* (L.) Sm. (McDOWALL *et al.*), *Lepidium densiflorum* Schrad. and *L. virginicum* L. (KJÆR *et al.* 1953 c). In several other species, glucotropaeolin has been reported on paperchromatographic evidence (SCHULTZ and GMELIN 1953). Some of the latter assignments undoubtedly need revision. It is interesting that benzyl *isothiocyanate* has been identified as a constituent of several plants not belonging to the family *Cruciferae*. Besides *Tropaeolaceae*, seeds of *Salvadora oleoides* Den. *(Salvadoraceae)* (PATEL *et al.*) and *Carica papaya* L. *(Caricaceae)* (ETTLINGER and HODGKINS 1956) as well as root material of *Moringa pterygosperma* Gaertn. *(Moringaceae)* (KURUP and RAO) have been reported as sources of the mustard oil.

The parent glucoside, glucotropaeolin, has recently been crystallized as its tetramethylammonium salt (ETTLINGER and LUNDEEN 1957). It is of interest that enzymic hydrolysis of the glucoside is often accompanied by the formation of considerable amounts of benzyl cyanide, $C_6H_5CH_2CN$, which may be further hydrolyzed to phenylacetic acid, known as an active growth-substance. The possible identity of the latter with a natural growth factor of seeds and roots of *Lepidium sativum* led SCHULTZ and GMELIN (1954 b) to suggest a possible metabolic relationship between *isothiocyanate* glucosides and certain growth-substances. More work, however, is needed to test this interesting hypothesis.

m-Methoxybenzyl *isothiocyanate*, (m)-CH_3O-$C_6H_4CH_2NCS$. From seeds of the American plant *Limnanthes douglasii* R.Br., belonging to the small family *Limnanthaceae*, ETTLINGER and LUNDEEN (1956a) isolated m-methoxybenzyl *isothiocyanate* subsequent to enzymic hydrolysis of a not further characterized glucoside contained in the seeds. The finding is remarkable in view of the rare occurrence in Nature of *meta*-disubstituted aromatic compounds.

p-Methoxybenzyl *isothiocyanate*, (p)-CH_3O-$C_6H_4CH_2NCS$. At about the same time, the existence in Nature of a glucoside *(glucoaubrietin)*, affording the isomeric p-methoxybenzyl *isothiocyanate* on enzymic hydrolysis, was revealed in this laboratory. Fresh plants of *Aubrietia deltoidea* D.C. *(Cruciferae)*, as well as of other *Aubrietia* species, served as sources of the new *isothiocyanate* (KJÆR *et al.* 1956a). Its natural occurrence may not be very surprising because of the long recognized existence in plants of sinalbin, furnishing p-hydroxybenzyl mustard oil on enzymic fission (V.B.c.). The biogenetic formation of glucoaubrietin by biological methylation of sinalbin appears likely, although no supporting experimental evidence can be adduced at the present.

β-Phenylethyl *isothiocyanate*, $C_6H_5CH_2CH_2NCS$. This volatile mustard oil is liberated on enzymic hydrolysis from the glucoside *gluconasturtiin*, demonstrated by GADAMER (1899 c) to be present in fresh plants of *Nasturtium officinale* R. Br. and *Barbaraea praecox* R. Br. Other sources of the isothiocyanate have been: root tissue of *Reseda odorata* L. (BERTRAM and WALBAUM) and horseradish (HEIDUSCHKA and ZWERGAL). It occurs frequently also in members of the genus *Brassica*, preferentially in the roots. Thus, KUNTZE (b) reported the presence of β-phenylethyl *isothiocyanate* in cortical root tissue of a certain variety of turnip (*B. rapa* var. *rapifera* Metzger) whereas STAHMANN *et al.* provided good evidence for its occurrence also in roots of *B. nigra*, *Sinapis alba* and *B. oleracea capitata* (cabbage). No correlation could be established between the content of mustard oil and the resistance to the clubroot organism (*Plasmodiophora brassicae* Wor.). Lately, GERRETSEN and HAAGSMA isolated a steam-volatile oily substance, "rapine", from cortical tissue of *B. rapa* tubers, possessing a high degree of antifungal activity. It appears likely that this product is identical with β-phenylethyl *isothiocyanate*. It is of interest that the latter is absent from seeds of *Brassica* species or present in traces only (JENSEN *et al.*). This fact merits further consideration and may be of importance for the understanding of the biogenesis of mustard oils. Recently, ANDRÉ and DELAVEAU have provided evidence for the occurrence of phenylethyl *isothiocyanate* as a minor constituent of the mustard oil mixture obtained upon destillation of enzymically hydrolyzed rape seed cake.

Ester *isothiocyanates*.

Recent work has revealed the presence in Nature of two glucosides both giving rise to *isothiocyanates* with ester structure upon enzymic hydrolysis. There is evidence to suggest that additional compounds of this general structure type are present in higher plants.

Methyl 4-*isothiocyanatobutyrate*, $CH_3OOC(CH_2)_3NCS$. The presence of an alkali-labile glucoside in several *Erysimum* species was recently established in this laboratory by paperchromatographic methods. The new glucoside, *glucoerypestrin*, was isolated from seeds of *Erysimum rupestre* D.C. in the form of a crystalline tetraacetate. Enzymic hydrolysis of the glucoside furnished equimolecular amounts of glucose and sulfate in addition to a mustard oil, whose identity as a methyl ester of 3-carboxypropyl *isothiocyanate* was deduced by degradation and confirmed by synthesis (KJÆR and GMELIN 1957a). Glucoerypestrin may derive biogenetically from a corresponding glucoside containing a free carboxyl group, although the acid mustard oil expected as a product of enzymic hydrolysis of the latter has not yet been recognized in Nature.

Thus far the known occurrence of glucoerypestrin is limited to species of the genus *Erysimum*, in every case in admixture with minor amounts of glucocheirolin, the glucoside long known as a characteristic constituent in species of the taxonomically related genus *Cheiranthus*.

3-Benzoyloxypropyl *isothiocyanate*, $C_6H_5CO\text{-}O\text{-}(CH_2)_3NCS$. This ester mustard oil, of a somewhat deviating structural type, constitutes the enzymic hydrolysis product of a glucoside of the traditional type, *glucomalcolmiin*, encountered in seeds of *Malcolmia maritima* (L.) R.Br. Its structure has been elucidated by degradation and confirmed by synthesis (KJÆR and GMELIN 1956g). The benzoate structure suggests a hydroxypropyl mustard oil glucoside as a conceivable biosynthetic precursor although the latter has not as yet been recognized in Nature.

The distribution of glucomalcolmiin is still insufficiently known, but it is noteworthy that this glucoside also occurs in admixture with glucocheirolin. Again, a combination possibly conditioned by the taxonomic similarity between the genera *Malcolmia* and *Cheiranthus*.

c) Non-volatile *isothiocyanates*.

The classification of naturally occurring *isothiocyanates* according to their volatility may appear rather arbitrary. Certainly, it does not imply any fundamental chemical difference between the two types but has been selected for methodical reasons only, as will appear from the following.

Although a few glucosides (sinalbin, glucocheirolin and glucoerysolin), containing non-volatile *iso*thiocyanates, have been recognized for many years, the great majority of natural mustard oils is still looked upon as a group of essential oils. Recent discoveries, however, have made modifications of this view necessary. In addition to the glucosides already mentioned, paper chromatography has revealed the existence of several others affording non-volatile *iso*thiocyanates and therefore escaping notice upon enzymic cleavage and distillation. The isolation of such mustard oils must be performed in other ways, *e.g.* by extraction.

Recent additions to this class include the *iso*thiocyanates containing a sulf-oxide-group (Table 1). The parent glucosides often occur in admixture with the corresponding sulfide-glucosides and hence have been discussed in a preceding section. It is very probable that the former arise from the latter by enzymic oxidation during transportation or storage in the living cells. Observed variations in the relative amounts of the two glucoside-types in various plant materials may reflect corresponding variations in the sulfide-oxidase activity of the tissues.

In this laboratory, additional non-volatile *iso*thiocyanates have been discovered. Structural work on the new compounds is in progress. It may well be that before long the non-volatile mustard oils have outnumbered the volatile, "classical" representatives.

p-**Hydroxybenzyl *iso*thiocyanate, (*p*)-HOC$_6$H$_4$CH$_2$NCS.** This mustard oil occurs as part of the glucoside *sinalbin*, first isolated from seeds of white mustard (*Sinapis alba* L.) (BOUTRON and ROBIQUET 1831, GADAMER 1897). Whereas other glucosides have been crystallized as potassium salts, crystalline sinalbin was obtained as a salt of the complicated organic base sinapine.

$$\text{HO}-\!\!\left\langle\underset{}{\bigcirc}\right\rangle\!\!-\text{CH}_2-\text{N}-\text{C}\underset{\diagdown \text{N}-\text{O}-\text{SO}_2-\text{O}^-\ \text{X}}{\overset{\diagup \text{S-Glucose}}{}}$$

Sinalbin

$$\text{X} = (\text{CH}_3)_3\overset{+}{\text{N}}-\text{CH}_2-\text{CH}_2-\text{O}-\text{OC}-\text{CH}=\text{CH}-\!\!\left\langle\underset{\text{OCH}_3}{\overset{\text{OCH}_3}{\bigcirc}}\right\rangle\!\!-\text{OH}$$

Sinapine

Enzymic hydrolysis affords glucose, sinapine sulfate and a non-volatile *iso*thiocyanate which has never been isolated in pure form. Its original identification as *p*-hydroxybenzyl *iso*thiocyanate (SALKOWSKI) rested on secondary evidence. The structure has recently been confirmed in a more direct way by formation of derivatives (KJÆR and RUBINSTEIN 1954b). The free mustard oil polymerizes spontaneously and is reasonably stable in very dilute solutions only.

Sinalbin is not a glucoside of extensive occurrence. Paperchromatographic studies in this laboratory have ascertained its presence in *Sinapis arvensis* L., *Lepidium campestre* (L.) R. Br. and probably various species of *Aubrietia* and *Bunias*.

d) Natural *iso*thiocyanates suffering spontaneous cyclization.

In 1938, HOPKINS described the isolation of a heterocyclic compound, 2-mercapto 5,5-dimethyloxazoline (II), from seed of the crucifer *Conringia orientalis* L. and suggested that (II) might be a secondary product, formed by cyclization of a parent *iso*thiocyanate (I). This view has proved untenable because synthetic (I) (KJÆR *et al.* 1953e) does not display any tendency to cyclization.

I   II   III

More recently, ASTWOOD *et al.* isolated the goitrogenic principle of root and seed of turnip and established its structure as *l*-5-vinyl-2-oxazolidinethione (III), recently confirmed by synthesis (ETTLINGER). The substance has also been found in certain other species of the genus *Brassica* and appears to be genuinely present as a water-soluble precursor, namely a glycoside, from which it arises by enzymic action (GREER *et al.*). Recently, RACISZEWSKI *et al.* isolated (III) from rapeseed oilmeal and again confirmed its identity by synthesis.

In this laboratory, the suggested precursor of (III) has since been shown to be an *iso*thiocyanate glucoside of the ordinary type (IV), possessing a hydroxy-group in the side chain (KJÆR *et al.* 1956 b).

$$CH_2\!=\!CH\!-\!CHOH\!-\!CH_2\!-\!C\!\underset{\underset{N-O-SO_2-O^-\ \ X^+}{IV^1}}{\overset{S\text{-Glucose}}{\diagdown}} \qquad\qquad CH_2\!=\!CH\!-\!CHOH\!-\!CH_2\!-\!NCS$$
$$V$$

More direct evidence has recently been adduced by GREER who succeeded in isolating the crystalline genuine glucoside, named *progoitrin*, as the sodium salt (IV, X = Na). Similar results were obtained by SCHULTZ and WAGNER (1956 c) who introduced the name *glukorapiferin* for (IV) and characterized it as a crystalline pentaacetate.

The evidence is based on paperchromatographic recognition of a glucoside, with the localisation expected from (IV), in seed extracts of the above-mentioned species. Such extracts exhibit uncharacteristic end-absorption only in ultra-violet light. Upon addition of myrosinase, 2-hydroxy-3-butenyl *iso*thiocyanate (V) is liberated. This cyclizes spontaneously to (III) as inferred from the ultra-violet absorption spectrum which now displays the intense band at *ca.* 240 mμ, diagnostic of (III). The enzymic reaction has further proved to be accompanied by the liberation of glucose and sulfuric acid.

In a similar way, the occurrence in seed of *Conringia orientalis* L. of a gluco-side, possessing the side-chain $(CH_3)_2C(OH)CH_2\!-\!$, has been established. Again, enzymic hydrolysis liberates the corresponding *iso*thiocyanate which spontane-ously cyclizes to the HOPKINS compound (II) (KJÆR *et al.* 1956 b).

Systematic investigations, at the present under way, have disclosed the existence of several additional *iso*thiocyanate glucosides of similar or closely related type. On enzymic hydrolysis they afford *iso*thiocyanates, again under-going spontaneous rearrangements to cyclic products the nature of which is now being studied. Very recently, one of these has been identified in this labora-tory (KJÆR and GMELIN 1957 b). Extracts of seeds of *Barbarea vulgaris* R.Br., as well as of other *Barbarea* species, afforded on enzymic hydrolysis glucose, sulfate and a heterocyclic compound, identified as (—)-5-phenyl-2-oxazolidine-thione (VI). This fact immediately establishes the chemical nature of *gluco-barbarin*, the parent glucoside, which undergoes hydrolysis to a 2-hydroxy-2-phenylethyl *iso*thiocyanate (VII), cyclizing spontaneously, in turn, to the ring compound (VI).

$$\begin{array}{cc} \underset{VI}{\overset{\displaystyle H_2C\!-\!\!-\!NH}{\underset{\displaystyle H}{\overset{C_6H_5\cdots}{\diagdown}}C\underset{O}{\diagup}CS}} \longleftarrow & \underset{VII}{\overset{\displaystyle CH_2NCS}{C_5H_5\!-\!\!-\!H\atop OH}} \end{array}$$

The identity of the latter has further been confirmed by synthesis in such a way as to establish its absolute configuration at the asymmetric carbon atom.

---

[1] This formula has been corrected so as to be in conformity with the revised general glucoside structure of ETTLINGER and LUNDEEN (1956 b).

Consequently, the natural derivatives, (VI) and (VII), possess the absolute configurations indicated in the formulae above.

e) *iso*Thiocyanates of questionable identity.

Some *iso*thiocyanates, structurally different from those discussed above, have been reported in the literature. They have not been included above, either because the structures are based on insufficient evidence or because their existence has been questioned through failure to repeat their isolation.

Thus, Heiduschka and Zwergal attributed the structure (I) to a volatile component of fresh radish roots. This should be considered as purely hypothetical, since it is not supported by any chemical evidence but based solely on analyses and determination of molecular weight. The same authors claim the isolation of a phenylpropyl *iso*thiocyanate from root material of horseradish. This finding also should be accepted with reservation and needs corroboration. From roots of kohlrabi *(Brassica oleracea gongylodes)* Zwergal (1952) isolated a volatile sulfur-compound which he considered to be 4,4-dimethyl-5-vinyl-thiooxazolidone (II), analogous to the goitrogenic factor of Astwood *et al.*

$$CH_3CH_2CH_2CH_2SCH=CHCH_2CH_2NCS$$

I

II

No experimental facts are presented, however, in support of the structure (II). Zwergal's compound is volatile with steam and does not form a silver salt, two properties which are not easily reconcilable with the structure suggested. More recently, Schultz and Wagner (1956c) preferred the structure $CH_3SOCH = CHCH_2CH_2CH_2NCS$ over (II) for the kohlrabi mustard oil. Experimental documentation is lacking but evidence seems to rest mainly on structural analogy to the nitrile $CH_3SOCH=CH—CH_2CH_2CH_2CN$, originally presented by Zwergal (1951) as the structure of the kohlrabi constituent until it was replaced by the formulation (II). The name *glucocaulorapin* has been proposed for the parent glucoside, isolated as a non-homogeneous tetraacetate from kohlrabi seeds (Schultz and Wagner 1955, 1956c).

The popular garden-flower subject *Hesperis matronalis* L. contain four unidentified glucosides, one of which *(glucomatronalin)* has been isolated as a crystalline hexa- or hepta-acetate. Experimental evidence suggests that the corresponding mustard oil, whose structure is still unknown, contains two or three acetylable groupings (Wagner 1956).

Puntambekar reported the isolation of phenyl *iso*thiocyanate from a glucoside present in seed kernels of the Indian *Euphorbia*-species *Putranjiva Roxburghii* Wall. and characterized the mustard oil as a thiourea derivative. It is the only record of the presence in Nature of phenyl *iso*thiocyanate and therefore deserves further attention.

f) *iso*Thiocyanates. Biogenesis, metabolism and phylogenesis.

At present, very little can be said about the biosynthesis and metabolism of mustard oils. The chemical similarity of the side-chains of the *iso*thiocyanates and those of the natural α-amino acids is a striking feature, suggesting a biochemical relationship. The carbon-skeletons of the former probably arise from the same two- and three-carbon fragments, which are involved in the biosynthesis of α-keto and α-amino acids. At a certain stage in the sequence of reactions a branch, ultimately leading to *iso*thiocyanates, may be visualized. The incorporation of sulfur probably occurs as one of the final steps. Carefully conducted experiments are needed to test these vague suggestions. No more is known concerning the metabolism and possible physiological rôle of the *iso*thiocyanates and their parent glucosides.

An extensive survey of the types of *iso*thiocyanates present in various plants may eventually prove to be of value also for the purpose of botanical classification. In this laboratory, the phylogenetic consequences of such knowledge are being considered at the present.

Literature.

AKABORI, S., and T. KANEKO: Über einen schwefelhaltigen Riechstoff von „Shoyu". Proc. Imp. Acad. (Tokyo) 12, 131 (1936). *Cf.* Chem. Zbl. 1936 II, 2391. — ANDRÉ, É., et P. DELAVEAU: Recherches chimiques sur la composition des essences sulfurées des graines de colza. Oléagineux 9, 591–600 (1954). — ASTWOOD, E. B., M. A. GREER and M. G. ETTLINGER: *l*-5-vinyl-2-thicöxazolidone, an antithyroid compound from yellow turnip and from *Brassica* seeds. J. of Biol. Chem. 181, 121–130 (1949).

BADDILEY, J., G. L. CANTONI and G. A. JAMIESON: Structural observations on "Active Methionine". J. Chem. Soc. 1955, 2662–2664. — BARRON, E. S. G.: Thiol groups of biological importance. Adv. Enzymol. 11, 201–266 (1951). — BERSIN, TH., A. MÜLLER u. E. STREHLER: Methylmethioninsulfoniumsalze. Arzneimittel-Forsch. 6, 174—176 (1956). — BERTRAM, J., u. H. WALBAUM: Über das Resedawurzelöl. J. prakt. Chem. [2] 50, 555–561 (1894). — BINKLEY, F.: Enzymatic cleavage of thioethers. J. of Biol. Chem. 186, 287–296 (1950). — BLACK, S., and N. G. WRIGHT: Enzymatic formation of glyceryl and phosphoglyceryl methylthiol esters. J. of Biol. Chem. 221, 171–180 (1956). — BLANKSMA, J. J.: Pharmac. Weekbl. 51, 1383 (1914). Quoted after STOLL and JUCKER 1955. — BODINUS: Extractum *Capsellae bursae pastoris* fluidum. Apoth.-Ztg 35, 183–184 (1920). — BOTTOMLEY, W., and D. E. WHITE: The chemistry of Western Australian plants. II. The essential oil of *Codonocarpus cotinifolius* (Desf.). Roy. Austral. Chem. Inst. J. a. Proc. 17, 31–32 (1950). Quoted after Chem. Abstr. 45, 820 (1951). — BOUTRON et ROBIQUET: Sur la semence de moutarde. J. Pharmacie Chim. [II] 17, 279–308 (1831). — BUSSY, A.: Untersuchungen über die Bildung des ätherischen Senföls. Liebigs Ann. 34, 223–230 (1840). — BYWOOD, R.: Thesis, University of Leeds, 1953. Quoted after CHALLENGER and HAYWARD 1954. — BYWOOD, R., F. CHALLENGER, D. LEAVER and M. I. WHITAKER: The evolution of dimethyl sulphide from bracken, "Horse's Tail" and other plants on treatment with sodium hydroxide. Biochemic. J. 48, XXX–XXXI (1951).

CANTONI, G. L.: S-Adenosyl-methionine; a new intermediate formed enzymatically from L-methionine and adenosinetriphosphate. J. of Biol. Chem. 204, 403–416 (1953). — CANTONI, G. L., and D. G. ANDERSON: Enzymatic cleavage of dimethylpropiothetin by *Polysiphonia lanosa.* J. of Biol. Chem. 222, 171–177 (1956). — (a) CAVALLITO, C. J., and J. BAILEY: Allicin, the antibacterial principle of *Allium sativum*. I. Isolation, physical properties and antibacterial action. J. Amer. Chem. Soc. 66, 1950–1951 (1944). — (b) CAVALLITO, C. J., J. S. BUCK and C. M. SUTER: Allicin, the antibacterial principle of *Allium sativum*. II. Determination of the chemical structure. J. Amer. Chem. Soc. 66, 1952–1954 (1944). — CAVALLITO, C. J., J. H. BAILEY and J. S. BUCK: Allicin, the antibacterial principle of *Allium sativum*. III. Its precursor and "Essential Oil of Garlic". J. Amer. Chem. Soc. 67, 1032–1033 (1945). — CHALLENGER, F.: Biological methylation. Adv. Enzymol. 12, 429–491 (1951). — The biological importance of organic compounds of sulphur. Endeavour 12, No 48 (1953). — CHALLENGER, F., and D. GREENWOOD: Sulphur compounds of the genus *Allium*. Biochemic. J. 44, 87–91 (1949). — CHALLENGER, F., and B. J. HAYWARD: The occurrence of a methylsulphonium derivative of methionine (α-aminodimethyl-γ-butyrothetin) in *Asparagus*. Chem. a. Ind. 1954, 729–730. — CHALLENGER, F., and Y. C. LIU: The elimination of methylthiol and dimethyl sulphide from methylthiol- and dimethyl-sulphonium compounds by moulds. Rec. Trav. chim. Pays-Bas (Amsterd.) 69, 334–342 (1950). — CHALLENGER, F., and M. I. SIMPSON: Studies on biological methylation. Part XII. A precursor of the dimethyl sulphide evolved by *Polysiphonia fastigiata*. Dimethyl-2-carboxyethylsulphonium hydroxide and its salts. J. Chem. Soc. 1948, 1591–1597.

DELAVEAU, P.: Recherches sur les sénevols des graines de *Brassica*, à l'aide de la chromatographie sur papier. I. Graines de *B. nigra* et *B. juncea*. Ann. pharmaceut. franc. 14, 765–769 (1956). — II. Graines de *B. rapa, B. napus* et *B. oleracea*. Ann. pharmaceut. franç. 14, 770–777 (1956).

ETTLINGER, M. G.: Synthesis of the natural antithyroid factor *l*-5-vinyl-2-thicöxazolidone. J. Amer. Chem. Soc. 72, 4792–4796 (1950). — ETTLINGER, M. G., and J. E. HODGKINS:

The mustard oil of rape seed, allylcarbinyl *iso*thiocyanate, and synthetic isomers. J. Amer. Chem. Soc. **77**, 1831–1836 (1955). — The mustard oil of *Papaya* seed. J. of Org. Chem. **21**, 204–205 (1956). — (a) ETTLINGER, M. G., and A. J. LUNDEEN: The mustard oil of *Limnanthes douglasii* seed, *m*-methoxybenzyl *iso*thiocyanate. J. Amer. Chem. Soc. **78**, 1952–1954 (1956). — (b) ETTLINGER, M. G., and A. J. LUNDEEN: The structure of sinigrin and sinalbin; an enzymatic rearrangement. J. Amer. Chem. Soc. **78**, 4172 (1956). — ETTLINGER, M. G., and A. J. LUNDEEN: First synthesis of a mustard oil glucoside; the enzymatic Lossen rearrangement. J. Amer. Chem. Soc. **79**, 1764 (1957).

FEIST, K.: Das ätherische Oel von *Cardamine amara* L. Apoth.-Ztg **20**, 832 (1905).

GADAMER, J.: Über die Bestandteile des schwarzen und des weißen Senfsamens. Arch. Pharmaz. **235**, 44–114 (1897). — (a) GADAMER, J.: Über das ätherische Oel von *Cochlearia officinalis*. Arch. Pharmaz. **237**, 92–105 (1899). — (b) GADAMER, J.: Das ätherische Oel von *Tropaeolum majus*. Arch. Pharmaz. **237**, 111–120 (1899). — (c) GADAMER, J.: Ueber ätherische Kressenöle und die ihnen zu Grunde liegenden Glukoside. Arch. Pharmaz. **237**, 507–521 (1899). — GERRETSEN, F. C., and N. HAAGSMA: Occurrence of antifungal substances in *Brassica rapa*, *Brassica oleracea* and *Beta vulgaris*. Nature (Lond.) **168**, 659–660 (1951). — GMELIN, R., G. HASENMAIER u. G. STRAUSS: Über das Vorkommen von Djenkolsäure und einer C—S-Lyase in den Samen von *Albizzia lophantha* Benth. *(Mimosaceae)*. Z. Naturforsch. **12** b (1957), in press. — GREER, M. A.: Isolation from Rutabaga seed of progoitrin (L-2-hydroxy-3-butenyl *iso*thiocyanate glucoside) the precursor of the naturally occurring antithyroid compound, goitrin (L-5-vinyl-2-thiooxazolidone). J. Amer. Chem. Soc. **78**, 1260 (1956). — GREER, M. A., M. G. ETTLINGER and E. B. ASTWOOD: Dietary factors in the pathogenesis of simple goiter. J. Clin. Endocrin. **9**, 1069–1079 (1949).

HAAGEN-SMIT, A. J., J. G. KIRCHNER, C. L. DEASY and A. N. PRATER: Chemical studies of pineapple (*Ananas sativus* Lindl.). II. Isolation and identification of a sulfur-containing ester in pineapple. J. Amer. Chem. Soc. **67**, 1651–1652 (1945). — HAAS, P.: The liberation of methyl sulphide by seaweed. Biochemic. J. **29**, 1298–1299 (1935). — HEIDUSCHKA, A., u. A. ZWERGAL: Beiträge zur Kenntnis der Geschmacksstoffe von Meerrettich und Rettich. J. prakt. Chem. [2] **132**, 201–208 (1931). — HINE, R., and D. ROGERS: The crystal and molecular structure of (+)-*S*-methyl-L-cysteine *S*-oxide; a standard of absolute configuration for asymmetric sulphur. Chem. a. Ind. **1956**, 1428–1430. — HOFMANN, A. W.: Synthese des ätherischen Oels der *Cochlearia officinalis*. Ber. dtsch. chem. Ges. **7**, 508–514 (1874). — HOPKINS, C. Y.: A sulphur-containing substance from the seed of *Conringia orientalis*. Canad. J. Res., Sect. B **16**, 341–344 (1938).

IVÁNOVICS, G., and S. HORVÁTH: Raphanin, an antibacterial principle of the radish *(Raphanus sativus)*. Nature (Lond.) **160**, 297–298 (1947).

JANSEN, E. F.: The isolation and identification of 2,2′-dithiol-*iso*butyric acid from *Asparagus*. J. of Biol. Chem. **176**, 657–664 (1948). — JENSEN, K. A., J. CONTI u. A. KJÆR: *iso*Thiocyanates. II. Volatile *iso*thiocyanates in seeds and roots of various *Brassicae*. Acta chem. scand. (Copenh.) **7**, 1267–1270 (1953). — JIROUSEK, L.: Zur Frage des Brassica-Faktors und des endemischen Kropfes. Endokrinologie **33**, 310–320 (1956). — Isolace a identifikace organických polysulfidů ze zelí a jejich vztah k Brassica Faktoru. Chem. Listy **50**, 1840—1847 (1956). — On the antithyreoidal substances in cabbage and other Brassica plants. Naturwiss. **43**, 328–329 (1956).

(a) KJÆR, A., and K. RUBINSTEIN: Paper chromatography of thioureas. Acta chem. scand. (Copenh.) **7**, 528–536 (1953). — (b) KJÆR, A., and J. CONTI: *iso*Thiocyanates. V. The occurrence of *iso*propyl *iso*thiocyanate in seeds and fresh plants of various *Cruciferae*. Acta chem. scand. (Copenh.) **7**, 1011–1012 (1953). — (c) KJÆR, A., J. CONTI and I. LARSEN: *iso*Thiocyanates. IV. A systematic investigation of the occurrence and chemical nature of volatile *iso*thiocyanates in seeds of various plants. Acta chem. scand. (Copenh.) **7**, 1276–1283 (1953). — (d) KJÆR, A., J. CONTI and K. A. JENSEN: *iso*Thiocyanates. III. The volatile *iso*thiocyanates in seeds of rape (*Brassica napus* L.). Acta chem. scand. (Copenh.) **7**, 1271–1275 (1953). — (e) KJÆR, A., K. RUBINSTEIN and K. A. JENSEN: Unsaturated five-carbon *iso*thiocyanates. Acta chem. scand. (Copenh.) **7**, 518–527 (1953). — (a) KJÆR, A., and I. LARSEN: *iso*Thiocyanates. IX. The occurrence of ethyl *iso*thiocyanate in nature. Acta chem. scand. (Copenh.) **8**, 699–701 (1954). — (b) KJÆR, A., and K. RUBINSTEIN: *iso*Thiocyanates. VIII. Synthesis of *p*-hydroxybenzyl *iso*thiocyanate and demonstration of its presence in the glucoside of white mustard (*Sinapis alba* L.). Acta chem. scand. (Copenh.) **8**, 598–602 (1954).— (a) KJÆR, A., R. GMELIN and I. LARSEN: *iso*Thiocyanates. XII. 3-Methylthiopropyl *iso*thiocyanate (Ibervirin), a new naturally occurring mustard oil. Acta chem. scand. (Copenh.) **9**, 1143–1147 (1955). — (b) KJÆR, A., and R. GMELIN: *iso*Thiocyanates. XI. 4-Methylthiobutyl *iso*thiocyanate, a new naturally occurring mustard oil. Acta chem. scand. (Copenh.) **9**, 542–544 (1955). — (c) KJÆR, A., I. LARSEN and R. GMELIN: *iso*Thiocyanates. XIV. 5-Methylthiopentyl *iso*thiocyanate, a new mustard oil present in nature as a glucoside (glucoberteroin). Acta chem. scand. (Copenh.) **9**, 1311–1316 (1955). — (d) KJÆR, A., R. GMELIN and I. LARSEN:

*iso*Thiocyanates. XIII. Methyl *iso*thiocyanate, a new naturally occurring mustard oil, present as glucoside (glucocapparin) in *Capparidaceae*. Acta chem. scand. (Copenh.) **9**, 857–858 (1955). — (a) KJÆR, A., R. GMELIN and R. BOE JENSEN: *iso*Thiocyanates. XV. *p*-Methoxybenzyl *iso*thiocyanate, a new natural mustard oil occurring as glucoside (glucoaubrietin) in *Aubrietia* species. Acta chem. scand. (Copenh.) **10**, 26–31 (1956). — (b) KJÆR, A., R. GMELIN and R. BOE JENSEN: *iso*Thiocyanates. XVI. Glucoconringiin, the natural precursor of 5,5-dimethyl-2-oxazolidinethione. Acta chem. scand (Copenh.) **10**, 432–438 (1956). — (c) KJÆR, A., and R. GMELIN: *iso*Thiocyanates. XVIII. Glucocapparin, a new crystalline *iso*thiocyanate glucoside. Acta chem. scand. (Copenh.) **10**, 335–336 (1956). — (d) KJÆR, A., and R. GMELIN: *iso*Thiocyanates. XIX. L(−)-5-Methylsulphinylpentyl *iso*thiocyanate, the aglucone of a new naturally occurring glucoside (glucoalyssin). Acta chem. scand. (Copenh.) **10**, 1100–1110 (1956). — (e) KJÆR, A., and R. BOE JENSEN: *iso*Thiocyanates. XX. 4-Pentenyl *iso*thiocyanate, a new mustard oil occurring as a glucoside (glucobrassicanapin) in Nature. Acta chem. scand. (Copenh.) **10**, 1365–1371 (1956). — (f) KJÆR, A., R. GMELIN and R. BOE JENSEN: *iso*Thiocyanates. XXI. (−)-10-Methylsulphinyldecyl *iso*thiocyanate, a new mustard oil present as a glucoside (glucocamelinin) in *Camelina* species. Acta chem. scand. (Copenh.) **10**, 1614–1619 (1956). — (g) KJÆR, A., and R. GMELIN: *iso*Thiocyanates. XXII. 3-Benzoyloxypropyl *iso*thiocyanate, present as a glucoside (glucomalcolmiin) in seeds of *Malcolmia maritima* (L.) R.Br. Acta chem. scand. (Copenh.) **10**, 1193–1195 (1956). — (h) KJÆR, A., and R. GMELIN: *iso*Thiocyanates XXIII. L(−)-9-Methylsulphinylnonyl *iso*thiocyanate, a new mustard oil present as a glucoside (glucoarabin) in *Arabis* species. Acta chem. scand. (Copenh.) **10**, 1358–1359 (1956). — (a) KJÆR, A., and R. GMELIN: *iso*Thiocyanates. XXV. Methyl 4-*iso*thiocyanatobutyrate, a new mustard oil present as a glucoside (glucoerypestrin) in *Erysimum* species. Acta chem. scand. (Copenh.) **11**, 577–578 (1957). — (b) KJÆR, A., and R. GMELIN: *iso*Thiocyanates. XXVIII. A new *iso*thiocyanate glucoside (glucobarbarin) furnishing (−)-5-phenyl-2-oxazolidinethione upon enzymic hydrolysis. Acta chem. scand. (Copenh.) **11** (1957), in press. — KOCZKA, I., and G. IVÁNOVICS: The antibacterial substance of radish seeds. Acta Univ. szeged, Chem. et Phys. **2**, 205–206 (1949). Quoted after: Chem. Abstr. **44**, 5538 (1950). — KOOLHAAS, D. R.: Das Vorkommen von Methylmercaptan in den Blättern der *Lasianthus*-Arten. Biochem. Z. **230**, 446–450 (1931). — (a) KUNTZE, M.: Das ätherische Öl von *Cardamine amara* L. Arch. Pharmaz. **245**, 657–659 (1907). — (b) KUNTZE, M.: Das ätherische Öl von *Brassica rapa* var. *rapifera* Metzger. Arch. Pharmaz. **245**, 660–661 (1907). — KURUP, P. A., and P. L. N. RAO: Antibiotic principle of *Moringa pterygosperma*. Part II. Chemical nature of pterygospermin. Indian J. Med. Res. **42**, 85–95 (1954).

LEAVER, D., and F. CHALLENGER: Studies on biological methylation. Part XVI. Natural sulphonium compounds. The alkyl methyl sulphides evolved from the urine of dogs by boiling alkali. J. Chem. Soc. Lond. **1957**, 39–46.

MANNICH, C., u. P. FRESENIUS: Über den Hauptbestandteil des ätherischen Öles der *Asa foetida*. Arch. Pharmaz. **274**, 461–472 (1936). — MATSUKAWA, T., H. KAWASAKI, T. IWATSU and S. YURUGI: Synthesis of allithiamine and its homologues. J. of Vitaminol. (Japan) **1**, 13–26 (1954). — McDOWALL, F. H., I. D. MORTON and A. K. R. McDOWELL: Land-cress taint in cream and butter. New Zealand J. Sci. Technol. Sect. A **28**, 305–311 (1947). — McRORIE, R. A., G. L. SUTHERLAND, M. S. LEWIS, A. D. BARTON, M. R. GLAZENER and W. SHIVE: Isolation and identification of a naturally occurring analog of methionine. J. Amer. Chem. Soc. **76**, 115–118 (1954). — MITSUHASHI, S.: Decomposition of thioether derivatives by bacteria. Jap. J. of Exper. Med. **20**, 211–222 (1949). Quoted after: Chem. Abstr. **44**, 3088 (1950). — MORRIS, C. J., and J. F. THOMPSON: The identification of L(+)-S-methylcysteine sulphoxide in plants. J. Amer. Chem. Soc. **78**, 1605–1608 (1956).

NAKAMURA, N.: Über das Vorkommen von Methylmercaptan in frischer Raphanuswurzel. Biochem. Z. **164**, 31–33 (1925). — NENCKI, M.: Über das Vorkommen von Methylmercaptan im menschlichen Harn nach Spargelgenuß. Arch. exper. Path. u. Pharmakol. **28**, 206–209 (1891). — NEUBERG, C., u. J. WAGNER: Über die Verschiedenheit der Sulfatase und Myrosinase VIII. Mitt. über Sulfatase. Biochem. Z. **174**, 457–463 (1926).

PATEL, C. K., S. N. IYER, J. J. SUDBOROUGH and H. E. WATSON: Über das Fett von *Salvadora oleoides:* Khakanfett. J. Indian Inst. Sci., Sect. A **9**, 117–132 (1926). Quoted after: Chem. Zbl. **1927 I**, 465. — PUNTAMBEKAR, S. V.: Mustard oils and mustard oil glucosides occurring in the seed kernels of *Putranjiva Roxburghii* Wall. Proc. Indian Acad. Sci., Sect. A **32**, 114–122 (1950).

RACISZEWSKI, Z. M., E. Y. SPENCER and L. W. TREVOY: Chemical studies of a goitrogenic factor in rapeseed oilmeal. Canad. J. Technol. **33**, 129–133 (1955). — RAGLAND, J. B., and J. L. LIVERMAN: S-Methyl-L-cysteine as a naturally occurring metabolite in *Neurospora crassa*. Arch. of Biochem. a. Biophysics **65**, 574–576 (1956).

SALKOWSKI, H.: Über einige Derivate der *p*-Oxyphenylessigsäure und das ätherische Öl des weißen Senfs. Ber. dtsch. chem. Ges. **22**, 2137–2144 (1889). — SANDBERG, M., and O. M. HOLLY: Note on myrosin. J. of Biol. Chem. **96**, 443–447 (1932). — SATOH, K., and

K. MAKINO: Structure of adenylthiomethylpentose. Nature (Lond.) **165**, 769–770 (1950). — SCHMID, H., u. P. KARRER: Über Inhaltsstoffe des Rettichs. I. Über Sulphoraphen, ein Senföl aus Rettichsamen *(Raphanus sativus* L. var. *alba)*. Helvet. chim. Acta **31**, 1017–1028 (1948). — SCHNEIDER, W.: Über Cheirolin, das Senföl des Goldlacksamens. Sein Abbau und Aufbau. Liebigs Ann. **375**, 207–254 (1910). — SCHNEIDER, W., u. H. KAUFMANN: Untersuchungen über Senföle. II. Erysolin, ein Sulfonsenföl aus *Erysimum Perofskianum*. Liebigs Ann. **392**, 1–15 (1912). — SCHNEIDER, W., u. L. A. SCHÜTZ: Untersuchungen über Senfölglykoside. II. Glucocheirolin. Ber. dtsch. chem. Ges. **46**, 2634–2640 (1913). – (a) SCHNEIDER, W., D. CLIBBENS, G. HÜLLWECK u. W. STEIBELT: Untersuchungen über Senföle: V. Thiourethane und Thiourethanäther aus einigen natürlich vorkommenden Senfölen. Ber. dtsch. chem. Ges. **47**, 1248–1269 (1914). — (b) SCHNEIDER, W., u. F. WREDE: Untersuchungen über Senfölglucoside. V. Zur Konstitution des Sinigrins. Ber. dtsch. chem. Ges. **47**, 2225–2229 (1914). — SCHULTZ, O.-E., u. R. GMELIN: Papierchromatographie der Senfölglucosid-Drogen. Z. Naturforsch. **7**b, 500–506 (1952). — Papierchromatographie senfölglucosidhaltiger Pflanzen. Z. Naturforsch. **8**b, 151–156 (1953). — (a) SCHULTZ, O.-E., u. R. GMELIN: Das Senfölglukosid „Glukoiberin" und der Bitterstoff „Ibamarin" von *Iberis amara* L. (Schleifenblume). Arch. Pharmaz. **287/59**, 404–411 (1954). — (b) SCHULTZ, O.-E., u. R. GMELIN: Das Senfölglukosid von *Tropaeolum majus* L. (Kapuzinerkresse) und Beziehungen der Senfölglukoside zu den Wuchsstoffen. Arch. Pharmaz. **287/59**, 342–350(1954).— SCHULTZ, O.-E., u. H. WAGNER: Kristallisierte Azetylderivate von nicht oder schwer kristallisierenden Senfölglukosiden. Arch. Pharmaz. **288/60**, 525–532 (1955). — (a) SCHULTZ, O.-E., u. W. WAGNER: Trennung der Senfölglucoside durch absteigende Papierchromatographie. Z. Naturforsch. **11**b, 73–78 (1956). — (b) SCHULTZ, O.-E., u. W. WAGNER: Glucoalyssin, ein neues Senfölglucosid aus *Alyssum*-Arten. Z. Naturforsch. **11**b, 417–419 (1956). — (c) SCHULTZ, O.-E., u. W. WAGNER: Senfölglukoside als genuine Muttersubstanzen der natürlich vorkommenden antithyreoiden Stoffen. Arch. Pharmaz. **289/61**, 597–604 (1956). — SEMMLER, F. W.: Über das ätherische Öl von *Allium ursinum* L. Liebigs Ann. **241**, 90–150 (1887). — Über schwefelhaltige ätherische Öle. *Asa foetida* Öl. Arch. Pharmaz. **229**, 1–31 (1891). — (a) SEMMLER, F. W.: Über das ätherische Öl des Knoblauchs *(Allium sativum)*. Arch. Pharmaz. **230**, 434–443 (1892). — (b) SEMMLER, F. W.: Das ätherische Öl der Küchenzwiebel *(Allium Cepa* L.). Arch. Pharmaz. **230**, 443–448 (1892). — SIMANDL, J., u. J. FRANC: Die Isolierung des Tetraäthylthiuramdisulphids aus dem Tintenmistpilz *(Coprinus atramentarius)*. Coll. čechoslov. Chem. Commun. **22**, 331–332 (1957). — STAHL, W. H., B. McQUE, G. R. MANDELS and R. G. H. SIU: Microbiological degradation of wool. I. Sulfur metabolism. Arch. of Biochem. **20**, 422–432 (1949). — STAHMANN, M. A., K. P. LINK and J. C. WALKER: Mustard oils in crucifers and their relation to clubroot. J. Agricult. Res. **67**, 49–63 (1943). — STOLL, A., u. E. JUCKER: Modern Methods of Plant Analysis, vol. IV. Berlin-Göttingen-Heidelberg: Springer 1955. — STOLL, A., R. MORF, A. RHEINER u. J. RENZ: Über Inhaltsstoffe aus *Petasites officinalis* Moench. I. Petasin und die Petasolester B und C. Experientia (Basel) **12**, 360–368 (1956). — STOLL, A., u. E. SEEBECK: Über Alliin, die genuine Muttersubstanz des Knoblauchöls. Experientia (Basel) **3**, 114–115 (1947). — Über Alliin, die genuine Muttersubstanz des Knoblauchöls. Helvet. chim. Acta **31**, 189–210 (1948). — Über den enzymatischen Abbau des Alliins und die Eigenschaften der Alliinase. Helvet. chim. Acta **32**, 197–205 (1949). — Chemical investigations on alliin, the specific principle of garlic. Adv. Enzymol. **11**, 377–400 (1951). — SYNGE, R. L. M., and J. C. WOOD: (+)-(S-Methyl-L-cysteine S-oxide) in cabbage. Biochemic. J. **64**, 252—259 (1956).

TER MEULEN, H.: Recherches expérimentales sur la nature des sucres de quelques glucosides. Rec. Trav. chim. Pays-Bas (Amsterd.) **24**, 475–483 (1905). — THOMPSON, J. F., C. J. MORRIS and R. M. ZACHARIUS: Isolation of (—)S-methyl-L-cysteine from beans *(Phaseolus vulgaris)*. Nature (Lond.) **178**, 593 (1956).

VIGNEAUD, V. DU: A trail of research. Ithaca, N. Y.: Cornell University Press 1952. — WAGNER, W.: Papierchromatographische Analyse der Senfölglucoside, präparative Darstellung ihrer Acetylderivate und ein Beitrag zu ihrer allgemeinen Struktur. Inaug.-Dissert. Tübingen 1956, p. 82. — WEYGAND, F., R. JUNK u. D. LEBER: Adenyl-thiomethylpentose. Hoppe-Seylers Z. **291**, 191–196 (1952). — WEYGAND, F., O. TRAUTH u. R. LÖWENFELD: Konstitutionsaufklärung des Thiozuckers der Adenylthiomethylpentose. Chem. Ber. **83**, 563–567 (1950). — WILL, H.: Untersuchungen über die Constitution des ätherischen Öls des schwarzen Senfs. Liebigs Ann. **52**, 1–51 (1844). — WOLFF, E. C., S. BLACK and P. F. DOWNEY: Enzymatic synthesis of S-methylcysteine. J. Amer. Chem. Soc. **78**, 5958 (1956).— WRIGHT, L. D., E. L. CRESSON, J. VALIANT, D. E. WOLF and K. FOLKERS: Biotin *l*-sulfoxide. III. The characterisation of biotin *l*-sulfoxide from a microbiological source. J. Amer. Chem. Soc. **76**, 4163–4166 (1954).

ZWERGAL, A.: Beitrag zur Kenntnis der Inhaltsstoffe des Kohlrabis. Pharmazie **6**, 245 (1951). — Der *Brassica*-Faktor und andere antithyreoide Stoffe als die Ursache der Kropfnoxe. Pharmazie **7**, 93–97 (1952).

Die Schwefelspezialisten
unter den Mikroorganismen.

Von

W. Schwartz.

Mit 7 Abbildungen.

1. Übersicht.

Der Kreislauf des Schwefels in der Natur bewegt sich in der Hauptsache zwischen Sulfaten, Schwefelwasserstoff und Sulfiden, elementarem Schwefel und Schwefel in organischen Verbindungen. Soweit Mikroorganismen beim Übergang der einen in die andere Form beteiligt sind, handelt es sich teils um unspezifische Vorgänge, teils um die Tätigkeit von Spezialisten. Der erste Fall betrifft die Assimilation des Schwefels aus Sulfaten, das Freiwerden von Merkaptanen, Methyldisulfid (STARKEY, SEGAL und MANAKER 1953) und Schwefelwasserstoff bei Fäulnisprozessen, sowie die gelegentlich als Nebenreaktion bei Bakterien vorkommende Reduktion von Sulfaten zu Schwefelwasserstoff und Oxydation von Schwefel im Boden durch Pilze und Bakterien. Die Wirksamkeit von Spezialisten zeigt sich vor allem im anorganischen Abschnitt des Schwefelkreislaufs in oxydativen und reduktiven Prozessen, die sich zwischen Sauerstoffverbindungen des Schwefels, elementarem Schwefel und Schwefelwasserstoff abspielen. Beteiligt sind die farblosen, roten und grünen „Schwefelbakterien" und die desulfurizierenden Bakterien. Wie weit bei der Umsetzung gewisser organischer schwefelhaltiger Verbindungen, z. B. der Thiophane und Thiophene, Schwefelspezialisten auftreten, ist noch unbekannt. Eine thiocyanatoxydierende Art hat ihren Platz bei der Gattung *Thiobacterium (Thiobacillus)* gefunden.

Die bis jetzt bekannt gewordenen Desulfurizierer sind sämtlich Eubakterien. Die „Schwefelbakterien" sind eine ökologisch gekennzeichnete Gruppe von Mikroorganismen, die in BERGEYs Manual of Determinative Bacteriology (BREED usw. 1948) an verschiedenen Stellen eingeordnet worden ist: *Thiospira* und *Thiospirillum* bei den *Pseudomonadaceae*, *Thiobacterium* als einzige Gattung der *Thiobacterieae (Thiobacilleae)*[1] bei den *Nitrobacteriaceae*; *Achromatium*, *Thiovulum* und *Macromonas* bei den *Achromatiaceae* als Anhang der *Chlamydobacteriales*; die *Beggiatoaceae* bei den *Chlamydobacteriales* und schließlich die roten und grünen Formen in verschiedenen Familien der *Rhodobacteriineae*.

Offensichtlich gehört nur ein relativ kleiner Teil der Gattungen zu den eigentlichen Bakterien. Zu den Oscillatoriaceen müssen die Beggiatoaceen gestellt werden[2]. Auch die grünen Formen dürften Cyanophyceen sein. Bei den Rhodobacteriineen sind lediglich nach physiologischen Gesichtspunkten Formen ohne engere taxonomische Beziehungen zusammengefaßt, die teils zu den Cyanophyceen, teils zu den Eubakterien und vielleicht in einigen Fällen zu den Protisten gehören — oder deren Zugehörigkeit zu einer der großen taxonomischen Gruppen noch gänzlich unsicher ist (BAHR und SCHWARTZ 1957).

---

[1] Vgl. Fußnote S. 97.
[2] Für *Beggiatoa* und *Thiotrix* vgl. z. B. BAHR und SCHWARTZ (1957).

Die zusammenfassende Bezeichnung „Schwefelbakterien" ist demnach irreführend und sollte durch die im folgenden gebrauchte Bezeichnung „Schwefelmikroben" ersetzt werden.

2. Sulphuretum.

In der Natur hat Schwefelwasserstoff wegen seiner Giftigkeit auslesende Wirkung. Wo im Süßwasser oder Meerwasser oder in Gewässern höherer Salzkonzentration und in ihren Ablagerungen ständig H_2S vorhanden ist, entwickeln sich die Lebensgemeinschaften des Sulphuretums (BAAS-BECKING 1925, GALLIHER 1933), deren charakteristische Vertreter Schwefelmikroben sind (Abb. 1).

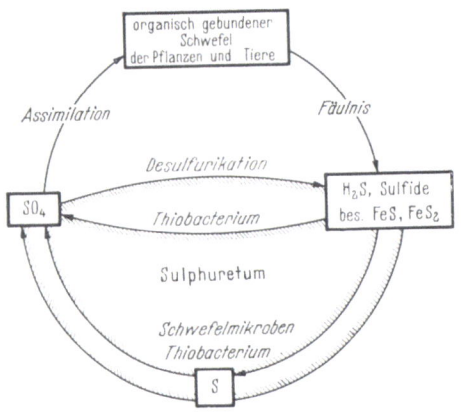

Abb. 1. Biologie des Schwefelkreislaufs und des Sulphuretums. ▭▭▭▭▭ Bereich des Sulphuretums.

Das wichtigste und ausgedehnteste Vorkommen von Sulphureten bieten sapropelische Ablagerungen. Stammt bei Gegenwart von Sulfaten H_2S aus der Desulfurikation, so kommen desulfurizierende Bakterien zu den Schwefelmikroben hinzu. Auch H_2S aus vulkanischen Prozessen kann, wie am Golf von Neapel, zur Entstehung mehr oder weniger ausgedehnter, über längere Zeit bestehender Sulphureten führen, während Eiweißfäulnis abgesehen von Abwässern, verunreinigten Wasserläufen und einzelnen besonderen Vorkommnissen, nur vorübergehend wirksam ist[1].

Trotz seiner Giftigkeit schließt H_2S jedoch nicht die Anwesenheit einiger anderer Arten von Mikroorganismen neben Schwefelmikroben, Desulfurizierern und Fäulnisbakterien aus. *Desulfovibrio* ist fast stets von bis jetzt wenig beachteten Eubakterien begleitet; ferner finden sich hier Oscillatoriaceen, Grünalgen, Diatomeen, Flagellaten, einzelne Ciliaten und Spirochaeten (BAVENDAMM 1924, POP 1936, CZURDA 1491, BÖCHER 1949). Einzelne Oscillatoriaceen speichern Schwefel; sie könnten in ähnlichen stoffwechselphysiologischen Beziehungen zu H_2S stehen wie die anaeroben

Abb. 2. Sulphuretum im Searles Lake, Cal. (Juni 1954). Temperatur der Salzsole etwa 54° C, $p_H > 9$. Desulfurikation im Sediment. Rotfärbung durch Schwefelpurpurmikroben und Purpurmikroben.

---

[1] In unseren Flüssen führen z. B. Vorgänge der sekundären Fäulnis von abgestorbenen *Sphaerotilus*-Massen (BAHR und SCHWARTZ 1956) zur vorübergehenden Ansiedlung von Schwefelmikroben.

Schwefelmikroben. SHTURM (1937) hat das gemeinsame Vorkommen von Desulfurizierern, Schwefelmikroben und anaeroben Cellulosevergärern von der Krim und der unteren Wolga beschrieben. Je nach Temperatur, H_2S-Konzentration, Salzgehalt, Reaktion wechselt die Zusammensetzung der Lebensgemeinschaften des Sulphuretums. Im kalifornischen Wüstengebiet findet sich im Searles Lake und im Owens Lake (GALE 1913) und in anderen teilweise ausgetrockneten Salzseen eine an hochkonzentrierte alkalische Salzsolen angepaßte Lebensgemeinschaft von Desulfurizierern, Schwefelpurpurmikroben, Purpurmikroben, halotoleranten und halophilen Eubakterien, Flagellaten und einigen

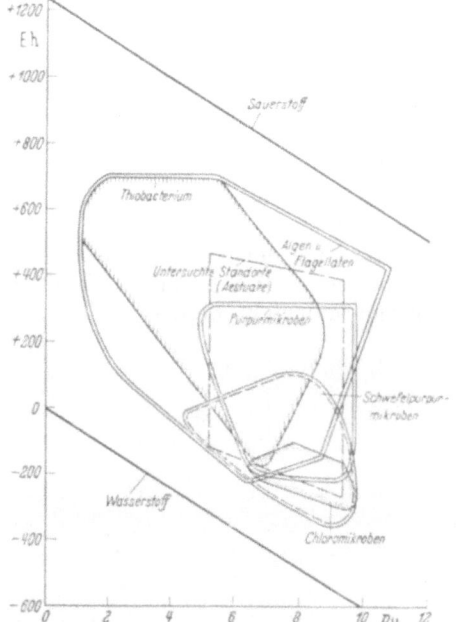

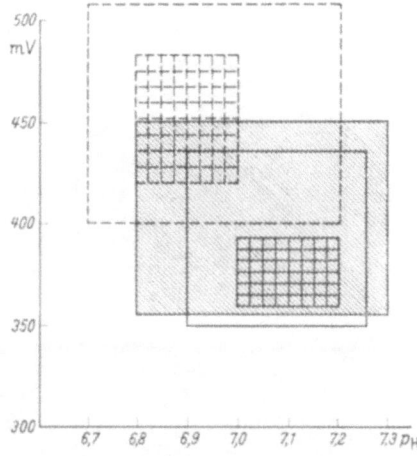

Abb. 3. Abhängigkeit der Standorte von *Desulfovibrio*, Schwefelmikroben, *Thiobacterium* und begleitenden Algen und Flagellaten von der Reaktion und vom Elektrodenpotential. Abszisse p_H, Ordinate Eh. Bereiche: Algen + Flagellaten, *Thiobacterium*, Purpurmikroben, Schwefelpurpurmikroben, Chloromikroben, untersuchte Standorte (Aestuare). (Aus BAAS-BECKING und WOOD 1955.)

Abb. 4. Bereiche von p_H und Elektrodenpotential (mV) für einige Beggiatoaceen (besonders *Begg. alba*, *Thiothrix nivea*) von verschiedenen Standorten. ---- Material aus Schwefelquellen (inneres Rechteck: optimaler Bereich); —— Material aus eutrophen Gewässern (inneres Rechteck: optimaler Bereich); ▨▨▨ Material aus Klärschlammablagerungen. (Aus BAHR und SCHWARTZ 1956.)

erst ungenügend untersuchten Mikroorganismen (BAAS-BECKING 1925, 1928, NEHRKORN und SCHWARTZ[1]) (Abb. 2).

　　Bei so verschiedenen Lebensbedingungen ist es nicht weiter verwunderlich, wenn es bei den beteiligten Arten zur Entstehung ökologischer Rassen kommen kann. Bei Desulfurizierern ist diese Erscheinung seit längerem bekannt. Sie findet sich ferner bei *Beggiatoa* in Beziehung zur Versorgung mit H_2S (BAHR und SCHWARTZ 1956). BAAS-BECKING und WOOD (1955) haben versucht, die schwer zu überblickenden ökologischen Verhältnisse in derartigen Ablagerungen aufzuklären. Sie haben zunächst in den Ablagerungen der Aestuarien die Standortbedingungen für die am Schwefelkreislauf beteiligten Mikroorganismen und die begleitenden Algen analysiert.

　　Nach p_H und Elektrodenpotential ließen sich durch Beobachtung am Standort und in Laboratoriumskulturen Areale abgrenzen, die sich teilweise überdecken und das gemeinsame Vorkommen von Vertretern verschiedener Gruppen verständlich machen (Abb. 3). BAHR und SCHWARTZ (1956) haben entsprechende Versuche mit Beggiatoaceen *(Beggiatoa, Thiothrix)* ausgeführt, die je nach der Herkunft von verschiedenen Standorten verschiedenes Verhalten zeigten (Abb. 4).

---

[1] Noch nicht veröffentlichte Untersuchungen.

3. Desulfurizierer.

Während Denitrifikation bei Bakterien häufig vorkommt, findet sich die energetisch ungünstigere Desulfurikation, die obligate Reduktion von Sulfaten zu H_2S als Grundlage des Betriebsstoffwechsels, nur bei wenigen Spezialisten[1].

Weitaus der wichtigste Vertreter ist die Gattung *Desulfovibrio* Kl. u. v. N, bei der auch sporenbildende Stämme (*Sporovibrio*, Starkey 1938) beobachtet worden sind. Hvid-Hansen (1951) und Beerens und Hvid-Hansen (1952) haben die desulfurizierende *Desulphori-stella hydrocarbonoclastica* beschrieben. Frisch aus Wasser und Schlammablagerungen isolierte Stämme einiger *Clostridium*-Arten sollen nach Prévot (1948, 1949) ebenfalls desulfurizieren. Beide Befunde bedürfen indessen der Bestätigung; nicht jede Reduktion von Sulfaten zu H_2S kann als Desulfurikation aufgefaßt werden.

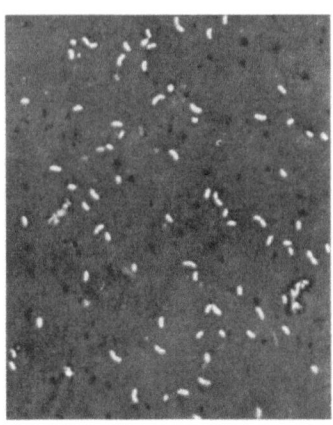

a

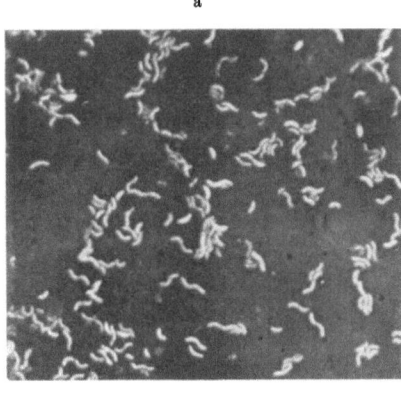

b

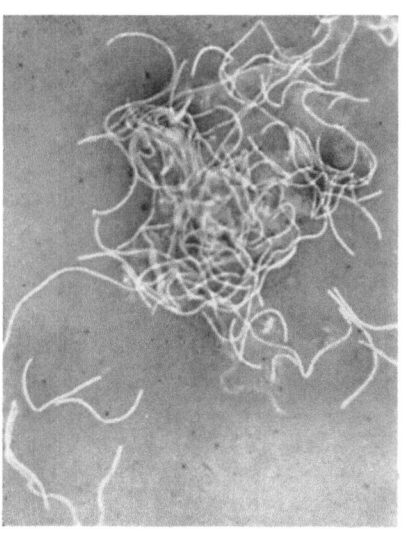

c

Abb. 5a—c. *Desulfovibrio desulfuricans.* a Kurze Vibrionen (30° C, 24 Std. alt, 850fach). b Vibrionen einzeln und in kurzen Ketten (45° C, 3 Tage alt, 850fach). c Fäden und Ketten von langen Vibrionen (von 30° auf 55° C übertragen, 17 Tage alt, 850fach). (Aus Starkey 1938.)

Die Gattung *Desulfovibrio* ist weit verbreitet. Aus Erdboden, rezenten und alten limnischen und marinen Ablagerungen, Erdöl, Salzwasser der Erdöl-Lagerstätten und von vielen anderen Stellen kann man *Desulfovibrio* isolieren. In den Reinkulturen ist das Bild wechselnd, je nach dem Stamm und nach den Kulturbedingungen. Man sieht in der Hauptsache kurze oder längere, schwach gebogene oder deutlich vibrionenförmige, bewegliche und unbewegliche Stäbchen, einzeln oder in Ketten zusammenhängend, die an Spirillen erinnern; auch Fäden treten auf (Abb. 5, 6). Die Abgrenzung einzelner Arten ist schwierig, Halophile und thermophile Stämme sind eher Rassen als eigene Arten.

---

[1] Zu den Energieverhältnissen vgl. Rippel-Baldes (1948).

Auch über die taxonomische Bewertung sporenbildender, C-autotropher und N-bindender Stämme besteht keine Klarheit, und ebensowenig befriedigt bis jetzt der Versuch, nach den verwertbaren H-Donatoren in Verbindung mit anderen physiologischen Merkmalen die einzelnen bis jetzt beschriebenen Arten[1] voneinander abzugrenzen, vielleicht mit Ausnahme von *D. rubentschicki*, der durch die größere Zahl verwertbarer Fettsäuren auffällt. Die Liste der als H-Donatoren in Frage kommenden Substanzen ist groß; sie umfaßt, abgesehen vom elementaren Wasserstoff, α-Aminosäuren Säureamide, Pepton, Alkohole (Äthyl-, Propyl-, Butylalkohol, Glycerin), Zucker (Glucose, Fructose, Galactose, Saccharose, Maltose, Lactose), fettsaure Salze (Acetat, Succinat, Lactat, Malat, Propionat, Butyrat, Valerianat, Palmitat, Stearat — die letzten fünf für *D. rubentschicki*); ferner werden genannt Kohlenwasserstoffe (rein und Gemische), Chitin und Fette.

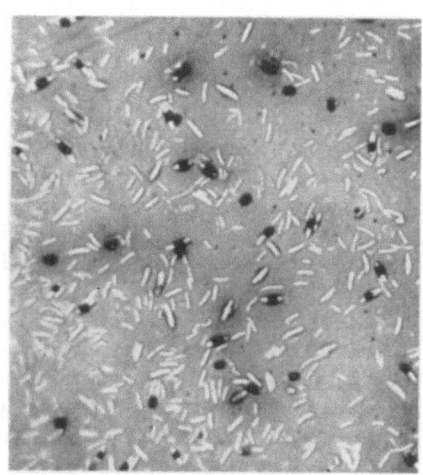

Die Anreicherung und Reinkultur von *Desulfovibrio* gelingt unter streng anaeroben Bedingungen z. B. in Medien mit fettsauren Salzen (Lactat-Medium nach STARKEY 1938), die durch Zugabe kleiner Mengen von Pepton oder Hefeextrakt verbessert werden können. Günstig wirken ferner hohe Einsaat und die Zugabe eines reduzierenden Agens (etwa 5 μmol/cm³ Na_2S, Cystein, Glutathion, Ascorbinsäure). Mit Ascorbinsäure ist Wachstum auch unter aeroben Bedingungen möglich (GROSSMANN und POSTGATE 1953). Schwierig ist die Abtrennung H_2S-resistenter Begleitbakterien, die zu *Desulfovibrio* in einem synergistischen Verhältnis stehen (POSTGATE 1953). Auch BAUMANN

Abb. 6. *Desulfovibrio desulfuricans*, Sporenbildung. Sporen teils noch in den Sporangien eingeschlossen, teils frei (45° C, 5 Tage alt, 850fach). (Aus STARKEY 1938.)

und DENK (1951) scheinen solche Beziehungen zu erwägen; sie vermuten die Produktion eines Stoffes mit biosartiger Wirkung durch *Desulfovibrio* selbst, der das Anwachsen einzelner Zellen erleichtert. CZURDA (1940) hat eine zweigliedrige Kultur eines *Desulfovibrio*-Stammes mit einem stäbchenförmigen Bakterium beschrieben, die in einer mineralischen Nährlösung ohne Wasserstoffgabe kräftig desulfurizierte; der Energiespender war unbekannt. Die Polymorphie, die man in Reinkulturen beobachtet, z. B. bei dem häufig untersuchten HILDENBOROUGH-Stamm, findet zum Teil darin ihre Erklärung, daß Gemische mit Begleitbakterien vorgelegen haben.

Die Produktion größerer oder kleinerer Mengen von H_2S bei ausreichender Versorgung mit Sulfat und einem Wasserstoffdonator hängt von zahlreichen Faktoren wie Herkunft der Stämme, Kulturbedingungen, Vorkultur und Alter der Kultur ab (BAARS 1930, MILLER 1949). Die H_2S-Toleranz ist hoch; es werden z. B. 2500 mg/l H_2S vertragen (MILLER 1949) und die H_2S-Produktion steigt weiter an, wenn H_2S, z. B. durch Zugabe von $CdCO_3$ zum Medium, als Sulfid gebunden wird (MILLER 1950). Hält man die Kultur während mehrerer Wochen in der logarithmischen Phase, so erreicht die SO_4-Reduktion mit Lactat etwa 3,5 mol ''SO_4 je mg Zelltrockengewicht und Stunde und bei Gegenwart von Wasserstoff anstelle von Lactat sogar 12,5 (BUTLIN und POSTGATE 1953). Auch an

---

[1] ZoBELL führt in BERGEYS Manual (BREED usw. 1948) 3 Arten auf: *D. desulfuricans* (BEIJ.) KL. und v. NIEL, *D. aestuarii* (v. DELD.) *comb. nov.* und *D. rubentschicki* (BAARS) *comb. nov.* und erwähnt einige weitere hierhin gehörende Formen als Anhang.

natürlichen Standorten können erhebliche Mengen von H_2S entstehen (Butlin 1949).

Die Verwendbarkeit von elementarem Wasserstoff als Donator zur Reduktion von Sulfat, Sulfit, Thiosulfat haben Stephenson und Stickland (1931) an reinen Kulturen nachgewiesen, nachdem die Möglichkeit schon lange vorher geäußert worden war, z. B. durch Nikitinsky (1907). Daß bei Gegenwart von Wasserstoff C-Autotrophie unter Verwertung von CO_2 oder Bicarbonat möglich ist, haben erst Wight und Starkey (1945) gezeigt. Fakultativ autotrophe Stämme scheinen häufig vorzukommen; bei Gegenwart von Wasserstoff und milchsauren oder traubensauren Salzen kann je nach den Eigenschaften des Stammes die Sulfatreduktion gleichzeitig autotroph und heterotroph erfolgen (Senez 1954). In marinen Sedimenten fanden Sisler und ZoBell (1950, 1951) zahlreiche hydrogenaseaktive Stämme, die Wasserstoff teils in einem Lactatmedium, teils in einem mineralischen Medium verarbeiten konnten. Sisler und ZoBell (1951) glauben auch die Verwertung von elementarem N als N-Quelle bei hydrogenasepositiven Stämmen an Veränderungen des Stickstoff/Argon-Verhältnisses im Gasgemisch über den Kulturen nachgewiesen zu haben.

Im einzelnen ist der Verlauf der Sauerstoffabgabe aus Sulfat, Thiosulfat usw. und die Beschaffenheit des beteiligten Enzyms „Desulfurikase" ebensowenig bekannt wie die Bedeutung des von Postgate (1954) nachgewiesenen reversiblen cytochromartigen Pigmentes.

Mit Wasserstoff können Sulfat, Sulfit, Thiosulfat und Tetrathionat und, allerdings sehr langsam, kolloider Schwefel in den zu erwartenden Verhältnissen reduziert werden. Die Sulfatreduktion scheint über Sulfit zu verlaufen. Sie wird durch Na-Selenat schon bei einem Selenat/Sulfat-Verhältnis von 0,02 gehemmt; Sulfit hebt die Hemmung auf (Postgate (1949). Auch Monofluorophosphat bewirkt eine durch Sulfit aufhebbare Hemmung von Wachstum und Sulfatreduktion mit Wasserstoff (Butlin und Postgate 1953), während Chromat beide Vorgänge nichtkompetitiv hemmt.

Mit Lactat und Sulfat verläuft der Stoffwechsel über Brenztraubensäure und führt zu Äthylalkohol und Essigsäure, wobei das Wachstum keineswegs immer dem Ausmaß der Sulfatreduktion entspricht (Senez 1951) und in einzelnen Fällen Wachstum mit Brenztraubensäure ohne Desulfurikation beobachtet worden ist (Postgate 1952). Niedere Fettsäuren (Essigsäure, Propionsäure, n-Buttersäure) konnten bei dem von Ghose und Wikén (1955) benutzten Stamm weder als C-Quelle noch als H-Donatoren verwandt werden; sie griffen vermutlich in die intracellulare Umwandlung von Brenztraubensäure in Essigsäure ein und blockierten die SH-Gruppen des an der Acetatbildung beteiligten Coenzyms A.

NO_3-Stickstoff wird nach Baumann und Denk (1951) als N-Quelle verwertet und unabhängig von der Desulfurikation zu NH_3 reduziert (Denk 1950).

4. Schwefelmikroben.

Bei den Schwefel und Schwefelverbindungen oxydierenden Mikroorganismen ist die Mannigfaltigkeit der Formen und der stoffwechselphysiologischen Vorgänge besonders groß. Wir treffen auf zwei stoffwechselphysiologische Hauptgruppen: Farblose aerobe Chemosynthetiker und rote und grüne anaerobe Photosynthetiker. Die zweite Gruppe (Schwefelpurpurmikroben: „*Thiorhodaceae*" und „*Chlorobacteriaceae*") ist durch zahlreiche Übergänge zwischen Autotrophie und Heterotrophie mit den Purpurmikroben (*„Athiorhodaceae"*) verbunden (Gest 1951), auf die hier nicht eingegangen werden kann. Auch bei der ersten Gruppe kommt Heterotrophie bei *Beggiatoa* vor (Cataldi 1940). Eine heterotrophe, offensichtlich

Thiothrix nahestehende Art haben HAROLD und STANIER (1955) untersucht und als *Leucothrix mucor* OERSTED identifiziert.

Die Mehrzahl der Schwefelmikroben ist noch nicht kultiviert worden und viele Gattungen sind erst ungenügend bekannt; die einzelnen Arten werden oft nur nach den Größenverhältnissen der Zellen voneinander unterschieden (Tabellen 1 bis 3). Versuche zur Erzielung von Kulturen haben sich in erster Linie auf die häufiger und in größeren Ansammlungen vorkommenden Gattungen wie *Beggiatoa*, *Thiothrix, Lamprocystis, Chroma-*

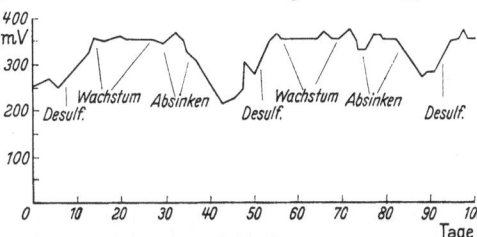

tium und besonders auf Purpurmikroben erstreckt. Roh- und Anreicherungskulturen sind im allgemeinen leicht zu gewinnen.

Die Reinkultur bei den farblosen aeroben Formen unter Anwendung von Gasgemischen (BAVENDAMM 1924, 1934) ist kompliziert, im Ergebnis durch die Herkunft des Materials von verschiedenen Standorten beeinflußt und

Abb. 7. Rohkultur von *Beggiatoa alba* mit *Desulfovibrio*. Periodischer Verlauf der Entwicklung von *Beggiatoa* (Wachstum und Absinken) während 100 Tagen im Zusammenhang mit Schwankungen des Elektrodenpotentials durch Verbrauch und Nachlieferung von H_2S. (Aus BAHR und SCHWARTZ 1956.)

Tabelle 1. *Übersicht über die Gattungen der farblosen Schwefelmikroben*[1].
(Nach BAVENDAMM 1924 und VAN NIEL in BERGEYs Manual 1948.)
Reinkulturen liegen anscheinend nur von *Beggiatoa-* und *Thiothrix*-Arten vor. Die Beggiatoaceen sind farblose hormogonale Cyanophyceen (BAHR und SCHWARTZ 1957). Die Achromatiaceen sind eine heterogene Gruppe von Schwefelmikroben, von denen *Thiospira* eine Pseudomonadacee ist und die Zugehörigkeit der übrigen Gattungen unbekannt ist.

Gattungen und Anzahl der Arten	Morphologie, Größenverhältnisse der Zellen, Beweglichkeit	Gattungen und Anzahl der Arten	Morphologie, Größenverhältnisse der Zellen, Beweglichkeit
Beggiatoaceae		*Achromatiaceae*	
Beggiatoa TREVISAN (6)	Durchmesser der Fäden 1 bis 55 μ. Kriechbewegung	*Achromatium* SCHEW. (1)	Zellen kuglig (7 μ Durchmesser) bis ovoid (bis $35 \times 100 \mu$). Beweglich, Geißeln 0. Einschlüsse von $CaCO_3$
Thiothrix WINOGR. (7)	Durchmesser der Fäden 0,3 bis 30 μ. Festgewachsen, Fadenbruchstücke zeigen Kriechbewegung	*Thiophysa* HINZE (2)	Zellen kuglig (Durchmesser 7—40 μ) bis ovoid (bis $18 \times 29 \mu$). Beweglich, Geißeln 0
Thioploca LAUTERB. (4)	Durchmesser der Fäden 1 bis 9 μ. Kriechbewegung	*Thiosphaerella* NADS. (1)	Zellen: $4,8 \times 6 \mu$. Beweglich, Geißeln 0
Thiospirillopsis UPHOF (1)	Durchmesser der Fäden 2 bis 3 μ. Kriechbewegung ähnlich *Spirulina*	*Thiovulum* HINZE (3)	Zellen: $4,9 \times 12,8$ bis $17 \times 18 \mu$. Beweglich, Geißeln + (peritrich, Geißelkranz)
		Thiospira WISL. (5)	Zellen (Spirillen) $(1,2$ bis $1,5) \times (6,6—50) \mu$. Beweglich, Geißeln + (polar)
		Macromonas UNTERM. und KOPPE (2)	Zellen bis $(8—14) \times (12$ bis $30) \mu$. Beweglich, Geißeln + (polar), Einschlüsse von $CaCO_3$

[1] Die Größenangaben der Zellen in den Tabellen 1—3 dienen lediglich als Anhaltspunkt für den bei den einzelnen Gattungen beobachteten Größenbereich, die Anzahl der Arten ist unsicher; auch in bezug auf die Gattungen bestehen in vielen Fällen noch Unsicherheiten.

Tabelle 2. *Übersicht über die Gattungen der roten Schwefelmikroben (Schwefelpurpurmikroben, „Thiorhodaceae").* (Nach Bavendamm 1924 und van Niel in Bergeys Manual 1948.) Die Zusammenfassung in Familien ist nach dem Fehlen oder nach der Beschaffenheit der am natürlichen Standort vorkommenden Zellkolonien erfolgt und besagt nichts über die Verwandtschaft der Gattungen, deren Zugehörigkeit zu größeren taxonomischen Gruppen, von wenigen Ausnahmen (z. B. *Thiospirillum*) abgesehen, ungewiß ist. Reinkulturen liegen erst von einzelnen Arten vor, z. B. von *Lamprocystis rosa-persicina* und *Chromatium warmingii.*

Gattungen und Anzahl der Arten	Morphologie, Größenverhältnisse der Zellen, Beweglichkeit	Gattungen und Anzahl der Arten	Morphologie, Größenverhältnisse der Zellen, Beweglichkeit
Thiocapsaceae		*Lamprocystaceae*	
Thiocystis Winogr. (2)	Gallertkolonien mit Schwärmstadien, Geißeln +. Durchmesser der Zellen 1—5 μ.	*Lamprocystis* Schroeter (5)	Gallertkolonien, hohlkuglig, netzförmig, mit Schwärmstadien. Geißeln + Zellen rund bis oval. Durchmesser 1,5 bis 7 μ
Thiocapsa Winogr. (2)	Gallertkolonien ähnlich *Aphanocapsa*, nicht schwärmend. Durchmesser der Zellen 1,5 bis 3 μ.		
		Thiopediaceae	
Thiosarcina Winogr. (1)	Paketförmige Kolonien ähnlich *Sarcina*, nicht schwärmend. Durchmesser der Zellen 2—2,5 μ	*Thiopedia* Winogr. (1)	Tafelförmige Kolonien ähnlich *Merismopedia*, jedoch mit Schwärmstadien. Durchmesser der Zellen 1—2 μ. Geißeln +
Amoebobacteriaceae		*Chromatiaceae*	
Amoebobacter Winogr. (3)	Familien amoeboid beweglich. Zellen kuglig. Durchmesser 0,5 μ oder gestreckt (1,5—35) × (2 bis 6) μ	*Chromatium* Perty (12)	Zellen gestreckt, oft gebogen. Durchmesser 1—8 μ. Länge 2—25 μ. Geißeln + (polar)
		Rhabdochromatium Winogr. (5)	Zellen stab- oder unregelmäßig spindelförmig. Durchmesser 2—8,5 μ, Länge 5—60 μ. Geißeln + (polar)
Thiothece Winogr. (1)	Gallertkolonien mit Schwärmstadien ähnlich der unbeweglichen *Aphanothece.* Zellen kuglig (4,2 μ) bis gestreckt	*Thiospirillum* Winogr. (5)	Zellen (Spirillen) 1,0 × (8 bis 18) μ bis (2,5—4) × (30—100) μ. Geißeln + (polar)
Thiodictyon Winogr. (1)	Netzförmige Kolonien ähnlich *Hydrodictyon* mit langsam beweglichen Schwärmstadien. Zellen 1,7 × 5 μ. Geißeln 0	*Rhodocapsaceae*	
		Rhodocapsa Molisch (1)	Zellen gestreckt in Schleimkapsel, bewegliche ohne Kapsel (?), Durchmesser (ohne Kapsel) 1,8—3,5 μ, Länge meist 10—20 μ
Thiopolycoccus Winogr. (1)	Dichte Kolonien, nicht schwärmend, ähnlich *Micrococcus*-Kolonien. Durchmesser der Zellen 1,2 μ	*Rhodothece* Molisch (1)	Zellen kuglig mit Gallerthülle, Durchmesser (ohne Hülle) 1,8—2,5 μ. Unbeweglich

daher nicht immer ohne weiteres reproduzierbar. *Beggiatoa alba* und wohl auch andere Formen lassen sich auch in zweigliedrigen Kulturen halten, bei denen H₂S durch Desulfurikation geliefert wird (Bahr und Schwartz 1956). *Beggiatoa alba* zeigt unter diesen Verhältnissen, besonders in den kräftiger sich entwickelnden Rohkulturen, einen periodischen Verlauf des Wachstums im Zusammenhang mit Schwankungen des Elektrodenpotentials (Abb. 7).

Die roten und grünen Formen sind leichter zugänglich (van Niel 1932) und können in Medien kultiviert werden, deren Zusammensetzung je nach Vorhandensein oder Fehlen von Licht, Sauerstoff oder Wasserstoff wechselt. Der Stickstoffbedarf kann aus Ammoniumsalzen oder Aminosäuren gedeckt werden. Purpur-

Tabelle 3. *Übersicht über die Gattungen der grünen Schwefelmikroben (Chloromikroben, ,,Chlorobacteriaceae'') und der Purpurmikroben (,,Athiorhodaceae''*). (Nach van Niel in Bergeys Manual 1948.) Für die Chloromikroben hat Pringsheim (1953) eine weitere Gattung *(Microchloris)* kurz beschrieben. Die Zugehörigkeit der auf Bakterien, Flagellaten, Amöben angetroffenen Chloromikroben zu besonderen Gattungen ist unsicher.

Gattungen und Anzahl der Arten	Morphologie, Größenverhältnisse der Zellen, Beweglichkeit	Gattungen und Anzahl der Arten	Morphologie, Größenverhältnisse der Zellen, Beweglichkeit
Chlorobacteriaceae			
Chlorobium Nads. (1)	Kuglige bis ovoide Zellen (Durchmesser 0,5—1 μ) in Gallertlagern. Nicht beweglich	*Chlorobacterium* Lauterb. (1)	Stäbchenförmige Zellen (0,5 × 2—5 μ) auf Amoeben und Flagellaten
Pelodictyon Lauterb. (3)	Stäbchenförmige Zellen in netzförmigen Gallertkolonien, Durchmesser 0,5 bis 1,5 μ, Länge 1—6 μ. Nicht beweglich	*Chlorochromatium* Lauterb. (1)	Cylindrische Zellen (0,5 bis 1) × (1,5—2,5) μ, eine stäbchenförmige Bakterienzelle mit polarer Geißel umhüllend
Clathrochloris Geitler (1)	Kuglige Zellen (Durchmesser 0,5—0,7 μ) in netzförmig durchbrochenen Gallertkolonien. Nicht beweglich	*Cylindrogloea* Perfiliev (1)	Stäbchenförmige Zellen (0,5—1) × (2—4) μ, umhüllen eine kapselbildende Bakterienzelle
Athiorhodaceae			
Rhodopseudomonas (Kl. u. v. N.) emend. v. Niel (4)	Zellen meist stäbchenförmig. Durchmesser 0,5 bis 4 μ, Länge 1—15 μ. Beweglich, Geißeln + (polar)	*Rhodospirillum* (Mol.) emend. v. Niel (2)	Spirillen, Durchmesser 0,5—1,5 μ, Länge 2 bis 50 μ, beweglich, Geißeln + (polar)

mikroben sind vitaminheterotroph, Schwefelpurpur- und Chloromikroben soweit bekannt vitaminautotroph, doch wird bei ihnen im synthetischen Medium durch Zugabe von Hefeextrakt oder Pepton das Wachstum erheblich verbessert, ohne daß hierfür Spurenelemente mit Sicherheit verantwortlich gemacht werden konnten.

5. *Thiobacterium* (*Thiobacillus* Beij.).

Da es sich bei den bis jetzt bekannt gewordenen Arten durchweg um Nichtsporenbildner handelt, scheint die Gattungsbezeichnung *Thiobacterium* angemessener zu sein[1]. Die Gattung ist gekennzeichnet durch kurze, gram-negative, bewegliche oder unbewegliche Stäbchen. Der Stoffwechsel basiert auf der aeroben, autotrophen Oxydation von Schwefel und Schwefelverbindungen sowie von Ferroverbindungen, jedoch haben Baas-Becking und Wood (1955) auch anaerobes Wachstum beobachtet. Denitrifikation kommt vor, desgleichen Verwertung von Thiocyanat (Tabelle 4) sowie C-heterotrophes Wachstum (*Th. novellus*, vielleicht auch Stämme von *Th. denitrificans*). Einige Arten sind angepaßt an recht eigenartige Standorte, besonders an das Leben bei hohen H-Konzentrationen, die bei der Oxydation von Schwefel und Schwefelverbindungen zu Schwefelsäure

---

[1] Die Auffassung, es handele sich bei Bezeichnungen wie ,,*Thiobacillus*'' und ,,*Lactobacillus*'' um Eigennamen, bei denen die sonst übliche Unterscheidung sporenbildender und nichtsporenbildender Formen als ,,*Bacillus*'' und ,,*Bacterium*'' keine Anwendung finden sollte, ist nicht zwingend und letzten Endes irreführend. Auch Rippel-Baldes (1955) ist dieser Ansicht. Benennungen wie *Thiobacterium* und *Lactobacterium* verdienen den Vorzug, wenn das Auftreten oder Fehlen von Endosporen in den Gattungsdiagnosen und Gattungsnamen der Bakterien berücksichtigt werden soll.

Tabelle 4. *Übersicht über die wichtigsten Arten der Gattung Thiobacterium (Thiobacillus* Beij.).
(Nach Starkey in Bergeys Manual 1948, Parker und Prisk 1953 u. a.)

Art	Vorkommen	Stoffwechsel	Bemerkungen
Th. thiooxidans Waksm. u. Joffe	Erde, Schlamm	oxydiert Thiosulfat, S, H_2S; H$\cdot$-Toleranz: $p_H < 1$	Tetrathionat wird, wenn es als einzige Schwefelverbindung im Medium enthalten ist, nicht angegriffen obgleich es bei der Oxydation von Thiosulfat vorübergehend auftritt
Th. thioparum Beij.	Wasser, Schlamm, Erde	oxydiert Thiosulfat, S; H$\cdot$-Toleranz: p_H etwa 3—4	
Th. novellum Starkey	Erde	oxydiert langsam Thiosulfat, H$\cdot$-Toleranz: gering (p_H etwa 5)	
Th. coproliticum Lipm. u. McLees	Isoliert aus Coproliten (Trias, Arizona)	oxydiert Thiosulfat, S	
Thiobact. X	Isoliert von korrodiertem Beton	oxydiert Thiosulfat, Tetrathionat, S, H_2S; H$\cdot$-Toleranz: p_H etwa 3,0	Parker (1947)
Th. denitrificans Beij.	Schlamm, Wasser, Erde, Torf	oxydiert Thiosulfat, Tetrathionat, S; aerob oder mit Nitrat anaerob	Baalsrud (1954) NO_3 wird zu N_2 denitrifiziert
Th. thiocyanoxidans	thiocyanathaltiges Brunnenwasser, Abflüsse von Kläranlagen	oxydiert Thiocyanat, Thiosulfat, S, Sulfid	Thiocyanat dient als N-, C- und S-Quelle (Happold usw. 1954, Youatt 1954)
Th. ferrooxidans	Isoliert aus sauren Wässern in Kohlengruben	oxydiert Thiosulfat, Ferrosulfat zu Ferrisulfat; H$\cdot$-Toleranz: p_H etwa 1—2	Temple und Colmer (1951), Leathen usw. (1953)

auftreten. Abgesehen von Erde, Torf, Schlamm und Wasser findet man *Thiobacterium*-Arten z. B. in Braunkohlenflözen mit Markasit und Pyrit (FeS_2) und in der Oxydationszone sulfidischer Kupfer- und Eisenerzlagerstätten. Anreicherung und Reinkultur gelingen in flüssigen Medien, bereiten jedoch häufig Schwierigkeiten beim Übergang auf feste Substrate.

6. Beziehung von Schwefelspezialisten zu geologischen und mineralogischen Prozessen.

Seit Ehrenberg (1854) und Nadson (1903) ist man auf die Beziehungen zwischen der Lebenstätigkeit von Mikroorganismen und geologischen Prozessen sowie Vorgängen der Mineralbildung aufmerksam geworden, die schließlich zur Entwicklung eines besonderen Arbeitsgebietes, Geomikrobiologie, geführt haben (Schwartz und Müller 1953, 1956). Soweit Mikroorganismen des Schwefelkreislaufs in Frage kommen, handelt es sich um die Beteiligung an der Fällung von kohlensaurem Kalk, an der Entstehung von Schwefellagerstätten und der Fällung und Oxydation sulfidischer Erze. Ein gemeinsames Merkmal vieler geomikrobiologischer Prozesse besteht darin, daß Mikroorganismen das zu erwartende Produkt, z. B. Schwefel oder FeS_2 (als wasserhaltiges FeS gefällt), in feinster

Verteilung liefern. Erst im Lauf diagenetischer Veränderungen, bei denen die Mitwirkung von Mikroorganismen gegenüber chemischen und physikalischen Faktoren meist immer mehr zurücktritt, finden Anreicherung, Kristallisation usw. statt.

Die Frage des Vorkommens fossiler Schwefelmikroben ist mehrfach erörtert worden, besonders von SCHNEIDERHÖHN (1949) bei Betrachtungen über die Genese der Kupfererze des Mansfelder Kupferschiefers. Es hat nicht an Kritik gefehlt. Die in Dünnschliffen als vererzte Bakterien angesprochenen, zum Teil in Gruppen zusammenliegenden kokken- und stäbchenförmigen Gebilde sind umstritten; ihre Entstehung kann auch in anderer Weise erklärt werden. Gebilde, die sich zu Beggiatoen, Spirillen, Vibrionen oder anderen morphologisch besser gekennzeichneten Formen in Beziehung setzen lassen, sind bisher anscheinend nicht gefunden worden. Andererseits hat schon BEIJERINCK (1895) auf eine beginnende Vererzung und das Auftreten von Schwefeleisen-Einschlüssen in toten Zellen von *Spirillum (Desulfovibrio) desulfuricans* hingewiesen.

Daß auch in früheren geologischen Epochen Sulphureten und den Sapropelen entsprechende Ablagerungen vorhanden waren, ist ohne Zweifel. Auf faulschlammartige Ablagerungen gehen z. B. viele Ölschiefer und Diatomite, der Mansfelder Kupferschiefer und der Posidonienschiefer (Lias ε) zurück. Marine Faulschlamme sind die Muttersedimente der heutigen Erdöle gewesen. Auch bei der Entstehung von Salzlagerstätten ist es stellenweise zur H_2S-Bildung, wahrscheinlich durch Desulfurikation (MÜLLER und SCHWARTZ 1953), desgleichen zur Fällung von Schwefel und zur Entstehung bituminöser Substanzen gekommen.

Das Vorkommen von Schwefelmikroben und Desulfurizierern in tieferen Schichten, nicht nur in Verbindung mit Erdöllagerstätten (Literatur bei SCHWARTZ und MÜLLER 1948, 1953, 1955/56) ist wiederholt nachgewiesen worden, so in letzter Zeit z. B. in schwefelhaltigem Gips und Mergel sizilianischer Schwefellagerstätten (Desulfurizierer, *Thiobacterium*, SCHWARTZ[1]) und im Wasser artesischer Brunnen bei Tripolis, das aus etwa 1200 m Tiefe stammt (BUTLIN 1953) und thermophile Desulfurizierer mitführt, wie ja überhaupt die Tiefengrenze der Biosphäre in der Erdrinde erst mit dem Überschreiten des Temperaturbereichs thermophiler Mikroorganismen erreicht wird. Bemerkenswert ist das Vorkommen von Schwefelpurpur- und Purpurmikroben im Salzwasser von Erdöllagerstätten aus Tiefen bis zu etwa 2000 m (ISSATSCHENKO 1940).

Schwefelspezialisten (Desulfurizierer) sind neben zahlreichen anderen Mikroorganismen und höheren Pflanzen an der Fällung von $CaCO_3$ und der Entstehung ausgedehnter Lager von dichtem, ungeschichtetem Kalkstein beteiligt, wie sie sich in Florida und bei den Bahamabänken finden (BAVENDAMM 1932).

Noch unklar, was die wirksamen Faktoren und die biochemischen Vorgänge betrifft, sind die Zusammenhänge, die zwischen der Lebenstätigkeit desulfurizierender Bakterien und der Entstehung von Erdölkohlenwasserstoffen immer wieder vermutet worden sind (BEERSTECHER 1954).

Die Entstehung syngenetischer sedimentärer Schwefellagerstätten ist mit der Lebenstätigkeit von Mikroorganismen des Schwefelkreislaufs in Verbindung gebracht worden. Ein rezenter Vorgang dieser Art könnte sich an einigen Seen der Cyrenaica abspielen, die von BUTLIN und POSTGATE (BUTLIN 1953, BUTLIN und POSTGATE 1954) untersucht worden sind. Der Vorgang der Schwefelabscheidung, soweit er sich biochemisch abspielt, beruht dort auf einer Metabiose von Desulfizierern, die H_2S liefern, mit Schwefelpurpur- und Chloromikroben, besonders *Chromatium* und *Chlorobium*, die H_2S oxydieren und intracellular bzw. extra-

---

[1] Noch nicht veröffentlicht.

cellular Schwefel anhäufen[1]). Eine biochemische Entstehung hat Hunt (1915) für die sizilianischen Schwefel-Gips-Lagerstätten angenommen; Murzaiev (1937) vertritt die gleiche Auffassung für einige sedimentäre russische Lagerstätten. Thode und seine Mitarbeiter (Tudge und Thode 1950, Macnamara und Thode 1951, Thode usw. 1951, 1953) haben durch Untersuchung der Isotopen-Verhältnisse bei C und besonders bei S zur Stützung der biochemischen Hypothese beigetragen und für die Schwefellagerstätten am Golf von Mexiko (Texas, Louisiana) ebenfalls eine Entstehung unter Beteiligung von Mikroorganismen wahrscheinlich gemacht.

Bei der Fällung von Metallsulfiden und ihrer Anreicherung in Sedimenten, die schließlich zur Entstehung sulfidischer Erze führen kann, besteht neben anderen, nicht-biologischen Prozessen die Möglichkeit für eine Mitwirkung der Desulfurizierer.

Was zunächst die mikrobiologischen Voraussetzungen anbetrifft, so sind sie zweifellos gegeben. Die Fällung von Schwefeleisen, FeS, als Vorstufe der stabileren FeS_2-Sulfide Pyrit und Markasit und seine Anhäufung in dunkelfarbigen rezenten Sedimenten ist weitverbreitet. Gegenüber geringen Mengen von Schwermetallverbindungen im Medium, die als Sulfide gefällt werden, ist *Desulfovibrio* im allgemeinen wenig empfindlich (Miller 1950). Die direkte Desulfurikation von Schwermetallsulfaten verläuft ebenfalls mit einer verhältnismäßig hohen Toleranz[2].

Für das Vorkommen von Markasit und Pyrit in Braunkohle sind unter anderem Desulfurizierer verantwortlich gemacht worden (Thiessen 1920, Edwards und Baker 1951). Ihre Mitwirkung bei der Fällung der sulfidischen Erze des Mansfelder Kupferschiefers und ähnlicher Lagerstätten ist von Trask (1925), Bastin (1926), Schneiderhöhn (1949) und vielen anderen diskutiert worden, und das Fehlen eindeutig erkennbarer Reste fossilisierter Bakterien ist kein Gegenargument.

Spezialisten des Schwefelkreislaufs sind auch an den entgegengesetzt gerichteten Prozessen beteiligt, die mit der Oxydation und Verwitterung sulfidischer Erze verbunden sind. Standorte sind z. B. die Sickerwässer in pyrithaltigen Schiefergruben (Glockenberg bei Goslar)[3] und vor allem in der Oxydationszone von Gruben sulfidischer Erze (Rammelsberg bei Goslar, Meggen/Westfalen, Arizona/Texas)[3], ferner in kieshaltigen Braunkohlenflözen.

An den Oxydationsprozessen, die zur Entstehung eines Gemisches von Schwermetallsulfaten und freier Säure führen (Leathen usw. 1953, Temple usw. 1954) sind, soweit es sich um biochemische Vorgänge handelt, *Thiobact. ferrooxidans* und *thiooxidans* beteiligt, von denen das erste Ferrosulfat zu Ferrisulfat und das zweite den bei der Oxydation durch Luftsauerstoff entstandenen Schwefel zu Schwefelsäure oxydiert (Temple und Delchamps 1953)[4].

Literatur.

Baalsrud, K. u. K. S.: Studies on *Thiobacillus denitrificans*. Arch. Mikrobiol. 20, 34—62 (1954). — Baars, J. K.: Over sulphaatreductie door bacterien. Diss. Delft 1930. — Baas-Becking, L. G. M.: Studies on sulphur bacteria. Ann. of Bot. 39, 613—650 (1925). — On organisms living in concentrated brine. Tijdschr. nederl. dierkd. Verigg, 3. Ser. Afl.1,

---

[1] Das Vorkommen einer Desulfurikation, die nur bis zum Schwefel führt, ist verschiedentlich erwähnt worden (vgl. z. B. Datta 1949); es bleibt jedoch abzuwarten, ob eine Bestätigung erfolgt.

[2] Noch nicht veröffentlicht (A. Müller).

[3] Noch nicht veröffentlicht (W. Schwartz).

[4] Einen ähnlichen Vorgang hat Ohle (1936) bei einem Tongrubenteich mit stark saurem Wasser über tertiärem Ton mit Schwefelkies-Einschlüssen beobachtet.

1928. — BAAS-BECKING, L. G. M., and E. J. F. WOOD: Biological processes in the estuarine environment. I., II. Ecology of the sulphur cycle. Proc. Kon. Ned. Akad. v. Wetensch., Ser. B 58, Nr 3, 160—181 (1955). — BAHR, H., u. W. SCHWARTZ: Untersuchungen zur Ökologie farbloser fädiger Schwefelmikroben. Biol. Zbl. **75**, 451—464 (1956). — Vergleichende cyctologische Untersuchungen an farblosen fädigen Schwefelmikroben und an hormogonalen Cyanophyceen. Biol. Zbl. (im Druck) **1957.** — BASTIN, E. S.: A hypothesis of bacterial influence in the genesis of certain sulfide ores. J. Geol. **34**, 773—792 (1926). — BAUMANN,A., u. V. DENK (CZURDA): Zur Physiologie der Sulfatreduktion. Arch. Mikrobiol. **15**, 283—307 (1951). — BAVENDAMM, W.: Die farblosen und roten Schwefelbakterien des Süß- und Salzwassers. Pflanzenforschungen, H. 2. Jena 1924. — Die mikrobiologische Kalkfällung in der tropischen See. Arch. Mikrobiol. **3**, 205—276 (1932). — Kultur der am Kreislauf des Schwefels beteiligten Bakterien. In ABDERHALDENS Handbuch der biologischen Arbeitsmethoden, Abt. 12/2, S. 483—546. 1934. — BEERENS, H., et N. HVID-HANSEN: Présence dans les eaux sulfureuses francaises de bactéries réductrices de sulfate et d'une variété protéolytique *d'Actinobacterium israeli: A. israeli* var. *liquefaciens* (n. sp.). C. r. Acad. Sci. Paris **234**, 480—482 (1952). — BEERSTECHER jr., E.: Petroleum Microbiology. Houston and New York: Elsevier Press Inc. 1954. — BEIJERINCK, M. W.: Über *Sp. desulfuricans* als Ursache der Sulfatreduktion. Zbl. Bakter. II 1, 1—9, 49—59, 104—114 (1895). — BREED, R. S., E. G. D. MURRAY and A. P. HITCHENS: BERGEY's Manual of Determinative bacteriology. Baltimore 1948. — BÖCHER, T. W.: Studies on the sapropelic flora of the lake Flynders with special reference to the *Oscillatoriaceae.* Kgl. danske Vidensk. Selsk., biol. Medd. **21**, Nr 1, 1—45 (1949). — BUTLIN, K. R.: Some malodorous activities of sulphate-reducing bacteria. Proc. Soc. Appl. Bacter. **1949**, 39—42. — The bacterial sulphur cycle. Research (Lond.) **6**, 184—191 (1953). — BUTLIN, K. R., and J. R. POSTGATE: Microbiological formation of sulphide and sulphur. Symposium microbial metabolism. Riass. IV. Congr. Internat. Microbiol. Rom 1953. — The microbiological formation of sulphur in Cyrenaican lakes. In J. L. CLOUDSLEY-THOMPSON, Biology of Deserts. London: Inst. of Biology 1954.

CATALDI, MARIA, S.: Aislamiento di *Beggiatoa alba* en cultivo puro. Rev. Inst. Bacter. (Buenos Aires) **9**, 393—423 (1940). — CZURDA, V.: Zur Kenntnis der bakteriellen Sulfatreduktion. I. Arch. Mikrobiol. **11**, 187—204 (1940). — Schwefelwasserstoff als ökologischer Faktor der Algen. Zbl. Bakter. II **103**, 285—311 (1941).

DATTA, S. C.: Sulphate reduction and production of elemental sulphur by bacteria. J. Sci. a. Industr. Res. (India) **5**, 28—30 (1946). — DENK (CZURDA), V.: Zur Frage der Ammonentstehung im Stoffkreislauf der Natur. Arch. Mikrobiol. **15**, 308—314 (1950).

EDWARDS, A. B., and G. BAKER: Some influence of supergene iron sulphides in relation of their environments of deposition. J. Sedim. Petrology **21**, 34—46 (1951). — EHRENBERG, C. G.: Mikrogeologie. Text und Atlas. Leipzig: L. Voss 1854.

GALE, H. S.: Salines in the Owens, Searles and Panamint basins, Southeastern California, U. S. Geol. Surv. Bull. **580**, 251—323 (1913). — GALLIHER, E. W.: The sulfur cycle in sediments. J. Sedim. Petrology **3**, 51—63 (1933). — GEST, H.: Metabolic patterns in photosynthetic bacteria. Bacter. Rev. **15**, 183—210 (1951). — GHOSE, T. K., and T. WIKÉN: Inhibition of bacterial sulphate-reduction in presence of short chain fatty acids. Physiol. Plantarum (Copenh.) 8, 116—135 (1955). — GROSSMANN, J. P., and J. R. POSTGATE: Cultivation of sulphate-reducing bacteria. Nature (Lond.) **171**, 600—602 (1953).

HAPPOLD, F. C., K. J. JOHNSTONE, H. C. ROGERS and J. B. YOUATT: The isolation and characteristics of an organism oxidizing thiocyanate. J. Gen. Microbiol. **10**, 261—266 (1954).— HAROLD, RUTH, and R. Y. STANIER: The genera *Leucothrix* and *Thiothrix.* Bacter. Rev. **19**, 49—64 (1955). — HUNT, W. F.: The origin of the sulphur deposits of Sicily. Econom. Geol. **10**, 543—579 (1915). — HVID-HANSEN, N.: Sulfate-reducing and hydrocarbon-producing bacteria in ground-water. Acta path. scand. (Kobenh.) **29**, 314—334 (1951).

ISSATSCHENKO, V.: The microorganisms of the lower limits of the biosphere. J. Bacter. **40**, 379—381 (1940).

LEATHEN, W. W., S. A. BRALEY sr. and L. D. McINTYRE: The role of bacteria in the formation of acid from certain sulfuritic constituents associated with bituminous coal. I. *Thiobacillus thiooxidans.* II. Ferrous iron oxidizing bacteria. Appl. Microbiol. **1**, 61—64, 65—68 (1953).

MACNAMARA, J., and H. G. THODE: The distribution of S^{34} in nature and the origin of native sulphur deposits. Research (Lond.) **4**, 582—583 (1951). — METZNER, P.: Zur Kenntnis der Morphologie und Bewegung von *Thiovulum Mülleri* (WARMING) LAUTERBORN. Biol. Zbl. **68**, 49—58 (1949). — MILLER, L. P.: Rapid formation of high concentrations of hydrogen sulfide by sulfate-reducing bacteria. Contrib. Boyce Thompson Inst. **15**, 437—465 (1949). — Tolerance of sulfate-reducing bacteria to hydrogen sulfide. Contrib. Boyce Thompson Inst. **16**, 78—83 (1950). — Formation of metal sulfides through the activities of sulfate-reducing bacteria. Contrib. Boyce Thompson Inst. **16**, 85—89 (1950). — MÜLLER, A., u. W. SCHWARTZ: Geomikrobiologische Untersuchungen. III. Über das Vorkommen von Mikroorganismen in

Salzlagerstätten. Z. dtsch. geol. Ges. **105**, 789—802 (1955). — Murzaiev, P. M.: Genesis of some sulphur deposits of the USSR. Econom. Geol. **32**, 69—103 (1937).

Nadson, S.: Les microorganismes comme facteurs géologiques. Petersburg: 1903. [Russisch.] — Niel, C. B. van: On the morphology and physiology of the purple and green sulphur bacteria. Arch. Mikrobiol. **3**, 1—112 (1932). — Nikitinsky, J.: Die anaerobe Bindung des Wasserstoffs durch Mikroorganismen. Zbl. Bakter. II **19**, 495—499 (1907).

Ohle, W.: Der schwefelsaure Tonteich bei Reinbeck. Arch. f. Hydrobiol. **30**, 604 (1936).

Parker, C. D.: Species of sulfur bacteria associated with the corrosion of concrete. Nature (Lond.) **159**, 439 (1947). — Parker, C. D., and J. Prisk: The oxidation of inorganic compounds of sulphur by various sulphur bacteria. J. Gen. Microbiol. 8, 344—364 (1953). — Pop, L. J. J.: The influence of hydrogen sulphide on growth and metabolism of green algae. Thesis Univ. Leiden 1936. Delft 1936. — Postgate, J. R.: Competitive inhibition of sulphate reduction by selenate. Nature (Lond.) **164**, 67—71 (1949). — Growth of sulphate-reducing bacteria in sulphate-free media. Research (Lond.) **5**, 189—190 (1952). — On the nutrition of *Desulphovibrio desulphuricans*: A correction. J. Gen. Microbiol. **9**, 440—444 (1953). — Presence of cyctochrome in an obligate anaerobe. Biochemic. J. **56**, XI—XII (1954). — Prévot, A. R.: Recherches sur la reduction des sulfates et des sulfites mineraux par les bactéries anaérobies. Ann. Inst. Pasteur **75**, 571—574 (1948). — Anaérobies réducteurs des sulfates et formation des pétroles. Ann. Inst. Pasteur **77**, 400—418 (1949). — Pringsheim, E. G.: Die Stellung der grünen Bakterien im System der Organismen. Arch. Mikrobiol. **19**, 353—364 (1953).

Rippel-Baldes, A.: Die Energieverhältnisse bei einigen Vorgängen anaerober Atmung. Biol. Zbl. **67**, 60—64 (1948). — Grundriß der Mikrobiologie. Berlin-Göttingen-Heidelberg: Springer 1955.

Schneiderhöhn, H.: Erzlagerstätten. Stuttgart: Piscator-Verlag 1949. — Schwartz, W., u. A. Müller: Erdölbakteriologie. Erdöl u. Kohle 1, 232—240 (1948). — Geomikrobiologie, Entwicklung und Stand eines neuen Forschungsgebietes. Erdöl u. Kohle **6**, 523—527 (1953). — Mikrobiologie des Erdöls. Wiss. Z. der Ernst-Moritz-Arndt-Univ. Greifswald Festjahrg. zur 500-Jahrfeier, Math.-Naturwiss. Reihe **5**, 281—288 (1955/56). — Senez, J. C.: Étude comparative de la croissance de *Sporovibrio desulphuricans* sur pyruvate et sur lactate de soude. Ann. Inst. Pasteur **80**, 395—408 (1951). — Concurrence of autotrophic and heterotrophic metabolism in growing and in resting cells of sulphate-reducing bacteria. J. Gen. Microbiol. **11**, VI—VII (1954). — Shturm: L. D.: Zum Studium der Mikroflora schwefelhaltiger Ablagerungen. Mikrobiologija USSR. **6**, 481—487 (1937). — Sisler, F. D., and Cl. E. ZoBell: Hydrogen utilizing sulfate-reducing bacteria in marine sediments. J. Bacter. **60**, 747—756 (1950); **62**, 117—127 (1951). — Nitrogen-fixation by sulfate-reducing bacteria indicated by Nitrogen/Argon ratios. Science (Lancaster, Pa.) **113**, 511—512 (1951). — Starkey, R.: A study of spore formation and other morphological characteristics of *Vibrio desulfuricans*. Arch. Mikrobiol. **9**, 268—304 (1938). — Starkey, R. L., W. Segal and R. A. Manaker: Sulfur products of the decomposition of methionine and cystine by microorganisms. Riass. IV. Congr. Internat. Microbiol. Rom **1**, 167—168 (1953). — Stephenson, Marjory, and L. H. Stickland: Hydrogenase II. The reduction of sulphate to sulphide by molecular hydrogen. Biochemic. J. **25**, 215—220 (1931).

Temple, K. L., and A. R. Colmer: The autotrophic oxidation of iron by a new bacterium: Thiobacillus ferrooxidans. J. Bacter. **62**, 605—611 (1951). — Temple, K. L., and E. W. Delchamps: Autotrophic bacteria and the formation of acid in bituminous coal mines. Appl. Microbiol. 1, 255—258 (1953). — Temple, K. L., and W. A. Koehler: Drainage from bituminous coal mines. West Virginia Bull. 1954. Eng. Exp. Sta. Res. Bull. 25. — Thiessen, R.: Occurence and origin of finely disseminated sulfur compounds in coal. Trans. Amer. Inst. Mining a. Metallurg. Engr. **63**, 913—931 (1920). — Thode, H. G., H. Kleerekoper and D. McElcheran: Isotope fractionation in the bacterial reduction of sulphate. Research (Lond.) **4**, 581—582 (1951). — Thode, H. G., R. K. Wanless and R. Wallouch: The origin of native sulphur deposits from isotope fractionation studies. Ann. Meeting Geol. Soc. of Amer. 1953. — Trask, P. D.: The origin of the ore of the Mansfeld Kupferschiefer, Germany. A Review of the current literature. Econom. Geol. **20**, 746—761 (1925). — Tudge, A. P., and H. G. Thode: Thermodynamic properties of isotop compounds of sulphur. Canad. J. of Res., Sect. B **28**, 567—578 (1950).

Wight, K. M., and R. L. Starkey: Utilization of hydrogen by sulfate-reducing bacteria and its significance in anaerobic corrosion. J. Bacter. **50**, 238 (1945).

Youatt, J. B.: Studies on the metabolism of *Thiobacillus thiocyanoxidans*. J. Gen. Microbiol. **11**, 139—149 (1954).

Le cycle du soufre dans la nature.

Par

J. M. Wiame.

Avec 4 figures.

I. Introduction.

La circulation du soufre dans la nature, et le caractère cyclique de celle-ci résulte pour une large part de l'activité des organismes vivants. Elle se manifeste par des transformations qui mettent en jeu des oxydations et des réductions des atomes du soufre contenus dans les différents composés soufrés. Le soufre est à son état d'oxydation maximum dans les sulfates et à son état minimum dans les sulfures. On exprime parfois ces transformations par les changements de valence de cet élément; ceci n'exprime toutefois que partiellement l'état de l'atome de soufre et une expression plus générale consiste à attribuer à cet atome ce que l'on appelle un *nombre d'oxydation* (LATIMER 1952, PAULING 1947). Les nombres d'oxydation des dérivés inorganiques du soufre sont les suivants: sulfate, $+6$ (S^{VI}); sulfite, $+4$ (S^{IV}); thiosulfate, $+2$ (S^{+II}); soufre élément, 0 (S^O); disulfure, -1 (S^{-I}); sulfure, -2 (S^{-II}). Il faut remarquer que dans le thiosulfate, les deux atomes de soufre n'ont pas le même nombre d'oxydation, $+2$ représente une moyenne; il en est de même des tétrathionates ($Na_2S_4O_6$) dont le nombre d'oxydation moyen est $+2.5$.

Le passage d'un état d'oxydation à un autre s'exprime par la différence algébrique des nombres d'oxydation des deux états, et elle correspond au nombre d'électrons mis en jeu lors de cette transformation. Par exemple, le passage de sulfure (S^{-II}) à sulfate (S^{+VI}) met en jeu 8 unités de nombre d'oxydation, équivalent aux nombres d'électrons (e^-) mis en jeu dans l'équation (I):

$$H_2S + 4\,H_2O \rightarrow H_2SO_4 + 8\,H^+ + 8\,e^-. \tag{I}$$

Cette équation exprime une transformation importante du cycle du soufre.

L'étude du cycle du soufre met en jeu des réactions très nombreuses qui ont été étudiées chez un grand nombre d'organismes. L'exposé qui suit restera dans un cadre général, s'attachant plus à une vue d'ensemble qu'à l'étude minutieuse des organismes ou du mécanisme intime (enzymatique p. ex.) des réactions. Fréquemment, des équations seront écrites, qui représentent un bilan de plusieurs réactions. Des études approfondies sont présentées dans les différents chapitres de ce traité auxquels nous renvoyons le lecteur.

Tous les organismes vivants mettent en jeu des composés soufrés mais ceci à des titres divers, et on peut distinguer ceux où le soufre intervient (ou peut intervenir) en quantité considérable, contrairement à ceux où il n'intervient massiquement qu'en minime quantité. Le terme de comparaison qui peut être pris est p. ex. le poids des organismes (ou le carbone mis en jeu dans leur croissance); dans les premiers, le nombre d'atomes de soufre mis en jeu est au moins de l'ordre de grandeur de ceux du carbone alors qu'il est beaucoup plus faible dans les autres. Au premier groupe appartiennent les microorganismes dits «du soufre», le second groupe comprend les autres microorganismes, les végétaux et les animaux.

Il est normal que ce soit le premier groupe qui ait été étudiée tout d'abord, et que ce soit à la suite de ces études que la notion du cycle du soufre soit apparue le plus clairement. C'est ce cycle restreint du soufre que nous présenterons dans la section II en y introduisant les travaux essentiels des deux pionniers de la microbiologie des sols et des eaux, WINOGRADSKY et BEIJERINCK.

L'activité des autres bactéries du soufre sera complétée dans la section III; elle illustrera les différentes modalités des transformations présentées dans le «Cycle restreint», elles résultent de nombreux travaux, tout particulièrement ceux de l'école de WAKSMAN. Nous placerons dans le même chapitre les transformations très intéressantes de la photosynthèse bactérienne liées aux composés soufrés dont l'explication a été donnée par VAN NIEL, et qui a ouvert la voie à une expression généralisée de la photosynthèse qui constitue à présent une des plus belles études de biochimie comparée.

En quittant les organismes spécialisés du soufre, nous généraliserons le cycle de cet élément (section IV) en y étudiant la participation des animaux, des végétaux et des microorganismes habituels.

Les composés soufrés constituent dans certains habitats l'élément principal qui règle leur population; ceci constitue une application à l'écologie des organismes (section V).

II. Le cycle du soufre simplifié: Beggiatoa et Desulfovibrio.

La transformation des sulfates (S^{VI}) en sulfures (S^{-II}) et le retour de ceux-ci en sulfates furent réalisées expérimentalement et comprises pour la première fois par WINOGRADSKY (1877) lors de ses remarquables études sur les sulfuraires du genre *Beggiatoa*. Ces organismes furent décrits en 1865 par COHN qui les observa sur le fond d'un aquarium d'eau de mer. Quoique les *Beggiatoa* soient classés parmi les bactéries, ils ont une taille qui les rapproche des cyanophycées; ce sont des filaments d'un diamètre variant avec les espèces de 1 à 50 μ. C'est grâce à cette particularité que WINOGRADSKY pût comprendre leur physiologie, sans toutefois se soumettre à la méthode des cultures pures. Les observations de WINOGRADSKY étaient faites en microcultures de petites touffes de ces organismes, examinées sous le microscope et irriguées par des milieux nutritifs bien définis. En irriguant en aérobiose la microculutre au moyen d'eau chargée d'H_2S, il s'accumule dans les filaments du soufre élémentaire, que CRAMER avait déjà identifié dans ces organismes en 1870. Dès que l'H_2S est absent, le soufre intracellulaire est oxydé et excrété sous forme de sulfate. Les *Beggiatoa* placés en anaérobiose n'accumulent pas de soufre et ne peuvent pas oxyder le soufre intracellulaire. WINOGRADSKY démontra ainsi que ces organismes que l'on rencontre dans les eaux sulfureuses, ne sont pas la cause de la production de ce gaz, mais sont la conséquence de la présence de celui-ci. La formation d'H_2S était indépendante des *Beggiatoa* et résultait de l'activité d'un autre organisme qui, lui, était capable de réduire les sulfates en sulfures. Ce second point, et la réalisation de l'ensemble du cycle, sont le mieux illustrés par les observations que WINOGRADSKY fit lors de la culture des *Beggiatoa* en aquarium. Etant donné l'importance qu'eurent les travaux de WINOGRADSKY en ce qui concerne la compréhension du cycle du soufre, le texte original de cette expérience sera repris (WINOGRADSKY 1887—1945):

. . .«*Beggiatoa* est répandu dans les marais, les étangs à fond limoneux, généralement dans les eaux contenant de la matière végétale en voie de décomposition.

Mais ces végétations n'attirent l'attention que dans les eaux chargées d'hydrogène sulfuré. Aussi, nos premières tentatives de produire des cultures brutes de *Beggiatoa* dans de l'eau d'étang, ne donnaient longtemps aucun résultat, avant que je n'aie l'idée d'y ajouter un peu de gypse, cette eau étant trop pauvre en sulfate. On procédait comme suit: on jetait des morceaux d'un rhizome de *Butomus* dans un vase profond de 50 centimètres, que l'on remplissait avec de l'eau additionnée de gypse. Une odeur d'H_2S était déjà perceptible au bout de 2 jours, elle s'intensifiait de plus en plus, mais ce n'est qu'au bout de quelques semaines que l'on y découvrait les filaments recherchés. Il fallait généralement deux mois pour obtenir des végétations notables sous forme de délicats filets blancs et de touffes suspendues sous la surface de l'eau ou tapissant les parois du vase. Celui-ci était tenu à l'obscurité pour prévenir le développement des algues vertes. Ces cultures brutes permettent déjà de s'assurer que les sulfuraires filamenteux ne sont pour rien dans la production d'H_2S, car ils n'apparaissent que tardivement quand le dégagement est en train ...; à mesure que l'eau s'appauvrit en H_2S; les filaments se retirent dans les couches plus profondes.» ...

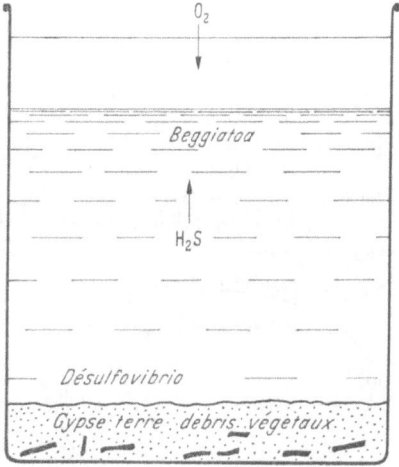

Fig. 1. Expérience de WINOGRADSKY.

WINOGRADSKY, par ses observations en cultures brutes ou en microcultures irriguées, observait toujours le mouvement des filaments, tel que ceux-ci soient dans une zône où à la fois se présentait simultanément une richesse adéquate en oxygène venant de l'air et en H_2S provenant de la zône anaérobique ou se réalise la réduction des sulfates par l'apport d'hydrogène des matières organiques et provoquée par des organismes différents des *Beggiatoa* (Fig. 1).

WINOGRADSKY concluait: «*L'énergétique des sulfuraires est basée sur l'oxydation du soufre qui est l'équivalent énergétique de l'acte respiratoire tel qu'on le connait chez les organismes qui vivent aux dépens de l'énergie accumulée dans la molécule organique.*»

Il restait toutefois un point capital à éclaircir. Lors d'expériences en microculture irriguée, WINOGRADSKY avait observé que les filaments de *Beggiatoa* croissaient en milieu strictement minéral (eau courante, H_2S et O_2); la matière organique devait donc dériver de carbone non organique. C'est par la découverte du chemoautotrophisme des *Nitrobacteriaceae* (1890) que WINOGRADSKY trouva simultanément une explication au mode de vie des sulfuraires; *ils assimilaient le CO_2 grâce à l'oxydation de l'H_2S ou du soufre.*

WINOGRADSKY avait été amené à supposer l'existence d'organismes réduisant les sulfates en H_2S. Il revint à BEIJERINCK de décrire et d'étudier l'organisme responsable de cette transformation, *Spirillum desulfuricans* (actuellement désigné sous le nom de *Desulfovibrio desulfuricans* par KLUYVER et VAN NIEL 1936). BEIJERINCK (1895) réalisa des cultures enrichies en ensemençant de terres ou de boues d'origines très diverses (marines ou d'eau douce) un milieu composé de 0.2% $FeSO_4$ $(NH_4)_2SO_4$ 6 H_2O, 0.2% de malate de potassium et 0.2% de peptone. La formation de sulfure de fer indique la production d'H_2S. Le *Desulfovibrio* ne fut isolé en culture pure qu'en 1903 par VAN DELDEN.

La signification biologique de la réduction des sulfates correspond à la possibilité pour *Desulfovibrio desulfuricans* de réaliser des oxydations anaérobiques de matières organiques à partir de la réserve d'oxygène des sulfates. Ce mécanisme d'oxydation est identique à celui de la dénitrification très répandue chez les bactéries, et de la transformation de l'acide carbonique en méthane élucidée par Barker (1936) et Barker et al. (1940). L'oxydation réalisée par les sulfates peut se porter sur l'hydrogène moléculaire et, comme l'ont montré Starkey et Wight (1945), ainsi que Butlin et Adams (1947), dans ces conditions, on peut démontrer le caractère chemosynthétique au moins de certaines souches de *Desulfovibrio desulfuricans*.

Nous pouvons résumer les transformations de composés sulfurés que nous avons rencontrés jusqu'à présent par le schéma qui représente le *cycle restreint du soufre* (Fig. 2).

Phase I   a) $SO_4^= + 4\,H_2 \rightarrow HS^- + 3\,H_2O + OH^-$ $A = +\ \ 27.000$ calgr.

b) $CO_2 + 2\,H_2 \rightarrow (HCOH)$ ₍bactéries₎ $+ H_2O$ $A = -\ 210.000$ calgr.

Phase II a) $2\,H_2S + O_2 \rightarrow 2\,S + 2\,H_2O$ $A = +\ 100.000$ calgr.

b) $CO_2 + 2\,H_2S \rightarrow (HCOH) + H_2O + 2\,S$ $A = -\ \ 12.700$ calgr.

Phase III a) $S + 3/2\,O_2 + H_2O \rightarrow SO_4^= + 2\,H^+$ $A = +\ 120.650$ calgr.

b) $3\,CO_2 + 2\,S + 5\,H_2O \rightarrow 3\,(HCOH) + 2\,SO_4^= + 4\,H^+$ $A = -\ \ 98.100$ calgr.

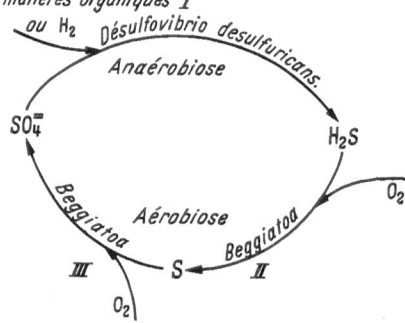

Fig. 2. Cycle restreint du soufre. La réduction (phase I) est conditionnée par la présence de sulfate, d'un donneur d'hydrogène et l'anaérobiose; *Desulfovibrio desulfuricans* est une bactérie anaérobique stricte. Les transforma tions II et III sont aérobiques. Ces transformations peuvent être exprimées par les trois couples d'équations ci dessus[1].

Dans tous les cas, les besoins énergétiques et matériels d'une vie autotrophique sont réalisés. Les réactions b correspondent à l'assimilation autotrophique du CO_2, l'affinité (A) de ces réactions est négative ou proche de zéro, chacune des réactions est couplée à une réaction à affinité positive, dont le rôle est énergétique. Ces réactions ne sont que des bilans de transformation et ne rendent pas compte du mécanisme de couplage. Au moins dans un cas, ce mécanisme a été éclairci (Vogler 1942, Vogler et Umbreit 1942, le Page et Umbreit 1943, le Page 1942). D'après les valeurs des affinités, il semble possible que la réalisation de l'assimilation de 1 mole de CO_2 n'implique pas plus que la participation d'une mole de composé soufré. Les expériences réalisées sur les organismes dont on a traité et sur ceux qui seront décrits dans la section suivante ont montré au contraire que les réactions du type a devaient se réaliser de nombreuses fois pour permettre la réalisation d'une réaction de type b. Ceci rend compte de la masse importante des composés soufrés qui est transformée par rapport à la masse cellulaire formée. Chez *Thiobacillus thiooxidans* réalisant la phase III, Waksman et Starkey (1922) ont montré que le rapport S oxydé/C assimilé était de 32, ce qui réduit le rendement énergétique à environ 6%.

---

[1] Les différentes équations sont accompagnées de la valeur de leur *affinité standard* (Prigogine et Defay 1944). L'affinité standard a une valeur égale à celle de $-\varDelta F$ ($F=$ énergie libre standard). Les valeurs de A ont été calculées à 298 K⁰, d'après les données des énergies libres de formation reprises de Latimer (1952) pour les composés inorganiques et de Parks et Huffman (1932) pour (HCOH) que l'on a pris $= 1/6$ de l'affinité de formation du glucose, soit $216.000/6 = 36.000$ cal. Les valeurs des affinités données ci-dessus ne sont pas exactement valables dans les conditions biologiques, étant donné qu'elles ne sont pas corrigées de l'*affinité de mélange* tenant compte de la concentration des substances.

III. Les microorganismes du soufre en general.

A. Le soufre dans la chemosynthèse.

Les *Beggiatoa* et *Desulfovibrio* qui furent les premiers microbes du soufre dont la physiologie fut élucidée, permettent d'illustrer le principe du cycle du soufre dans la nature. D'autres mircoorganismes existent qui se rattachent plus ou moins à la physiologie de ces deux premiers; leurs particularités peuvent toutefois être intéressantes dans la discussion de leur activité dans les divers habitats.

a) La réduction des sulfates est réalisée spécifiquement par *Desulfovibrio*, quoique le phénomène de réduction apparaisse très répandu. Différentes espèces ont été décrites, parfois caractéristiques de leur habitat. Le *D. desulfuricans*, l'espèce étudiée par BEIJERINCK et isolée par VAN DELDEN, est très répandue dans les terres, égoûts et des eaux douces, alors que *D. aesturii* (VAN DELDEN 1904) est une espèce marine. Alors que BAARS (1930) signale que cette espèce peut être acclimatée aux eaux douces, RITTENBERG (1941) considère l'halophilie comme un caractère stable. Certaines souches sont thermophiles (ELION 1924), ne croissant pas à moins de 30 à 40⁰ C avec un optimum de 55⁰ C. STARKEY (1938) a décrit un organisme qui peut croître à 65⁰ C produisant des spores. Cet organisme, qui pourrait également être obtenu par acclimatation aux hautes températures est pléomorphique et sa forme sporulée est plus proche du genre *Bacillus* que *Desulfovibrio*, et dans ce cas, les bactéries sont fréquemment gram positives et en forme de bâtonnets. Des variétés (ou espèces) ont été distinguées sur la base des donneurs d'hydrogène. Des *Desulfovibrio* sont capables également de réduire (BAARS 1930, BUTLIN, ADAMS, THOMAS 1949) les sulfites, les thiosulfates et moins activement le soufre élémentaire. La possibilité de croissance autotrophe n'est probablement pas commune à toutes les souches de *Desulfovibrio*, mais BUTLIN et al. ont montré que les souches pouvant employer l'hydrogène comme agent réducteur, pouvaient utiliser l'hydrogène résultant de l'immersion de fer métallique dans le milieu, ce qui constitue une donnée importante qui explique le rôle de ces bactéries dans la corrosion anaérobique des récipients ferreux, en particulier des tuyauteries. Nous reprendrons dans la section V le rôle des bactéries réduisant les sulfates dans d'autres habitats.

b) L'oxydation: l'oxydation $S^{-II} \rightarrow S^O \rightarrow S^{VI}$ qui se produit chez les *Beggiatoa* se retrouve avec diverses modalités chez d'autres microorganismes. Ceux-ci diffèrent considérablement entre eux; il existe tout d'abord des organismes de grande taille et formant des filaments multicellulaires qui se présentent sous forme de masses blanches, volumineuses (baregines, glairines). Ce sont avec les *Beggiatoa*, les *Thiotrix*, *Thiospirillopsis* et *Thioploca* formant la famille des *Beggiatoaceae*. Un autre groupe, formant la famille des *Achromatiaceae* (MASSART 1902) comprend des genres qui n'ont pas pu être cultivés en cultures pures; ils se retrouvent en général dans les eaux sulfureuses et contiennent des globules de soufre. Ces organismes généralement globuleux peuvent atteindre une taille allant de 5 à 100 microns.

Il existe d'autre part des organismes de la taille du micron, le genre *Thiobacillus*, placé dans la famille des *Nitrobacteriaceae* et qui sont des bactéries dans le sens habituel *(Eubactériales)*.

Les différents genres de *Beggiatoaceae* se distinguent par leur taille ou leur possibilité de se fixer sur les surfaces *(Thiotrix)*. Il existe des espèces marines (les *Beggiatoa* de grande taille), des espèces croissant dans les eaux douces ou marines, comme *Beggiatoa alba* qui est l'organisme le plus répandu et qui a été utilisé par WINOGRADSKY. Dans l'ensemble, ces organismes sont ubiquistes et une espèce telle que *Thiospirillopsis floridana* qui fut découverte par UPHOF (1927) et qui paraissait localisée aux sources sulfureuses de Wekiwa Springs et Palm Springs en Floride, a été retrouvée dans un aquarium d'eau de mer à Pacific Grove-Californie (BERGEY's Manual).

Les *Thiobacillus* sont tous spécialisés dans l'oxydation des composés soufrés et, d'une façon générale, il a été démontré que ces organismes étaient des autotrophes, stricts ou facultatifs. Dans ce groupe, l'organisme le plus remarquable est le *Thiobacillus thiooxidans* découvert par WAKSMAN et JOFFE (1922). STARKEY (1925, 1935) en a complété l'étude physiologique. Les propriétés surprenantes de cet organisme apparaissent dans le procédé d'enrichissement qui permet de l'obtenir. Il suffit d'introduire un peu de terre ou de boue dans un liquide contenant les quelques sels minéraux habituels des milieux microbiologiques et de la fleur de soufre, avec un accès abondant d'air, pour qu'il se développe dans le milieu une très forte acidité sulfurique après des temps variables (p. ex. 2 à 4 semaines). L'acide sulfurique peut ainsi atteindre une concentration 1.5 molaire. Cet organisme est le plus résistant à l'acidité qui soit connu; il est encore viable à 0.5 M H_2SO_4, son p_H optimum est entre p_H 3 et 4. La nature de cette résistance ne semble pas encore avoir été établie. La possibilité d'oxyder le soufre élémentaire, insoluble dans le milieu, est également très inattendue; il existe toutefois d'autres *Thiobacillus* capables d'utiliser le soufre et les animaux peuvent l'oxyder, mais le *Th. thiooxydans* est particulièrement actif, le soufre élémentaire constitue son aliment énergétique de choix. UMBREIT, VOGEL et VOGLER (1942) ont donné des arguments qui permettent de penser que le soufre élémentaire pénétrerait dans les cellules par l'intermédiaire de gouttelettes de lipides non saturés, dans lesquelles le soufre est soluble. Cette opinion n'est pas partagée par KNAYSI (1943).

Le *Th. thiooxidans* peut également utiliser les sulfures et les thiosulfates comme source de soufre réduit. Les propriétés de cette bactérie impliquent des habitats spécialisés (section V) quoiqu'on la retrouve dans la plupart des sols ordinaires, comme le prouve l'aisance avec laquelle on l'obtient par enrichissement. Le *Th. thioparus*, contrairement à *Th. thiooxidans*, n'utilise que lentement le soufre élémentaire. Il oxyde particulièrement bien l'hydrogène sulfuré, comme les *Beggiatoa*, mais le soufre n'est pas emmagasiné dans la cellule, il est excrété dans le milieu extérieur. Cette bactérie pourrait être à l'origine de dépôts de soufre sédimentaire rencontré dans certains terrains de Sicile (voir section V); comme *Beggiatoa*, il dépendrait de la réduction anaérobique de sulfate par des *Desulfovibrio*. *Th. thioparus* a été isolé d'eau de mer par NATHANSOHN (1902) et retrouvé par BEIJERINCK (1904) et STARKEY (1935); il est très répandu. Les thiosulfates sont également utilisés comme source de soufre réduit; dans ce cas, l'équation génératrice d'énergie est:

$$5\ Na_2S_2O_3 + H_2O + 4\ O_2 = 5\ Na_2SO_4 + H_2SO_4 + 4\ S \quad A = +\ 500.700\ calgr.$$

Thiobacillus novellus isolé par STARKEY (1935) oxyde particulièrement bien le thiosulfate, son p_H optimum est de 8 à 9, la croissance est arrêtée à p_H 5. Cet organisme est un autotrophe facultatif. Les réactions productrices d'énergie et d'assimilation sont:

$$S_2O_3^= + H_2O + 2\ O_2 = 2\ SO_4^= + 2\ H^+ \qquad A = +\ 174.000\ calgr.$$
$$S_2O_3^= + 3\ H_2O + 2\ CO_2 = 2\ (HCOH) + 2\ SO_4^= \quad A = -\ 56.000\ calgr.$$

Tous les organismes précédents qui dérivent leurs besoins énergétiques de l'oxydation des composés soufrés sont aérobies stricts. BEIJERINCK (1904) a cependant pu montrer qu'une oxydation bactérienne anaérobique du soufre pouvait se présenter dans la nature. C'est ce qui se passe chez *Thiobacillus denitrificans* où les nitrates sont l'agent oxydant. On a donc le couple de réactions :

$$6\ KNO_3 + 5\ S + 2\ H_2O = 3\ K_2SO_4 + 2\ H_2SO_4 + 3\ N_2 \quad A = +\ 612.000\ cal.$$

$$3\ CO_2 + 2\ S + 5\ H_2O = 3\ (HCOH) + 2\ SO_4^= + 4\ H^+ \quad A = -\ 98.100\ cal.$$

Cette bactérie, comme l'a montré LIESKE (1912) peut également utiliser le thiosulfate. Dans ce cas, la fixation de 1 g. de carbone nécessite l'oxydation de 100 g. de thiosulfate. Ceci permet d'évaluer à 8.7% le rendement énergétique du couplage des deux réactions.

HAPPOLD et KEY (1937) au cours d'une étude de l'évolution des eaux résiduaires des usines à gaz, ont montré que les thiocyanates disparaissaient à la suite de l'attaque par des organismes tels que celui qu'ils ont isolé en culture pure et qui peut croître avec le thiocyanate comme seule source de carbone. Le thiocyanate est oxydé en sulfate suivant l'équation :

$$NH_4CNS + 2\ H_2O + 2\ O_2 = (NH_4)_2SO_4 + CO_2.$$

B. Le rôle photochemosynthétique des composés soufrés.

La participation du soufre comme élément prédominant du métabolisme d'un groupe de microorganismes pigmentés (rouges et verts) fut bien établie par WINOGRADSKY (1888) ; toutefois, celui-ci rapprochait ces organismes des sulfuraires incolores, sans nier le rôle important du pigment et la nécessité de la lumière dans le développement de ces organismes. Les travaux d'ENGELMAN (1888), de MOLISCH (1907) et de BUDER (1909) apparemment contradictoires, ont contribué à montrer que l'on avait affaire à un type de physiologie inhabituel, par exemple que l'action photochimique de la lumière, si elle existait, n'entraînait pas la libération d'oxygène. Comme l'a montré VAN NIEL dans la suite de ses travaux sur les bactéries photosynthétiques, la difficulté résultait non seulement du fait de leur physiologie particulière, mais également de l'hétérogénéité du groupe. La nécessité d'obtenir des cultures pures était particulièrement importante dans ce cas. C'est VAN NIEL (1931—1944) qui réalisa successivement la culture pure et l'étude des trois groupes de bactéries photosynthétiques, les *Thiorhodaceae*, les *Chlorobacteriaceae* et les *Athiorhodaceae*. Les deux premières familles sont spécialisées dans l'utilisation des composés soufrés. La physiologie de ces organismes peut être résumée comme suit : ils peuvent croître sur des milieux minéraux contenant du bicarbonate et de l'hydrogène sulfuré dans des conditions strictement anaérobiques, à la condition d'être illuminés. L'hydrogène sulfuré est cependant oxydé. Cette oxydation ne peut donc se faire dans ce cas que par la réduction d'un seul composé, le CO_2, qui est transformé en substance cellulaire ; on a donc la réaction (b) de la phase II du cycle du soufre déjà rencontré à la p. 106.

$$CO_2 + 2\ H_2S = (HCOH) + H_2O + 2\ S \quad A = -\ 12.700\ calgr.$$

Cette réaction isolée, quoique satisfaisante matériellement, est cependant impossible d'après les données thermodynamiques ; son affinité est négative (—12.700). En anaérobiose, elle ne peut pas être compensée par la réaction a, phase II (v. p. 106) et c'est ici qu'apparait toute l'importance de la réaction photochimique qui rend le bilan d'affinité positif. L'étude quantitative du bilan de croissance, effectuée par VAN NIEL, est en très bon accord avec la réaction chimique précédente. Ceci entraîne comme conséquence un rendement matériel presque intégral entre le CO_2 assimilé et l'hydrogène sulfuré utilisé, ce qui est très différent de ce que l'on obtient avec les bactéries aérobiques du soufre non photosynthétiques. Si le rendement matériel de croissance est excellent, le rendement énergétique global, comprenant l'utilisation de l'énergie lumineuse est beaucoup plus réduit (Larsen 1953); il ne dépasse pas 8.5%, alors que la photosynthèse chez les plantes vertes est au minimum de 30% et pourrait atteindre 90% d'après les résultats de Warburg et Burk (1950). L'action des bactéries photosynthétiques sur les composés soufrés n'est pas uniforme et nous donnerons sucinctement les différentes caractéristiques des différents groupes. Ces bactéries sont de tailles très variables. Toutes contiennent de la bactériochlorophylle, mais les *Thiorhodaceae* (Molisch 1907) possèdent en plus des pigments caroténoïdes, que l'on ne trouve pas dans les *Chlorobacteriaceae* (Geitler et Pascher 1925, Bavendamm 1936). Les deux familles dont la classification a été revue par van Niel sont décrites dans le Manuel de Bergey (Breed, Murray et Hitchkens 1948). Elles contiennent de nombreux genres déjà décrits par Winogradsky (1888), les genres *Chromatium* (rouge) et *Chlorobium* (verts) jouent un rôle important dans le cycle du soufre des lacs de Cyrénaïque (Butlin et Postgate 1953). En présence d'hydrogène sulfuré, les *Thiorhodaceae* accumulent du soufre intracellulaire, ce qui donne aux cultures un aspect rose crayeux. En absence d'H_2S, le soufre disparaît, il est transformé en sulfate et les bactéries deviennent rouge vif. On a donc dans ce cas les phases II et III du cycle du soufre (v. p. 106).

Les *Thiorhodaceae* sont capables d'utiliser la plupart des composés soufrés réduits. Le développement de ces organismes est favorisé par des teneurs relativement faibles en H_2S. Dans les eaux riches en H_2S, ce sont les *Chlorobacteriaceae* qui se développent de préférence. Parmi des organismes dont plusieurs genres ont été décrits sur la base de la simple observation, seuls les *Chlorobium*, déjà observés par Nadson (1912) ont été cultivés en culture pure et étudiés en 1931 par van Niel *(Chlorobium limicola)*. Une nouvelle espèce, *Chlorobium thiosulfatophilum*, a été découverte, cultivée en culture pure et étudiée par Larsen (1950, 1952, 1953). Ces deux espèces se distinguent entre autres par les types de composés sulfurés utilisables.

Chlorobium limicola utilise le mieux H_2S. Cette bactérie peut aussi croître avec S et H_2 comme réducteurs. *Chl. thiosulfatophilum* peut en plus utiliser le thiosulfate et le tétrathionate. Chez *Chl. thiosulfatophilum*, le produit final d'oxydation est toujours H_2SO_4, alors que chez *Chl. limicola*, en présence d'H_2S, on a à la fois du soufre et H_2SO_4.

IV. Le cycle du soufre généralisé.
La participation des animaux, des végétaux et des microorganismes non spécialisés.

Chez les animaux, les plantes et la grande majorité des microorganismes, les transformations des composés soufrés sont beaucoup moins massives que chez les organismes qui ont été étudiés jusqu'à présent. Le rôle du soufre n'y est cependant pas moins important, mais il a une signification biologique différente. Alors que dans les organismes «du soufre», les transformations entrent pour une large part dans le bilan énergétique et matériel de l'organisme, dans les organismes non spécialisés, les composés soufrés interviennent essentiellement comme éléments d'architecture des protéines ou comme éléments de catalyse. Ici aussi cependant, une large part de la signification de ces réactions est due à des transformations d'oxydo-réduction du soufre. L'étude des transformations du bilan du soufre et par conséquent la connaissance de ces transformations, requiert d'autres méthodes d'investigations telles que: l'étude de la composition, de la nutrition, ainsi que l'étude enzymatique des transformations.

A. Généralités et réduction.

Le soufre cellulaire est essentiellement concentré sous forme réduite dans les 3 amino-acides, cystéine, cystine et méthionine. Les plantes tirent normalement leur soufre des sulfates du sol. Dans les composés soufrés organiques tels que la cystéine et la méthionine, le soufre peut être considéré avec le nombre d'oxydation —2. Dans la cystine, il est —1 (comme dans le cas des disulfures Na_2S_2). On peut donc considérer au moins en bilan, que les plantes effectuent essentiellement la phase I du cycle du soufre présenté à la page 106.

Les animaux, au contraire, sont incapables d'effectuer la réduction des sulfates. Le sulfate injecté sous forme du $S^{35}O_4^=$ est éliminé entièrement dans les urines; il ne participe pas à l'élaboration des constituants soufrés (TARVER et SCHMIDT 1939). Les animaux exigent des amino-acides soufrés dans leur régime alimentaire. Lorsque leur régime contient un excès de ceux-ci, ils l'éliminent en grande partie sous forme de sulfate dans les urines, et en moindre quantité sous forme de taurine (NH_2—CH_2—CH_2—SO_3H) à l'état d'acide taurocholique éliminé par les voies biliaires. Les animaux effectuent donc normalement l'oxydation $S^{-II} \rightarrow S^{+VI}$ * et participent donc aux phases II et III du cycle du soufre.

Pour les bactéries, les cas sont divers, certaines croissent avec les sulfates comme seule source de soufre (*Enterobacteriaceae*, *Pseudomonadaceae*, *Nitrobacteriaceae*, etc.) et elles effectuent donc la réduction comme les plantes. Cependant, comme c'est le cas par exemple de *Proteus vulgaris*, de *Escherichia coli* et de *Propionibacterium pentosaceum* (TARR 1933 a, 1933 b, 1934, DESNUELLE et FROMAGEOT 1939, DESNUELLE 1939), en présence de protéines contenant des amino-acides soufrés, il se produit une attaque de l'excédent de ces amino-acides et une production d'H_2S dans le milieu qui constitue une des caractéristiques de la putréfaction. La libération du soufre non oxydé est évidemment particulièrement nette en anaérobiose. La situation des bactéries est encore moins schématique, car elles peuvent souvent utiliser l'H_2S en remplacement des sulfates. Des bactéries particulièrement exigeantes (p. ex. *Lactobacteriaceae*) nécessitent comme les animaux un apport d'amino-acides soufrés.

Certaines bactéries sont capables d'oxyder le thiosulfate en polythionate (principalement tétrathionate), ces organismes sont hétérotrophes (STARKEY 1935 a, 1935 b) contrairement à ce que pensait TRAUTWEIN qui les a découverts (1921, 1924). Des *Pseudomonas* effectuent cette transformation. Dans certaines conditions, en présence d'hydrate de carbone comme réducteur, certaines souches de *Salmonella* effectuent la réaction inverse.

L'oxydation procéderait suivant l'équation:

$$2\, S_2O_3^= + H_2O + {}^1\!/_2\, O_2 = S_4O_6^= + 2\, OH^- \qquad A = + 15.000\,.$$

Cette oxydation est chimiquement comparable à celle des sulfures en disulfures (ou cystéine en cystine).

La réduction, avec un donneur d'hydrogène AH_2, peut s'écrire:

$$S_4O_6^= + AH_2 = 2\, S_2O_3^= + 2\, H^+ + A\,.$$

La signification biologique de ces transformations n'est pas connue, mais suggère que les stades thiosulfates, tétrathionates ne sont pas étrangers au métabolisme normal du soufre.

Ce qui vient d'être écrit ne constitue que les grandes lignes des transformations. Sans entrer dans le détail des nombreuses interconversions intracellulaires du soufre, nous essayerons de résoudre les réactions globales en leurs parties pour saisir les composés intermédiaires du cycle du soufre les plus significatifs.

---

* Le soufre de la taurine peut être considéré à l'état + IV.

LAMPEN *et al.* (1947) ont pu établir un certain nombre des stades inter-médiaires de la réduction des sulfates en amino-acides soufrés chez *Escherichia coli*. Par traitement à la lumière ultra-violette, ces auteurs ont produit de nombreux mutants qui ne possédaient plus la propriété de pouvoir croître avec les sulfates comme source de soufre. Le tableau 1 nous donne les exigences apparues dans

Tableau 1. *Utilisation des composés soufrés par les différents mutants d'Escherichia coli.*

Souche nº	Source du soufre							
	Na₂SO₄ (1)	Na₂SO₃ (2)	Na₂S₂O₃ Na₂S₂O₄ Na₂S (3)	Thio-glycolate (4)	l. cystine ou l. cystéine (5)	Cysta-thionine (6)	d. l. Homo-cystine (7)	d. l. méthio-nine (8)
E. coli typique	+	+	+	+	+	+	+	+
1251—171	—	+	+	+	+	+	+	+
932—230	—	+	+	+	+	+	+	±
1950—230	—	—	+	+	+	+	+	±
508—462	—	—	—	—	+	+	+	+
255—468	—	—	—	—	+	+	+	+
532—171	—	—	—	—	—	—	+	+
754 M—171	—	—	—	—	—	—	+	+
282—460	—	—	—	—	—	—	+	+
495—460	—	—	—	—	—	—	+	+
654—228	—	—	—	—	—	—	±	+
3—301	—	—	—	—	—	—	±	+
1—344	—	—	—	—	—	—	—	+
1—273—384	—	—	—	—	—	—	—	+

+ = croissance;   — = non croissance;   ± = croissance faible.

ces diverses souches. L'examen du tableau nous montre que les exigences se distribuent suivant une série de substances soufrées correspondant, dans les 3 premières colônnes, à des stades de plus en plus réduits et ensuite, à des trans-formations d'amino-acides n'impliquant plus de changement d'état d'oxydation (colonnes 4, 5, 6, 7). Une souche qui, par exemple, ne peut pas croître sur Na₂S, ne croît jamais sur un sulfite ou un sulfate. Ces résultats sont interprêtés suivant la méthode classique de l'école de BEADLE (voir à ce sujet TATUM 1949). Les souches ayant acquis des exigences ont perdu la possibilité génétique de produire *un enzyme* faisant partie de la chaîne métabolique allant du sulfate à la méthionine. Ceci permet d'échelonner la série des intermédiaires et conduit à admettre le schéma de la Fig. 3.

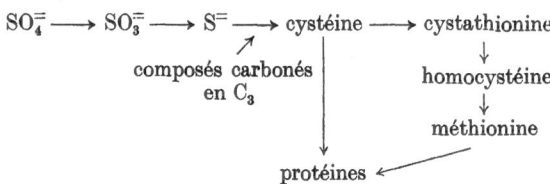

Fig. 3. La réduction des composés soufrés chez Escherichia coli.

On retrouve donc l'hydrogène sulfuré comme un des stades du cycle (Phase I du cycle) toutefois, il n'est pas libéré dans le milieu il est transformé immédiate-ment en amino-acides. Des études analogues sur *Ophiostoma multiannulatum* (FRIES 1945), sur *Aerobacter aerogenes*, DAGLEY, DAWES and MORISSON (1949) ont confirmé ces données, avec des modalités au niveau des conversions d'amino-acides (voir aussi BOLTON, COWIE, SANDS 1952). En utilisant la même méthode sur des mutants d'*Aspergillus nidulans*, HOCKENHULL (1949) propose la série de réactions:

$$SO_4^= \longrightarrow SO_3^= \longrightarrow (H_2SO_2) \longrightarrow H_2S_2O_3 \longrightarrow$$
$$\text{sulfoxylate} \qquad \text{thiosulfate}$$

$$HSO_3—S—CH_2—CH(NH_2)—COOH \longrightarrow HS—CH_2—CH(NH_2)—COOH$$
$$\text{serine thiosulfate} \qquad\qquad\qquad \text{cystéine}$$

dans ce cas, la synthèse ne passe pas par l'hydrogène sulfuré, mais par le thio-sulfate qui pourrait être formé par la dimérisation d'un composé tel que l'acide sulfoxylique qui ensuite serait incorporé aux composés carbonés sous forme de serine-thiosulfate donnant la cystéine (voir aussi ROBERTS *et al.* 1955).

Les résultats obtenus par PHINNEY (1948) avec *Neurospora crassa* diffèrent des précédents, par le couplage direct du sulfate en un composé organique, l'acide cystéique, qui serait ensuite réduit par une voie qui est l'inverse d'une forme d'oxydation de la cystéine présente dans les tissus animaux. La réduction des sulfates chez les plantes supérieures est traitée dans ce volume par THOMAS (v. p. 37).

Le soufre élémentaire n'apparaît pas dans le schéma précédent; il ne s'accumule pas et normalement, les organismes ne l'assimilent pas. On a cependant signalé que la levure peut réduire le soufre au sulfure. La réduction du soufre pourrait se produire suivant la réaction $2\,RSH + S = R—S—S—R + H_2S$ (GUTHRIE 1938, BERSIN 1950), RSH étant un sulfure cellulaire dont la forme oxydée est réduite physiologiquement.

B. Oxydation.

Le bilan du métabolisme du soufre chez les animaux est l'inverse de celui des plantes (revue: FROMAGEOT 1947—1955). Ils excrètent le soufre principale-ment sous forme de sulfate (libre ou estérifié à des phénols) et de taurocholate. On a pu retrouver également en petites quantités dans les urines d'autres sub-stances soufrées; ce sont: des thiosulfates (FROMAGEOT et ROYER 1945), du sulfocyanure, du taurocarbamate, de la cystine (LEFEVRE et RANGIER 1937). Certaines formes d'excrétion (cystinurie) sont pathologiques.

L'hydrogène sulfuré, à l'inverse des sulfates, est aisément absorbé par l'animal; il peut se former par l'attaque microbienne du contenu intestinal (ZÖRKEN-DÖRFER 1931). DZIEWIATKOWSKY (1945) a montré que H_2S^{35} se retrouve essen-tiellement sous forme de sulfate. Le soufre libre colloïdal peut être oxydé en sulfate par l'homme à raison de 500 à 750 mg. par jour (GREENGARD et WOOLLEY 1940); il en est de même des thiosulfates (PIRIE 1934, ZÖRKENDÖRFER 1935).

L'oxydation du soufre organique réduit en soufre oxydé (sulfate ou taurine) a fait l'objet de travaux récents en ce qui concerne les tissus animaux (FROMA-GEOT 1955), *Proteus vulgaris* (KEARNEY et SINGER 1953a, 1953b) et *Micro-sporum gypseum*, un champignon spécialisé dans l'attaque de la kératine, qui dégrade la laine et cause des mycoses de la peau et des poils (STAHL, McQUE, MANDEL et SIA 1949).

En principe, l'oxydation du soufre organique pourrait se réaliser par diffé-rentes voies, suivant l'état sous lequel le soufre quitte la molécule organique.

Le soufre de la cystéine, de l'homocystéine et de la méthionine (après sa transformation en cystéine par la voie inverse de la synthèse qui a été décrite) peut-être détaché avant toute oxydation, sous forme d'H_2S. Ce processus se rencontre à la fois dans les bactéries et les tissus animaux (revue: FROMAGEOT 1951). Le cas opposé serait réalisé par l'oxydation du soufre restant attaché sous la forme organique et donnant $R—SO_3—H$. C'est la forme sous laquelle on retrouve le soufre de la taurine (dans l'acide taurocholique). Ce type d'oxydation implique des stades intermédiaires que l'on peut figurer par la suite de réactions:

Cystéine $\longrightarrow$ cystéine sulfénique $\longrightarrow$ cystéine sulfinique $\longrightarrow$

R—SH   R—SOH   R—SO$_2$H

acide cystéique $\xrightarrow[-CO_2]{\hspace{2cm}}$ taurine

R—SO$_3$H   H$_2$N—CH$_2$—CH$_2$—SO$_3$H

BERNHEIM et BERNHEIM (1939) et MEDES (1939) ont montré en effet que les tissus animaux broyés transforment effectivement la cystéine en acide cystéique et que les mêmes préparations oxydaient également l'acide cystéine sulfinique. VIRTUE et DOSTER-VIRTUE (1939) ont observé une augmentation de la production d'acide taurocholique par injection simultanée d'acide cholique et d'acide cystéique. La décarboxydation enzymatique de l'acide cystéique a été démontrée par BLASCHKO (1942) et par MEDES et FLOYD (1942). Ces derniers ont également établi que la cystine est oxydée en taurine, le disulfoxyde de cystine étant un intermédiaire. L'acide cystéique injecté aux animaux n'augmente pas le taux de sulfate, donc l'acide cystéique n'est pas la voie majeure d'oxydation du soufre organique.

A ces deux processus extrêmes d'oxydation, il faut ajouter le cas où le soufre est détaché de la molécule organique à un stade intermédiaire d'oxydation. Ce type d'oxydation apparaît très important à la suite des travaux récents de l'école de FROMAGEOT (revue 1955) et qui concerne à la fois les tissus animaux et *Proteus vulgaris* (KEARNEY et SINGER 1953a, 1953b). Le fait essentiel qui ressort de ces travaux réside dans l'importance de l'acide cystéine-sulfinique et de sa transformation en acide β-sulfinylpyruvique et la libération à ce stade du soufre sous la forme d'H$_2$SO$_3$.

Ce mécanisme d'oxydation avait été partiellement prévu par PIRIE (1934); c'est également celui qui fut supposé par STAHL *et al.* (1949) lors de l'attaque de laine (contenant 12% de cystine) par *Microsporum gypseum* qui transforme complètement la cystine en sulfate.

Les études enzymatiques entreprises dans les laboratoires de FROMAGEOT ont en plus montré une troisième voie de transformation qui résulte de la décarboxylation de l'acide cystéine-sulfinique en hypotaurine CH$_2$(NH$_2$)—CH$_2$—SO$_2$H (aminoéthanesulfinique) qui pourrait être le principal intermédiaire vers la taurine. Les mécanismes précédents sont vraisemblablement la voie principale d'oxydation du soufre organique en sulfate.

L'importance de l'acide β-sulfinylpyruvique est encore accrue par le fait que ce composé semble être un intermédiaire dans la réduction S$^{VI} \rightarrow$ S^{-II} aussi bien que dans le processus inverse d'oxydation. CHAPEVILLE et FROMAGEOT (1954) ont en effet démontré la réversibilité de la réaction:

$$CH_3—CO—COOH + SO_2 \rightleftharpoons HO_2S—CH_2—CO—COOH.$$

Les débris végétaux et animaux seront minéralisés par l'action microbienne. Le résultat de cette minéralisation dépendra essentiellement des conditions; un organisme tel que *Proteus vulgaris*, en présence d'un excés de matières organiques et dans des conditions partiellement anaérobiques provoquera finalement une production d'H$_2$S; le même organisme peut cependant aussi, en aérobiose, oxyder les amino-acides soufrés jusqu'au stade de sulfate. Le premier stade de la minéralisation peut-être une proteolyse ordinaire *(Proteus)* ou une attaque des scléroprotéines qui doivent leur résistance à la présence de nombreuse liaisons disulfures dans leur structure *(Microsporum)*.

C. Le cycle général.

Les principales données présentées ci-avant sont reprises dans le schéma général donné à la Fig. 4. Seuls les composés soufrés ont été indiqués. Lorsqu'une transformation importante n'a pas été analysée quant à ses produits intermédiaires, elle est indiquée par une seule réaction, p. ex. $SO_4^= \rightarrow H_2S$ par *Desulfovibrio*, ce qui n'implique pas que cette transformation est essentiellement différente des autres transformations partant du même produit en aboutissant au même résultat.

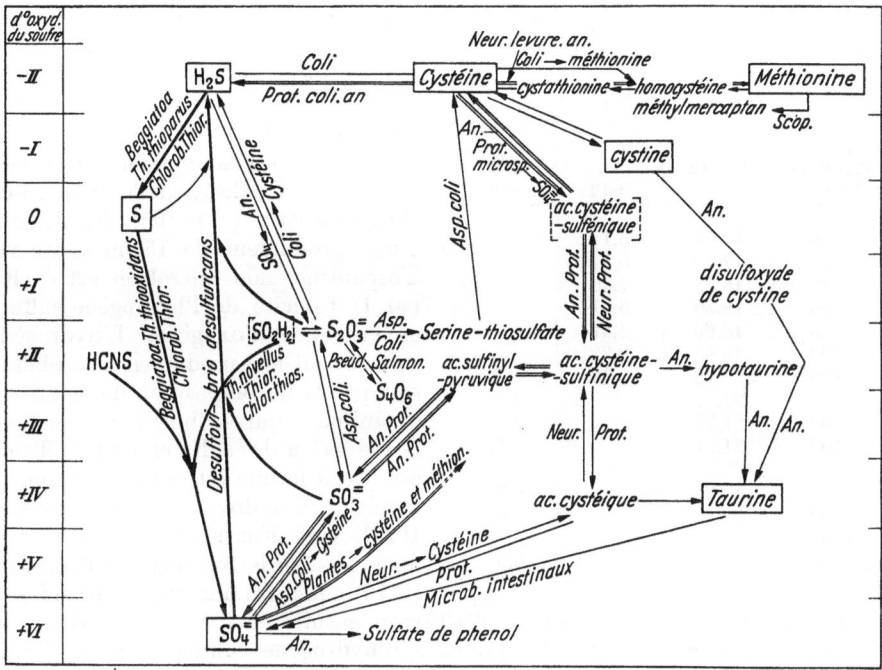

Fig. 4. Les différents produits soufrés sont classés suivant le degré d'oxydation du soufre (nombre d'oxydation. v. p. 103). Les substances soufrées qui s'accumulent dans le milieu extérieur ou dans les cellules sont encadrées. Les substances indiquées entre parenthèses pointillées sont hypothétiques. ⟶ Transformations où le soufre jour un rôle massique important; ⟹ voies principales; ⟶ voies secondaires ou hypothétiques. Abréviations: An.: Animaux; Asp.: *Aspergillus*; Chlor.: *Chlorobium*; Coli: *Escherichia coli*; Microsp.: *Microsporum gypseum*; Neur.: *Neurospora*; Prot.: *Proteus vulgaris*; Pseud.: *Pseudomonas*; Scop.: *Scopulariopsis brevicaule*; Th.: *Thiorhodaceae*.

V. Le rôle écologique du soufre.

En général, les composés soufrés ne constituent pas un facteur écologique important. Les sulfates sont très répandus dans les eaux douces, les eaux marines et les sols, et constituent la source normale de soufre nécessaire à la croissance de la flore et ensuite de la faune de cet habitat.

Toutefois, dans certains cas, le soufre joue un rôle déterminant, s'il se présente sous une forme particulière. En quantité très faible, il pourrait aussi être un facteur limitant du développement biologique.

La mer Noire et certains fjords Norvégiens sont des habitats particuliers en ce qui concerne le soufre (SVERDRUP; JOHNSON et FLEMING 1946). Ces masses d'eau marine sont séparées des eaux océaniques par des détroits fort resserrés et d'une profondeur faible. Alors que la mer Noire a une profondeur de 2.100 m, le Bosphore n'est profond que de 40 à 90 m. Dans ses conditions, il y a peu d'échange libre avec les eaux méditerranéennes. Alors que les eaux océaniques ont une composition relativement constante, au moins à partir d'une certaine profondeur, la mer Noire constitue un système fermé comparable aux eaux stagnantes et il y apparaît une stratification des constituants. Les substances les plus remarquables à cet égard sont l'oxygène, l'hydrogène sulfuré et les sels. La salinité de la Méditerrannée ne varie que dans d'étroites limites (de 38.4 à 38.8⁰/₀₀); la salinité de la mer Noire passe de 17.6⁰/₀₀ à la surface, à 22.27⁰/₀₀ à 1.000 m. L'oxygène est présent dans les eaux océaniques même aux plus grandes profondeurs et dans la Méditerrannée, sa concentration est d'environ 4 ml/l; au contraire, dans la mer Noire, l'oxygène décroît et pratiquement disparaît vers 150 mètres. Vers cette profondeur apparaît l'hydrogène sulfuré. Ces variations sont données au tableau 2, résultat de sondages de la Station Thor 172, août 10, 1910.

Tableau 2. *Conditions hydrographiques dans la mer Noire.* (Sverdrup et al. 1946.)

Profondeur (mètres)	Salinité ⁰/₀₀	O₂ ml/l	H₂S ml/l
0	17.59	5.14	
10	17.59	5.14	
25	18.22	7.40	
50	18.30	6.71	
75	18.69	4.41	
100	19.65	2.33	
150	20.75	0.17	
200	21.29		0.9
300	21.71		2.34
400	21.91		4.17
600	22.16		4.96
800	22.21		6.06
1000	22.27		6.04
1500			6.17
2000			6.5

La mer Noire est un habitat très semblable à celui réalisé dans l'expérience de Winogradsky (v. p. 105). En dessous d'une profondeur de 150 m toute vie d'organisme non microbien est exclue par la toxicité de l'hydrogène sulfuré et l'absence d'oxygène. L'hydrogène sulfuré peut-être produit en anaérobiose, soit par la décomposition des matières organiques qui sédimentent, soit par la réduction des sulfates (qui d'ailleurs est liée à la matière organique comme donneur d'hydrogène) (Issatchenko 1924). Une situation comparable existe dans les fjords Norvégiens; Gaarden (1916) a montré que dans le Mönefjord, à 60 m, on ne détecte pas d'H₂S; il s'accroît ensuite et atteint respectivement à 100 et 200 m, 1 et 2 ml/l. La présence d'hydrogène sulfuré dans ces fjords entraîne parfois la mort des organismes supérieurs vivant dans les eaux de surface, lorsque les vents amènent dans les fjords une masse d'eau océanique saline dense qui relève provisoirement le niveau des eaux sulfureuses.

Les marais et les bords de mer à pente très faible, les baies resserrées ainsi que les canaux à courant lent de Hollande, de Venise, offrent en général des conditions propres au développement des organismes producteurs d'H₂S. Butlin (1953) signale qu'à la côte Sud-Ouest Africaine, proche de la ville de Swakopmund, à la suite du développement de bancs de boues déposées sur un fond de gypse, il se produit des éruptions d'H₂S qui pénètrent à plus de 100 km vers l'intérieur des terres et corrodent les surfaces métalliques, et des milliers de poissons morts couvrent la plage.

La production microbiologique d'hydrogène sulfuré cause régulièrement la destruction des conduites de fer (Kühr, v. Wolzagen et van der Vlugt 1934, Bunker 1938). Cette corrosion est liée à la présence de sulfates et de conditions anaérobiques. Elle est favorisée par la présence de matières organiques. Depuis les ravaux de Starkey et Wight (1934), ainsi que de Butlin et al. (1947, 1949), on sait que le donneur d'hydrogène peut-être le fer en présence d'eau.

La présence d'H₂S entraîne l'apparition des organismes utilisateurs de cette substance, y compris les *Thiorhodaceae* qui forment parfois des masses rouges que Warming (1876) observa le long des côtes danoises et dans les eaux sales qui entourent Copenhague. Ces bactéries seraient aussi la cause de la baie rouge d'Odessa (JEGUNOV 1898) et des «mers de sang» des côtes de l'Holstein (GIETZEN 1931). La présence des sulfuraires dans les eaux stagnantes a été signalée par WINOGRADSKY, ces organismes forment des masses importantes dans les eaux de sortie des sources sulfureuses non captées.

L'activité des organismes du soufre se manifeste parfois par le dépôt de sédiments soufrés. Les boues formant le fond des eaux sulfureuses sont riches en sulfure ferreux. Une partie des dépôts de soufre de Sicile semble également être d'origine biologique. Ces couches de soufre allant de 1 à 30 m sont inter-stratifiées d'argile bitumeuse, de gypse et de tripoli (provenant de radiolaires et d'éponges). Ces dépôts sont accompagnés de gypse, ce qui suggère la trans-formation:

$$SO_4^{=} \longrightarrow S^{=} \longrightarrow S \qquad \text{(HUNT 1915)}.$$

Une étude très intéressante à cet égard a été réalisée récemment par BUTLIN et POSTGATE dans les lacs de Cyrénaïque (1953 a, b, c) de la région de El Agheila, principalement le lac Ain-ez-Zania. La présence de soufre colloïdal donne à la surface du lac une teinte bleue laiteuse. En bordure, sous l'eau peu profonde, il existe une bande d'une masse rouge, de consistance gélatineuse. L'eau de ce lac est saline (2 %) et saturée de sulfate de calcium, qui provient de gypse qu'on retrouve sur le fond du lac; les matières organiques sont en faible quantité. Au fond du lac, l'eau contient 108 mg H_2S/l et à la surface 20 mg/l. Les eaux du lac contiennent des *Desulfovibrio*, (des cultures d'enrichissement ont toujours donné ces organismes). Selon toute évidence, les bactéries réductrices de sulfate jouent un rôle primordial dans la production d'H_2S. La masse gélatineuse rouge, endessous de laquelle se trouvait parfois une matière verte, est composée de *Chromatium* (bactérie photosynthétique rouge), de *Chlorobium* (bactérie photo-synthétique verte) et de soufre. Ces organismes nécessitent de la lumière et se développent uniquement à faible profondeur. Du soufre se deposé dans les lacs et forme une couche atteignant 15 cm, qui est récoltée par des habitants de la région. Des trois lacs de la région, on tire environ 200 tonnes de soufre annuellement. Les auteurs ont reproduit en culture mixte de *Chlorobium* et de *Desulfovibrio*, le cycle du soufre qui se produit dans ce lac, grâce à la fois à la présence de sulfate et à la forte luminosité. Il existe toutefois un point à éclaircir, qui concerne le donneur d'hydrogène nécessaire à la réduction des sulfates; la matière organique en solution est peu importante, mais il se peut qu'un apport constant existe du fait que les *Chromatium* et *Chlorobium* sont autotrophes et seraient aussi des donneurs de matière organique pour les *Desulfovibrio* réduc-teurs de sulfate.

La présence du soufre élémentaire crée également un habitat très spécifique pour le *Thiobacillus thiooxidans*. Les sols traités par le soufre comme fongicide sont un habitat d'enrichissement pour cet organisme et l'acide sulfurique produit peut provoquer une acidification du sol qui entrave le développement de certains actinomycètes comme l'*Actinomyces scaber*, parasite de la pomme de terre (FOSTER 1951). *Th. thiooxydans* est responsable de l'acidification des terrains voisinant les mines de soufre et c'est vraisemblablement à son action acidifiante qu'était dû l'emploi d'engrais composés de phosphate insoluble, de soufre et de terre. Cet organisme serait également la cause de la déterioration des joints de canalisa-tion contenant du soufre. Quoique *Th. thiooxidans* se développe dans ces habitats très particuliers, c'est un organisme très répandu.

La production d'acide sulfurique par des bactéries du type *Thiobacillus* explique la corrosion des constructions en béton. L'attaque des egoûts étudiée par PARKER (1954, 1951) procède en plusieurs étapes: production d'H_2S et ensuite oxydation de celui-ci via des composés soufrés tels que les thiosulfates, polythionates, soufre élémentaire, ce dernier étant transformé en acide sulfurique par une bactérie proche de *Thiobacillus thiooxidans (Th. concretivorus)*.

Les eaux et sols sont généralement riches en sulfates qui constituent, avec les chlorures, les substances qui ont une tendance à s'accumuler en solution à la suite de l'altération des roches.

Le soufre est en général présent dans l'eau de mer à raison de 25 mg atome/l, alors que l'azote et le phosphore sont en quantités beaucoup moindres ($N < 0.05$; $P < 0.003$).

Les eaux douces sont également largement pourvues en sulfates. On a décrit des cas cependant où les sulfates sont en quantités très réduites et où ils pourraient jouer un rôle de facteur limitant du développement biologique (BEAUCHAMP 1946, 1953). C'est le cas de certains lacs africains. Le lac Tanganyka ne contient que $4/10^6$ de sulfate, le lac Victoria 0.8 à $1.0/10^6$, le lac George $0.5/10^6$, le lac Mweru ne contient que des traces indosables. Cette faible quantité de sulfate est dûe au faible apport d'eau qui entre dans ces lacs, à leur utilisation par le plancton et, finalement, à la précipitation de sulfure insoluble qui interrompt le cycle du soufre.

BEAUCHAMP signale également que certains sols africains sont très pauvres en sulfates et qu'une amélioration des cultures pourrait résulter de l'emploi d'engrais sulfatés.

Bibliographie.

BAARS, J. K.: Cité par BULTIN *et al.*, 1949. Over Sulfaatreductie door bacteriën. Diss. Delft 1930. — BARKER, H. A.: On the biochemistry of the methane fermentation. Arch. Mikrobiol. **7**, 404 (1936). — BARKER, H. A., S. RUBEN and M. D. KAMEN: The reduction of radioactive carbon dioxide by methane producing bacteria. Proc. Nat. Acad. Sci. U.S.A. **26**, 426 (1940). — BAVENDAMM: *Chlorobacteria*. Erg. Biol. **13**, 49 (1936). — BEAUCHAMP, R. S. A.: Lake Tanganyika. Nature (Lond.) **157**, 183 (1946). — Sulphates in african inland waters. Nature (Lond.) **171**, 769 (1953). — BEIJERINCK, N. M.: Über *Spirillum desulfuricans* als Ursache von Sulfatreduktion. Zbl. Bakter. II **1**, 1, 49, 104 (1895). — Über die Bakterien, welche sich im Dunkel mit Kohlensäure als Kohlenstoffquelle ernähren können. Zbl. Bakter. II **2**, 593 (1904). — BENECKE, W.: Bakteriologie des Meeres. In E. ABDERHALDENS Handbuch der biologischen Arbeitsmethoden, Bd. 5, Abt. IX. Berlin 1933. — BERNHEIM, F., and M. L. C. BERNHEIM: The effect of titanum on the oxidation of sulfhydryl groups by various tissues. J. of Biol. Chem. **127**, 695 (1939). — BERSIN, TH.: Die Phytochemie des Schwefels. Adv. Enzymol. **10**, 223 (1950). — BLASCHKO, E. T., D. B. COWIE and M. K. SAND: The metabolic fate of sulfate sulfur. J. Bacter. **63**, 309 (1952). — BLASCHKO, H.: Biochemic. J. **36**, 571 (1942). — BOLTON, E. T., D. B. COWIE and M. K. SANDS: Sulfur metabolism in *Escherichia coli*. J. Bacter. **63**, 309 (1952). — BREED, R. S., E. G. D. MURRAY and A. P. HITCHKENS: BERGEY's manual of determinative bacteriology. 6st edit. Baltimore: Williams & Wilkins Co. 1948. — BUDER, J.: Zur Bakteriologie des Bakteriopurpurins und der Purpurbakterien. Jb. wiss. Bot. **58**, 525 (1919). — BUNKER, H. J.: Microbiological experiments in anaerobic corrosion. J. Soc. Chem. Industr. (Lond.) **58**, 93 (1938). — BUTLIN, K. R.: The bacterial sulphur cycle. Research (Lond.) **6**, 184 (1953). — BUTLIN, K. R., and M. E. ADAMS: Autotrophic growth of sulphate reducing bacteria. Nature (Lond.) **160**, 154 (1947). — BUTLIN, K. R., M. E. ADAMS and M. THOMAS: The isolation and cultivation of sulphate reducing bacteria. J. Gen. Microbiol. **3**, 46 (1949). — BUTLIN, K. R., and J. R. POSTGATE: Microbiological formation of sulphide and sulphur. Symposium on microbial metabolism. Roma, Istituto Superiore di Sanita 1953. — The microbiological formation of sulphur in Cyrenaican Lakes. In: Biology of desert, J. L. CLOUDSLEY-THOMPSON, Institute of Biology, London 1953.

CHALLENGER, F.: Biological methylation. Adv. Enzymol. **12**, 429 (1951). — CHAPEVILLE, F., and P. FROMAGEOT: La formation enzymatique de l'acide cystéine-sulfinique à partir de sulfite. Biochim. et Biophysica Acta **14**, 415 (1954). — COHN, F.: Zwei neue *Beggiatoen*. Hedwigia **4**, 81 (1865). — CRAMER: Cité par WINOGRADSKY 1949.

DELDEN, A. van: Beitrag zur Kenntnis der Sulfatreduktion durch Bakterien. Zbl. Bakter. II **11**, 31, 113 (1903). — DESNUELLE, P.: Dégradation anaérobie de la cystéine par *B. coli*. III. Spécifité optique de la cystéinase. Enzymologia (Den Haag) **6**, 387 (1939). —

DESNUELLE, P., u. C. FROMAGEOT: La décomposition anaérobie de la cystéine par *Bacterium coli*. I. Existence d'une cystéinase, ferment d'adaptation. Enzymologia (Den Haag) **6**, 80 (1939). — DZIEWIATKOWSKY, D. D.: Fate of ingested sulfide sulfur, labeled with radioactive sulfur, in the rat. J. of Biol. Chem. **161**, 723 (1945).

ELION, L.: A thermophilic sulphate reducing bacterium. Zbl. Bakter. II **63**, 58 (1925). — ENGELMANN, W.: Die Purpurbakterien und ihre Beziehungen zum Licht. Bot. Ztg **46**, 662 (1888).

FOSTER, J. W.: Autotrophic assimilation of carbon dioxide. Dans: C. W. WERKAN and P. WILSON, Bacterial Physiology, p. 362. New York: Academic Press 1951. — FRIES, N.: X-Ray induced mutations in the physiology of *Ophiostoma*. Nature (Lond.) **155**, 757 (1945). — FROMAGEOT, C.: Oxidation of organic sulfur in animals. Adv. Enzymol. **7**, 369 (1951). — Desulfhydrase. Dans: SUMNER and MYRBÄCK: The Enzymes, vol. I, part 2. New York: Academic Press 1951. — The metabolism of sulfur and its relation to general metabolism. Harvey Lect. **1955**. — FROMAGEOT, C., u. A. ROYER: La présence constante du thiosulfate dans l'urine des animaux et sa signification physiologique. Enzymologia (Den Haag) **11**, 361 (1945). — FROMAGEOT, C., E. WOOKEY u. P. CHAIX: Sur la dégradation anaérobique de la cystéine par la désulfurase du foie. Enzymologia (Den Haag) **9**, 198 (1940).

GAARDEN: Cité par W. BENECKE 1933. — GEITLER, L., u. A. PASCHER: *Cyanochloridneac, Chlorobactc;iaceae*. Die Süßwasserflora Deutschlands, Österreichs und der Schweiz. Bd. 12, S. 451. Jena 1925. — GIETZEN: Untersuchungen über marine *Thiorhodazeen*. Zbl. Bakter. II **83**, 183 (1941). — GREENGARD, H., and J. R. WOOLLEY: Studies on colloïdal sulfur, polysulfide mixture. Absorption and oxidation after oral administration. J. of Biol. Chem. **132**, 83 (1940). — GREENSTEIN, J. B., and F. M. LEUTHARDT: Degradation of cystine peptides by tissues. I. Exocystine desulfurase and dehydropeptidase in rat liver extracts. J. Nat. Canc. Inst. **5**, 209 (1944). — GUTHRIE, J. D.: Availability of sulfate during synthesis of glutathione to potatoes treated with ethylene chlorhydrin. Contrib. Boyce Thompson Inst. **9**, 233 (1938).

HAPPOLD, F. C., and A. KEY: The bacterial purification of gas-works liquors. Biochemic. J. **31**, 1323 (1937). — HOCKENHULL, D. J. D.: The sulfur metabolism of mold fungi. Biochim. et Biophysica Acta **3**, 326 (1949). — HUNT, F. W.: The sulfur deposits of Sicily. Econ. Geol. **10**, 543 (1915).

ISSATCHENKO, B. L.: Sur la fermentation sulfhydrique dans la mer Noire. C. r. Acad. Sci. Paris **178**, 2204 (1924).

JEGUNOV, M.: Cité par BENECKE 1933. Zbl. Bakter. II **4**, 257 (1898).

KEARNEY, E. B., and T. P. SINGER: The oxidation of cysteinsulfinic and cysteic acid in *Proteus vulgaris*. Biochim. et Biophysica Acta **11**, 270 (1953a). — Enzymic transformations of L-cysteine-sulfinic acid. Biochim. et Biophysica Acta **11**, 276 (1953b). — KLUYVER, A. J., and C. B. VAN NIEL: Prospects for a natural system of classification of bacteria. Zbl. Bakter. II **94**, 369 (1936). — KNAYSI, G.: A cytological and microchemical study of *Thiobacillus thiooxidans*. J. Bacter. **46**, 451 (1943). — KÜHR, C. A. H., v. WOLZAGEN und I. S. VAN DER VLUGT: Cité par BUTLIN *et al.* 1949. The graphitization of cast iron as an electrochemical process in anaerobic soils. Water **18**, 147 (1934).

LAMPEN, J. O., R. R. ROEPKE and M. J. JONES: Studies on the sulfur metabolism of *Escherichia coli*. III. Mutant strains of *E. coli* unable to utilize sulfate for their complete sulfur requirements. Arch. of Biochem. **13**, 55 (1947). — LARSEN, H.: On the culture and general physiology of the green sulfur bacteria. J. Bacter. **64**, 187 (1952). — On the microbiology and biochemistry of the photosynthetic green sulfur bacteria. Kongl. norske Vidensk. Selsk., Skr. **1953**, Nr 1. — LARSEN, H., C. S. YOCUM and C. B. VAN NIEL: On the energetics of the photosynthesis in green sulfur bacteria. J. Gen. Physiol. **36**, 161 (1952). — LATIMER, W. H.: Oxidation potentials. New York: Prentice Hall 1952. — LEFEVRE, C., et M. RANGIER: Contribution à l'étude de la répartition du soufre organique urinaire. Bull. Soc. Chim. biol. Paris **19**, 1697, 1711 (1937). — LE PAGE, G. A.: The biochemistry of autotrophic bacteria. The metabolism of *Thiobacillus thiooxidans* in the absence of oxidizable sulphur. Arch. of Biochem. **1**, 255 (1942). — LE PAGE, G. A., and W. W. UMBREIT: Phosphorylated carbohydrates esters in autotrophic bacteria. J. of Biol. Chem. **147**, 263 (1943). — LIESKE, R.: Untersuchungen über die Physiologie denitrifizierender Schwefelbakterien. Ber. dtsch. bot. Ges. **36**, 12 (1912).

MASSART, J.: Rec. Inst. Bot. Univ. Bruxelles **5**, 251 (1902). — MEDES, G.: Metabolism of sulphur. VIII. Oxidation of the sulphur-containing amino-acids by enzymes from the liver of the albino rat. Biochemic. J. **33**, 1559 (1939). — MEDES, G., and N. FLOYD: Metabolism of sulphur. Cysteic acid. Biochemic. J. **36**, 836 (1942). — MOLISCH, H.: Die Purpurbakterien. Jena: Gustav Fischer 1907.

NADSON, G. A.: Mikrobiologische Studien. I. *Chlorobium limicola* Nads., ein grüner Mikroorganismus mit inaktivem Chlorophyll. Bull. Imp. Bot. Garden St. Petersburg **12**, 55 (1912). — NATHANSOHN, A.: Über eine neue Gruppe von Schwefelbakterien und ihren

Stoffwechsel. Mitt. zool. Stat. Neapel 15, 655 (1902). — NIEL, C. B. VAN: On the morphology and physiology of the purple and green sulfur bacteria. Arch. Microbiol. 3, 1 (1931). — The culture, general physiology, morphology and classification of the non-sulfur purple and brown bacteria. Bacter. Rev. 8, 1 (1944).

PARKER, C. C.: Mechanics of corrosion of concrete sewers by hydrogen sulphide. Sewage Industr. Wastes 23, 1477 (1951). — PARKER, C. D.: The corrosion of concrete. I. The isolation of a species of bacterium associated with the corrosion of concrete exposed to atmospheres containing hydrogen sulphide. Austral. J. Exper. Biol. a. Med. Sci. 23, 81—90 (1945). — PARKS, C. S., and H. M. HUFFMAN: The free energies of some organic compounds. New York: Chem. Catalog. Co. 1932. — PAULING, L.: General Chemistry. San Francisco: Freeman 1947. — PHINNEY, B. O.: Genetics 33, 624 (1948). — PIRIE, N. W.: The formation of sulphate from cysteine and methionine by tissues in vitro. Biochemic. J. 28, 305 (1934). — POLLOCK, M. R., and R. KNOX: Bacterial reduction of tetrathionate. Biochemic. J. 37, 476 (1943). — PRIGOGINE, I., et R. DEFAY: Thermodynamique chimique, Tome I. Liège: Desoer 1944.

RITTENBERG, S. C.: Cité par BUTLIN 1949, Studies on marine sulfate-reducing bacteria. Thesis Univ. California 1941. — ROBERTS, R. B., P. H. ABELSON, D. B. COWIE, E. T. BOLTON and R. J. BRITTEN: Studies of biosynthesis in Escherichia coli. Washington, D.C., Carnegie Institution of Washington 1955.

STAHL, W. H., B. MCQUE, G. R. MANDEL and R. G. U. SIA: Studies on the microbiological degradation of wool. I. Sulfur metabolism. Arch. of Biochem. 20, 422 (1949). — STARKEY, R. L.: Concerning the physiology of Thiobacillus thiooxidans an autotrophic bacterium oxidizing sulphur under acid conditions. J. Bacter. 10, 135, 165 (1915). — Isolation of some bacteria which oxidize thiosulphate. Soil Sci. 39, 197 (1935a). — Products of the oxidation of thiosulphate in mineral media. J. Gen. Physiol. 18, 325 (1935b). — Spore formation by the sulphate-reducing vibrio. Kon. Ned. Akad. Wetensch. 41, 422 (1938). — STARKEY, R. L., and K. M. WIGHT: Anaerobic corrosion of iron in soil. New York: American Gas Association 1945. — SVERDRUP, H. U., M. W. JOHNSON and R. H. FLEMING: The oceans. New York: Prentice-Hall 1946.

TARR, H. L. A.: The anaerobic decomposition of l-cystine by washed cells of Proteus vulgaris. Biochemic. J. 27, 759 (1933a). — The enzymic formation of hydrogen sulphide by certain heterotrophic bacteria. Biochemic. J. 27, 1869 (1933); 28, 192 (1934). — TARVER, H. L. A., and C. L. A. SCHMIDT: The conversion of methionine to cystine. Experiments with radioactive sulfur. J. of Biol. Chem. 130, 67 (1939). — TATUM, E. L.: Amino-acid metabolism in mutant strains of microorganisms. Federat. Proc. 8, 511 (1949). — TAYLOR, C. B., and G. H. HUTCHINSON: Corrosion of concrete caused by sulphur-oxidizing bacteria. J. Soc. Chem. Industr. (Lond.) 66, 54 (1947). — TRAUTWEIN, K.: Beiträge zur Physiologie und Morphologie der Thionsäurebakterien (OMELIANSKI). Zbl. Bakter. II 53, 513 (1921). — Die Physiologie und Morphologie der fakultativ autotrophen Thionsäurebakterien unter heterotrophen Ernährungsbedingungen. Zbl. Bakter. II 61, 1 (1924).

UMBREIT, W. W., H. R. VOGEL and K. G. VOGLER: The significance of fat in sulphur oxidation by Thiobacillus thiooxidans. J. Bacter. 43, 141 (1942). — UPHOF: 1927. Cité par BENECKE 1933.

VIRTUE, R. W., and M. E. DOSTER-VIRTUE: Studies on the production of taurocholic acid in the dog. J. of Biol. Chem. 127, 431 (1932). — VOGLER, K. G.: Studies on the metabolism of autotrophic bacteria. J. Gen. Physiol. 26, 103, 109 (1942). — VOGLER, K. G., and W. W. UMBREIT: Metabolism of autotrophic bacteria. J. Gen. Physiol. 26, 157, 159 (1942).

WAKSMAN, S. A., and J. S. JOFFE: Microorganisms concerned in the oxidation of the soil. II. Thiobacillus thiooxidans a new sulphur-oxidizing organism isolated from soil. J. Bacter. 7, 239 (1922). — WAKSMAN, S. A., and R. L. STARKEY: On the growth and respiration of sulphur oxidizing bacteria. J. Gen. Physiol. 5, 285 (1922). — WARBURG, O., and D. BURK: The maximum efficiency of photosynthesis. Arch. of Biochem. 25, 410 (1950). — WARMING: Cité par BENECKE 1933. — WINOGRADSKY, S.: Über Schwefelbacterien. Beiträge zur Morphologie und Physiologie der Bakterien. Leipzig 1888. — Recherches sur les organismes de la nitrification. Ann. Inst. Pasteur 4, 213 (1890). — Microbiologie du sol. Paris: Masson & Cie. 1949. — WOHLGEMUTH, J.: Über die Herkunft der schwefelhaltigen Stoffwechselprodukte im tierischen Organismus. Z. physiol. Chem. 43, 469 (1905).

ZÖRKENDÖRFER, W.: Über die Sulfit- und Sulfidbindung aus Sulfaten im Darm und ihr Anteil an der abführenden Wirkung schwefelsaurer Salze. Arch. exper. Path. u. Pharmakol. 161, 437 (1931). — Über die Ausscheidung des Thiosulfates und seine Bestimmung im Harn. Biochem. Z. 278, 191 (1935).

Der Stoffwechsel der P-haltigen Verbindungen. Übersicht.

Von

P. Schwarze.

Mit 1 Abbildung.

Als Pfeffer kurz vor der Jahrhundertwende sein Lehrbuch der Pflanzenphysiologie schrieb, war über die Rolle des Phosphors im Stoffwechsel noch wenig bekannt. „Dieser ist mit Rücksicht auf die Verkettung mit Proteinstoffen unter allen Umständen unentbehrlich. Jedoch ist unbekannt, inwieweit der Phosphor noch anderweitig Bedeutung im Organismus gewinnt" (Band I, S. 422). Bereits 1840 hatte Liebig gezeigt, daß der Phosphor dem Boden entnommen wird, und Sachs und Knop führten 1860 den Nachweis, daß er zu den unentbehrlichen Elementen gehört. Ein Vierteljahrhundert nach Pfeffers Lehrbuch erschien die „Biochemie der Pflanzen" von Czapek und wenig später das Lehrbuch der chemischen Physiologie von Kostytschew. Czapek behandelt die große Zahl der Arbeiten, die sich mit den damals bekannten P-Verbindungen (Phosphatiden, Phosphoproteinen, Nucleinsäuren und Phytin), den an ihrem Umsatz beteiligten Enzymen und mit der Aufnahme und Wanderung der Phosphorsäure befassen. Obwohl viele Ergebnisse dieser Arbeiten heute überholt sind, da sie mit unzulänglichen Methoden gewonnen wurden, ist Czapeks dreibändiges Werk als Fundgrube der älteren Literatur auch jetzt noch für den Physiologen von großem Nutzen. Gemessen an unserem heutigen Wissen sind zu dieser Zeit die Einblicke in den Stoffwechsel der P-Verbindungen noch gering und ist kaum etwas über ihre Bedeutung und Funktion im Zellstoffwechsel bekannt. Kostytschew bemerkt, daß Phosphor als Orthophosphorsäure dem Boden entnommen wird und „keine großen Umwandlungen" im Pflanzenkörper erfährt. Es ist bekannt, daß bei der Gärung Hexosediphosphorsäure entsteht und Stärke ein phosphorhaltiger Körper ist; „einige Verfasser nehmen an, daß solche Ester in den Pflanzen allgemein verbreitet sind". Im Phytin wird das erste Produkt der P-Assimilation vermutet. „Die hervorragende Bedeutung des Phosphors besteht folglich darin, daß er in den wichtigsten Plasmakolloiden enthalten ist und in Form von Phosphorestern des Zuckers eine wichtige Rolle bei der alkoholischen Gärung spielt ...". Daß heute dem P-Stoffwechsel ein wesentlicher Raum innerhalb des Handbuches der Pflanzenphysiologie zuerkannt werden muß und zudem keiner der Bände, die sich mit dem Stoffwechsel beschäftigen, auf die Behandlung wenigstens von Teilfragen des P-Stoffwechsels verzichten kann, bringt sehr überzeugend die in den seither vergangenen 25 Jahren erzielten Fortschritte zum Ausdruck.

Die Phosphorsäure ist, wie wir heute wissen, aus mehreren Gründen von fundamentaler Bedeutung für den Stoffwechsel. Die Anlagerung des Phosphorsäurerestes an organische Verbindungen erleichtert oder ermöglicht erst deren Umsetzung durch die Zelle. Beispiele dafür sind die Phosphorsäureester, wie sie im

Stoffwechsel der Kohlenhydrate, bei der alkoholischen Gärung, bei der Glykolyse, beim aeroben Kohlenhydratabbau und auch bei der Bildung der Kohlenhydrate im Chemo- und Photosyntheseprozeß entstehen. Eine Reihe der den Stoffwechsel katalysierenden Enzyme ist selbst P-haltig. Ein oder mehrere Phosphorsäurereste sind obligate Bausteine ihrer Coenzyme, z. B. in den Riboflavin- und Pyridin-nucleosiden, in der Cocarboxylase, im Coenzym A und im Pyridoxalphosphat. Da der enzymatische Vorgang mit einer Bindung zwischen Substrat und Enzym einsetzt, ist zu vermuten, daß in beiden Fällen der Wirkungsmechanismus der Phosphorsäure derselbe ist, nur wird dieser im ersten Fall durch das Substrat, im zweiten Fall durch das Enzym in den zerfallsbereiten Komplex eingebracht. Eine andere wichtige Funktion der Phosphorsäure besteht darin, daß mit Hilfe der Phosphatgruppe die bei bestimmten chemischen Umsetzungen freiwerdende Energie aufgefangen und zur Speisung der verschiedensten energiebedürftigen Reaktionen benutzt werden kann.

Die Speicherung der Phosphorsäure (III B.). Phosphor wird von der Pflanze als Orthophosphat, also wie der Schwefel in oxydierter Form aufgenommen. (Näheres in den Beiträgen von Robertson und Stiles, über die P-Quellen des Bodens und der Gewässer in den Beiträgen von Wiklander und Gessner in Band IV). Im Gegensatz zum Schwefel wird er im Stoffwechsel nicht reduziert, sondern erfüllt seine vielfältigen Funktionen immer als Ortho- oder Pyro-phosphorsäure. Am Stoffwechsel beteiligt sich die Phosphorsäure durch die Ausbildung von Phosphatestern, deren alkoholische Komponenten den verschiedensten Typen organischer Körper angehören. Diese Phosphorsäureester sind sehr umsatzbereit; durch Enzyme werden sie wieder zerlegt, wobei der Phosphatrest als anorganische Phosphorsäure in Freiheit gesetzt oder, was häufiger und physiologisch bedeutungsvoller ist, auf einen anderen organischen Körper übertragen wird.

P-Angebot und P-Bedarf sind naturgemäß aus äußeren und inneren Gründen nicht immer aufeinander abgestimmt. Verarmung des Bodens an Phosphorsäure löst die Symptome des P-Mangels aus, Angebot und Aufnahme über den Bedarf hinaus können die Pflanze zur Speicherung von Phosphorsäure veranlassen. Als Speicherformen werden die schon vor längerer Zeit in Pflanzen aufgefundenen Polyphosphate und das Phytin aufgefaßt.

Polyphosphate, zuerst von Liebermann (1888) in der Hefe entdeckt, finden sich in vielen Mikroorganismen, bestimmten Bakterien, verschiedenen Formen der Hefe und anderen Pilzen, Algen und wahrscheinlich auch in Protozoen. Es besteht heute nach den Untersuchungen von Wiame die Auffassung, daß die als Volutin (vgl. den Beitrag von Steffen[1] in Band I) bezeichneten Granula im wesentlichen aus Polymetaphosphaten bestehen. In der Hefe bilden sie sich bei reichlicher Nährstoff- und Phosphatzufuhr, um bei Energie- und Phosphatmangel wieder abzunehmen, was in der Tat dafür spricht, daß es sich um Phosphat- und Energiereserven handelt (Hoffmann-Ostenhof und Weigert 1952). Stich (1953) berichtete, daß *Acetabularia* Polymetaphosphate während der Photosynthese bildet.

Phytin ist das Calcium-Magnesiumsalz der Inosit-hexaphosphorsäure. Die Häufung der Phosphorsäure in seinem Molekül legte die Vermutung nahe, daß es sich um eine Reserveform des Phosphors handelt. Der hohe Phytingehalt vieler Samen, besonders der Cerealien, und das gleichzeitige Vorkommen der Phytase, die bei der Samenkeimung die Phosphorsäure abspaltet und für den

---

[1] Steffen, K.: Einschlüsse.

Stoffwechsel verfügbar macht, stützen diese Vermutung. ALBAUM und UMBREIT (1943) stellten fest, daß bei der Keimung von Hafer die Aufspaltung des Phytins mit der Bildung von Glycerinphosphorsäuren, Fructosediphosphat und Adenosintriphosphat konform geht. Im Getreidekorn findet sich Phytin in großer Konzentration in der Aleuronschicht und im Scutellum. Aber auch in anderen Pflanzenteilen, in Rhizomen, Knollen, Blättern und Pollenkörnern der verschiedensten Familien, ist es nachgewiesen worden. Die alkoholische Komponente des Phosphorsäureesters Phytin ist der Myoinosit (Mesoinosit, Bios I), ein cyclischer Alkohol, der in allen darauf untersuchten Pflanzen festgestellt wurde. Trotz der engen konstitutionellen Beziehungen zur d-Glucose war es noch nicht möglich, seine Biosynthese aufzuklären (vgl. den Beitrag von BALLOU[1] in Band X dieses Handbuches).

Phosphatide (III C; ausführlichere Behandlung in Band VII[2]). Die Phosphatide sind wesentliche Bestandteile des Protoplasmas. Sie wurden als Bauelemente des Kernes, der Plastiden, der Mitochondrien und Sphärosomen erkannt und fehlen auch nicht im strukturlosen Hyaloplasma der Zelle, wie in den Beiträgen von SERRA[3], GRANICK[4] und STEFFEN[5] im Band I dieses Handbuches dargelegt wird. Die Plasmaoberflächen bestehen zum großen Teil aus Phosphatiden (Beitrag von SEIFRIZ[6], Band I dieses Handbuches), und auch in Vacuolen und in den gelegentlich im Plasma auftretenden Lipoidtropfen (Beitrag STEFFEN[7], Band I) hat man diese Stoffe festgestellt.

Im Gegensatz zur Mehrzahl der P-Verbindungen, die als Bestandteile von Enzymen, als Energiespeicher und Energiespender und als reaktionsfähige Kohlenhydratester Träger der Dynamik des Stoffwechsels sind, müssen in den Phosphatiden wohl in erster Linie Bauelemente der Plasmastrukturen, insbesondere der die Reaktionsräume begrenzenden Lamellen und Membranen, an denen der Stoffwechsel abläuft, gesehen werden.

Obligate Bauelemente aller Phosphatide sind Fettsäuren und Orthophosphorsäure. Die „klassischen" Phosphatide Lecithin und Kephalin enthalten außerdem Glycerin und die Basen Cholin bzw. Colamin. Beide sind in tierischen und pflanzlichen Organismen aller Organisationsstufen anzutreffen. Noch nicht aufgefunden wurden in Pflanzen Diaminophosphatide vom Typ des Sphingomyelins, und nur in zwei Fällen wird über das Vorkommen von Acetalphosphatiden oder Plasmalogenen berichtet. Letztere wurden im Rohphosphatid der Sojabohne und im Olivenöl festgestellt.

Phosphatide ungewöhnlicher und oft sehr komplexer Zusammensetzung kommen in Bakterien vor. Bei Mycobakterien finden sich fast N-freie, aus Fettsäuren und phosphorylierten Kohlenhydraten aufgebaute Phosphatide, deren Kohlenhydratkomponente aus Mannose und Inosit, in anderen Fällen aus Mannose und Glycerin oder aus Inosit und Glycerin zusammengesetzt ist. Mehrfach untersucht wurde mit widersprechenden Ergebnissen die abnorm gebaute Phosphatidfraktion des Diphtheriebacteriums, unter deren Hydrolyseprodukten unter anderen Corynolsäure, eine mehrfach methylierte und hydroxylierte Fettsäure mit ungewöhnlich langer C-Kette ermittelt wurde. Die Phosphatide anderer Bakterien, z. B. von Salmonella-Arten, enthalten an Stelle von Cholin und

---

[1] BALLOU, C. E.: Inositol and related compounds (cyclitols).
[2] LOVERN, J. A.: The phosphatids and glycolipoids.
[3] SERRA, I. A.: Physical chemistry of the nucleus.
[4] GRANICK, S.: Plastid structure, development and inheritance.
[5] STEFFEN, K.: Chondriosomen und Mikrosomen (Sphärosomen).
[6] SEIFRIZ, W.: The physical chemistry of cytoplasma.
[7] STEFFEN, K.: Einschlüsse.

Äthanolamin Aminosäuren. Bemerkenswert ist der in *Micrococcus pyogenes* in großer Menge vorliegende Phospholipid-proteinkomplex.

Hefe, *Saccharomyces cerevisiae* und *Torula utilis*, die beide eingehend untersucht wurden, enthalten Lecithin und Kephalin. Auch bei einer Reihe anderer Pilze, z. B. *Aspergillus sydowi*, scheinen normal gebaute Phosphatide vorzukommen oder wenigstens vorzuherrschen, lediglich bei *Penicillium chrysogenum* wurde unter den Hydrolyseprodukten der Kephalinfraktion Inosit gefunden.

In Samen, aber auch in anderen Geweben höherer Pflanzen kommen überwiegend Lecithine und Kephaline vor, von denen sich letztere in einigen Fällen als sehr komplex erwiesen haben. Die Sojabohne enthält mindestens zwei inosithaltige Kephaline; im Hydrolysat eines dieser Kephaline wurden außerdem Galaktose, Mannose und Arabinose festgestellt. In einem Kephalin der Erdnuß war die N-Base ein N-Glykosylderivat des Colamins, weitere Komponenten waren Inosit und die Zucker Arabinose und Galaktose. Die Phosphatidsäuren, über deren Vorkommen eine Reihe von Autoren berichten, sind sicherlich Artefakte, welche durch autolytische Vorgänge während der Extraktion entstehen.

Schon diese wenigen Beispiele zeigen eine bedeutende Variabilität des chemischen Aufbaues der Phosphatide, ein Faktum, das ebenfalls als Hinweis auf die Rolle der Phosphatide als strukturbildendes Element aufgefaßt werden kann; denn ein und dieselbe Struktur kann aus ähnlichen Baustoffen errichtet werden, wenn es auch nicht ausgeschlossen ist, daß Beziehungen zwischen den verschiedenen Stoffwechselvorgängen und dem chemischen Aufbau der sie beherbergenden Strukturen bestehen.

Über den Phosphatidumsatz während der Entwicklung der Pflanze ist kaum etwas bekannt. In den verschiedensten Geweben wurden Enzyme festgestellt, die Phosphatide angreifen, Cholin aus Lecithin abspalten und die Phosphatidsäuren weiter in Glycerin, Fettsäuren und Phosphorsäure zerlegen. Bei Bakterien wurden interessante Unterschiede in der Reihenfolge der Abspaltung der einzelnen Komponenten festgestellt, wobei die verschiedensten Zwischenprodukte auftreten können. Interessant ist zweifellos die Beobachtung, daß bei der Keimung die Neutralfette sehr rasch in Kohlenhydrate umgewandelt werden, während sich der Gehalt an Phosphatiden verdoppelt. Eine Deutung dieses Befundes ist zur Zeit noch nicht möglich. Im ganzen gesehen sind die Erkenntnisse über den Bau und den Stoffwechsel der pflanzlichen Phosphatide noch recht begrenzt, was vor allem darauf beruht, daß die Isolierung der Phosphatide aus den Geweben Schwierigkeiten bereitet. Meist sind die Phosphatide mit Proteinen oder Kohlenhydraten und vielleicht auch anderen Stoffen fest verbunden. Um diese Bindung zu lösen, sind Maßnahmen erforderlich, die häufig auch die Phosphatide in unkontrollierbarer Weise verändern, sei es, daß Bestandteile abgespalten oder nicht zum eigentlichen Phosphatid gehörige anhaftende Verbindungen mitextrahiert werden. Hinzu kommt, daß Phosphatide die Löslichkeit der verschiedensten Stoffe in unkontrollierbarer Weise verändern, was ebenfalls die Reindarstellung sehr erschwert.

Die Phosphorsäureester des Kohlenhydratstoffwechsels (III D). Die von den englischen Biochemikern Harden und Young zu Beginn des Jahrhunderts bekanntgegebene Beobachtung, daß phosphorylierte Zucker am Kohlenhydratumsatz teilnehmen, hat eine Fülle von Untersuchungen ausgelöst, die diesen Zweig des Stoffwechsels weitgehend klargelegt haben. Eines der wesentlichen Ergebnisse ist die Erkenntnis, daß die Organismen im Laufe ihrer Evolution mehrere Möglichkeiten des Kohlenhydratabbaus entwickelt haben, die auf verschiedene Organismenarten verteilt sind oder in ein und demselben Organismus nebeneinander oder zu verschiedenen Zeiten ablaufen können. Bei all diesen

Arbeiten haben Mikroorganismen eine wesentliche Rolle gespielt, und viele der grundlegenden Erkenntnisse sind gerade an ihnen gewonnen worden. Sie waren bahnbrechend für die Untersuchung des tierischen Stoffwechsels, und die Erkenntnisse an Mikroorganismen und Tieren kamen dem Studium des Kohlenhydratstoffwechsels der höheren Pflanze zugute, das wegen der diffizileren Untersuchungsmethodik mit den Arbeiten an jenen anderen Lebewesen nicht Schritt halten konnte.

Die Befunde an Mikroorganismen werden im vorliegenden Band nur überblicksmäßig wiedergegeben (UMBREIT), da ihre ausführliche Behandlung dem Band XII dieses Handbuches, „Atmung, einschließlich Gärungen und Säurestoffwechsel", vorbehalten bleiben mußte. In der eingehenderen Darstellung der höheren Pflanzen (ALBAUM) liegt das Gewicht auf der Beschreibung der an ihrem Kohlenhydratumsatz beteiligten phosphorylierten Zucker und der Enzyme, welche Bildung und Umsetzung dieser Ester vermitteln.

Die Untersuchungen an Mikroorganismen haben ergeben, daß zwei wesentlich voneinander verschiedene Abbauwege anzutreffen sind: die nach dem Embden-Meyerhof-Schema vor sich gehende Glykolyse, gekennzeichnet durch die Reaktionsfolge Fructose-6-phosphat $\rightarrow$ Fructose-1,6-diphosphat $\rightarrow$ Triosephosphat, und der erst in neuerer Zeit näher untersuchte oxydative Abbau, für den die Oxydation von Glucose-6-phosphat zu 6-Phosphogluconsäure charakteristisch ist. Alle anderen Reaktionen sind nicht spezifisch für die eine oder andere Form des Abbaues. Glykolyse findet bei bestimmten Milchsäurebakterien und anaeroben und aeroben Sporenbildnern statt. Ferner sind dazu die Gewebe der Tiere und der höheren Pflanzen befähigt.

Wesentlich häufiger konnte bei Mikroorganismen der oxydative Abbau nachgewiesen werden. Bei einer kleinen Gruppe wird die Phosphogluconsäure über 2-Keto-3-desoxy-6-phosphogluconsäure in Brenztraubensäure und Glycerinaldehyd überführt, während es bei der Mehrzahl der Mikroorganismen zu einer Oxydation der 6-Phosphogluconsäure zu 3-Ketophosphogluconsäure kommt, aus der schließlich durch CO_2-Abspaltung Ribulose-5-phosphat entsteht. Dieser Ester ist eine Schlüsselsubstanz des Kohlenhydratstoffwechsels und kann auf die verschiedenste Weise weiter verarbeitet werden. Als Umwandlungsprodukte des Ribulosephosphats wurden erkannt: Ribosephosphat, Glycerinaldehyd-3-phosphat, aktiver Glykolaldehyd (eine Verbindung des Glykolaldehyds mit Thiaminpyrophosphat), Heptulosediphosphat, Fructose-6-phosphat, L-Erythrulosephosphat und Heptulosediphosphat, das für eine Vorstufe der Shikimisäure gehalten wird. Dieser oxydative Abbau über Pentosen, auch Pentosecyclus genannt, kommt außer bei Mikroorganismen auch in tierischen Geweben und in höheren Pflanzen vor (BEEVERS und GIBBS 1954). Hier wurde er z. B. in Gersten-, Zuckerrüben-, Tabak- und Spinatblättern, in der Mohrrübe und in Erbsen- und Maisinternodien nachgewiesen. Eine Reihe von Beobachtungen spricht dafür, daß in jungen Geweben, vor allem in Meristemen, die Glykolyse, in älteren Geweben die direkte Oxydation vorherrscht. In jedem Fall beginnt der Zuckerabbau mit einer Phosphorylierung und bilden sich bis zum Anfall der Brenztraubensäure, die entsprechend der Organisation der Lebewesen in mannigfacher Weise verarbeitet wird, eine ganze Reihe phosphorylierter Zwischenprodukte.

Die den Kohlenhydratabbau einleitende Phosphorylierung vermittelt das Adenosintriphosphat (ATP), dessen endständiger Phosphorsäurerest zusammen mit der in der Pyrophosphatbindung gelegenen Energie auf den Zucker übertragen wird. Die Erzeugung neuer energiereicher Verbindungen ist neben der Bereitstellung von C_3-Bausteinen für die verschiedensten Synthesen Sinn und Ziel des Kohlenhydratabbaues. Bis zur Brenztraubensäure werden im Fall der

Glykolyse nur drei Moleküle ATP, bei vollständiger Verbrennung bis zu CO_2 und Wasser aber etwa 36 Moleküle ATP je Molekül Hexose gebildet.

Seit 1944 sind Untersuchungen über die Beteiligung von Phosphorylierungsvorgängen an der Photosynthese im Gang. Anregungen dazu gaben Befunde an *Thiobacillus thiooxydans*. Vogler erkannte, daß bei diesem Mikroorganismus der energieliefernde Vorgang, die Schwefeloxydation, von der CO_2-Reduktion zeitlich zu trennen ist und die Übertragung der Energie vom energieliefernden auf den energieverbrauchenden Prozeß mit einer reversiblen Phosphatbindung einhergeht. Einen analogen Mechanismus der Energieübertragung nahmen Emerson, Stauffer und Umbreit (1944) auch für die Photosynthese an. Kandler stellte 1950 Phosphatbindung durch *Chlorella* bei Belichtung fest, und Strehler (1953) wies Adenosintriphosphat als Photosyntheseprodukt nach. Ein weiterer wichtiger Schritt in der Erforschung der Phosphorylierungsvorgänge bei der Photosynthese ist der Nachweis des Phosphorylierungsvermögens isolierter Spinatchloroplasten und *Spirogyra*-Chloroplastenfragmente durch Arnon und Mitarbeiter (1954) bzw. Thomas und Haans (1955). In einer Reihe von Untersuchungen haben Simonis und Mitarbeiter die lichtabhängige Phosphorylierung analysiert. Die Gesamtheit der vorliegenden Befunde veranlaßt sie, vier verschiedene Arten der Beeinflussung des Phosphatstoffwechsels durch das Licht zu unterscheiden:

1. Die Photosynthese-Phosphorylierung, unter der die mit den Primärvorgängen der Photosynthese zusammenhängenden Phosphorylierungsprozesse verstanden werden,

2. die Beeinflussung des Phosphatstoffwechsels, die im Verlauf der Folgeprozesse der Photosynthese auftritt,

3. die „photosensibilisierte Phosphorylierung" und

4. die wahrscheinlich lichtabhängige Phosphataufnahme.

Unter der „photosensibilisierten Phosphorylierung" verstehen Simonis und Ehrenberg (1957) die von ihnen nachgewiesene Förderung der Phosphateinlagerung durch Licht bei chlorophyllfreien, nicht zur Photosynthese befähigten Objekten (Wurzeln von Gerstenkeimlingen, Hefe). Näheres über diesen Typ der Phosphorylierung wissen wir noch nicht, es ist jedoch damit zu rechnen, daß sie am komplexen Vorgang der lichtabhängigen Phosphorylierung grüner Pflanzen beteiligt ist.

Neben Adenosintri- und Adenosindiphosphat sind weitere P-Verbindungen an der Photo- und ebenso an der Chemosynthese beteiligt. Es handelt sich um dieselben Typen von Verbindungen und reversiblen Reaktionen, wie sie auch beim Kohlenhydratabbau vorkommen, wenn die Photosynthese auch nicht, wie zunächst vermutet wurde, der Umkehr des glykolytischen Abbaues und des Citronensäurecyclus gleichkommt. Insbesondere auf Grund von Versuchen mit radioaktivem CO_2 ($^{14}CO_2$) haben Calvin und Mitarbeiter (s. Calvin 1956) ermitteln können, welche C-Verbindungen an der Photosynthese beteiligt sind. Nach kurzfristiger Darbietung von radioaktivem CO_2 wurde untersucht, welche Verbindungen radioaktiv sind und wie sich die Aktivität auf die einzelnen C-Atome verteilt. Die Befunde gaben die Möglichkeit, ein Schema des Photosynthesecyclus zu entwickeln (Abb. 1). Es enthält eine stattliche Zahl von phosphorylierten C-Verbindungen mit 3, 5, 6 und 7 C-Atomen. Unter Anlagerung an Ribulosediphosphat, einer Carboxylierungs- oder CO_2-Fixierungsreaktion, tritt CO_2 in den Photosynthesecyclus ein. Alle Reaktionen des Cyclus konnten in vitro durchgeführt und die daran beteiligten Fermente nachgewiesen oder isoliert werden. Der Ablauf des Cyclus erfordert reduzierendes Agens und

Adenosintriphosphat (ATP), und zwar 4 bzw. 3 Moleküle für den Einbau von 1 Molekül CO_2 in Triosephosphat. Die Energie für die Bildung des reduzierenden Agens (TPNH oder DPNH aus TPN bzw. DPN), die vermutlich über das Liponsäuresystem erfolgt, und von ATP aus ADP wird von der photochemischen Reaktion geliefert (Näheres s. Band V, Abschnitt III).

P-haltige Coenzyme (III E). Vor 50 Jahren fanden HARDEN und YOUNG, daß die Gärkraft von Hefeextrakt durch Phosphat gesteigert wird und die Gärung nur vor sich geht, wenn neben der thermolabilen Zymase, wie zunächst der Komplex der Hefefermente genannt wurde, ein dialysabler und hitzestabiler Faktor, die Cozymase, zugegen ist. Da, wie sich später herausstellte, viele

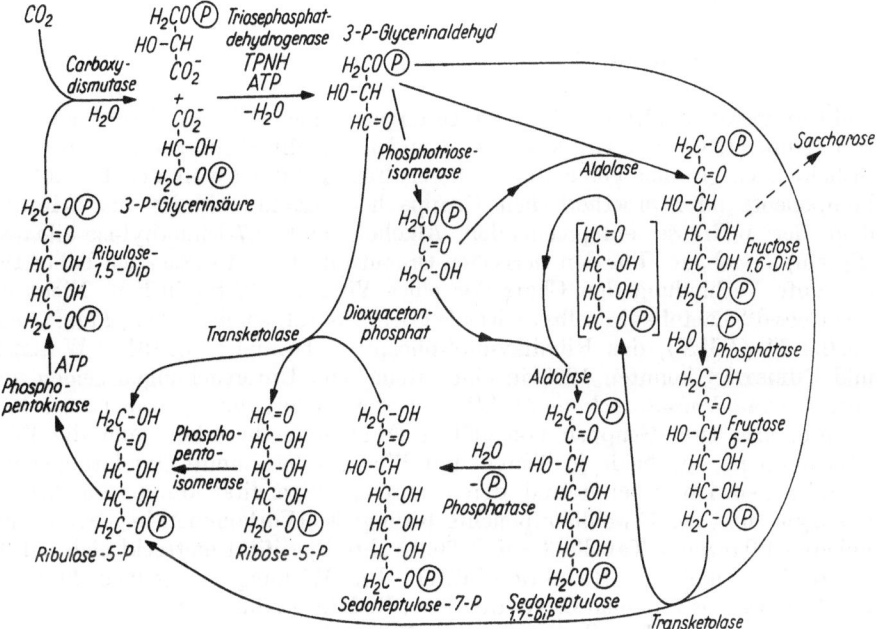

Abb. 1. Schema des Photosynthesecyclus. [Nach M. CALVIN: Angew. Chem. **68**, 253 (1956).]

Enzyme aus zwei Komponenten bestehen, wurde der Nichtproteinanteil allgemein als Coenzym und die Proteinkomponente als Apoenzym bezeichnet. Die Isolierung der Cozymase und die Aufklärung ihrer Struktur ist an die Namen SCHLENK und v. EULER (1936) und WARBURG und CHRISTIAN (1936) geknüpft. Letztere entdeckten einen zweiten ganz ähnlichen Faktor in Pferdeerythrocyten. Beide Coenzyme sind Nicotinsäureamid-adenin-dinucleotide und unterscheiden sich nur dadurch voneinander, daß die Cozymase zwei und das Warburgsche Coenzym drei Moleküle Phosphorsäure enthält (Formeln s. S. 178). Nach WARBURG und CHRISTIAN (1936) werden sie als *Diphosphopyridinnucleotid (DPN) und Triphosphopyridinnucleotid (TPN)* bezeichnet. In den Pyridinnucleotiden entfaltet das Vitamin Nicotinsäure (oder Nicotinsäureamid) seine physiologische Wirkung. Sie kommen wahrscheinlich in allen pflanzlichen und tierischen Organismen vor und können als die prosthetischen Gruppen der Pyridinenzyme, einer Gruppe von Dehydrogenasen, aufgefaßt werden. Die Wirkungsspezifität, die Fähigkeit, Wasserstoff aufzunehmen und wieder abzugeben, ist an das Coenzym, die Substratspezifität an die Eiweißkomponente gebunden. Der chemische Wirkungsmechanismus der Pyridin-Coenzyme besteht in der reversiblen Reduktion des

Pyridinringes. Bisher sind etwa 50 verschiedene Pyridindehydrasen aufgefunden worden, die an den Dehydrierungen des Kohlenhydratabbaues teilnehmen, aber auch eine wichtige Rolle bei der Photosynthese spielen. Hier werden sie als Acceptoren für den bei der Photolyse des Wassers entstehenden Wasserstoff eingesetzt. Über den Stoffwechsel der Pyridinnucleotide, einschließlich Biosynthese und Abbau der Nicotinsäure, liegen eine Reihe von Befunden vor, über die in diesem Band berichtet wird. Ihre Funktion im Photosyntheseprozeß wird im Band V, ihre Rolle beim Kohlenhydratabbau im Band XII[1] behandelt.

Auf das *Thiaminpyrophosphat*, das in der Thiazolkomponente Schwefel enthält, wurde bereits in der Übersicht über die S-haltigen Verbindungen (S. 3) kurz eingegangen. Der Beitrag von O'Brien in diesem Band befaßt sich mit dem Stoffwechsel (Bildung und Abbau) dieses Coenzyms.

Weitere oxydierende Enzyme, deren prosthetische Gruppen Phosphorsäure enthalten, sind die gelben Fermente. Ihre Entdeckung verdanken wir Warburg und Christian (1932); grundlegende Untersuchungen über ihre Struktur wurden von Kuhn (1935) und Karrer (1935) und ihren Mitarbeitern ausgeführt. Die rötlich- oder grünlich-gelbe Farbe dieser Enzyme rührt von der Lactoflavinkomponente ihrer prosthetischen Gruppe her. Lactoflavin ist aus D-Ribitol, dem der D-Ribose entsprechenden Alkohol, und 6,7-Dimethyl-iso-alloxazin (Flavin) aufgebaut. Für den tierischen Organismus besitzt diese auch Riboflavin genannte Verbindung den Charakter eines Vitamins (Vitamin B_2). Nicht das „Nucleosid" Lactoflavin selbst tritt als Coenzym auf, sondern das „*Flavinmononucleotid*" *(FMN)*, das Riboflavin-5'-phosphat (Formel s. S. 191). Warburg und Christian konnten 1938 in einer Reihe von Untersuchungen zeigen, daß auch *Flavin-adenin-dinucleotid (FAD)* als Coenzym vorkommt (Formel auf S. 191). Es gibt also zwei Gruppen von gelben Fermenten, die FMN- und die FAD-Flavoproteine, die beide in Tieren und Pflanzen bestimmte Oxydationen und Reduktionen katalysieren und deren jeweilige Spezifität durch die Art des Coenzyms und der Eiweißkomponente bedingt ist (Vorkommen in niederen und höheren Pflanzen s. Tabelle 3 auf S. 200). Ihre Spezifität erstreckt sich auf das Oxydations- und das Reduktionsmittel. Die Wirkung der gelben Fermente beruht auf der reversiblen Oxydation bzw. Reduktion ihrer Wirkgruppe, genauer der N-Atome des Iso-alloxazinringes. Nach Untersuchungen von Theorell und Nygaard (1954) vermittelt die Phosphorsäure die Bindung des Coenzyms an das Apoenzym, indem der Phosphatrest mit NH_2-Gruppen der Eiweißkomponente reagiert. Die Funktion der gelben Fermente bei der Atmung der Pflanzen wird im Band XII[2] diskutiert.

An vermutlich allen biologischen Acetylierungen ist das Coenzym A beteiligt, dessen Struktur nach wesentlichen Vorarbeiten durch Lynen u. a. Gregory, Novelli und Lipmann (1952) aufklären konnten (Formel s. S. 211). Als Bausteine wurden Pantothensäure, Adenin, Ribose, β-Mercaptoäthylamin (Thioäthanolamin), Pyrophosphorsäure und Phosphorsäure festgestellt. Letztere läßt sich durch eine Phosphomonoesterase abspalten, und eine Nucleotidpyrophosphatase zerlegt das Coenzym in Diphosphoadenosin und Phosphopantethein, die beide enzymatisch unwirksam sind. Das aus Pantothensäure und Thioäthanolamin aufgebaute Pantethein erwies sich als der schon länger bekannte, als LB-Faktor (LBF) bezeichnete Wuchsstoff für *Lactobacillus bulgaricus*. Von wesentlicher Bedeutung für die Wirkungsweise des Coenzyms ist die SH-Gruppe der Thioäthanolkomponente, bei deren Acetylierung Acetylcoenzym A, das aktivierte Acetat (Lynen und

---

[1] Hasse, K.: Pyridin-Dehydrasen.
[2] Meeuse, B. J. D.: The flavin enzymes.

Mitarbeiter 1951), gebildet wird. Die Thioesterbindung ist energiereich (Co A-S $\sim$ COCH$_3$); ihre freie Energie entspricht größenordnungsmäßig der der Pyrophosphatbindung. Wie ATP die Rolle eines Phosphatdonators bei Phosphorylierungen spielt, kommt dem Acetylcoenzym A die Funktion eines Acetyldonators bei physiologischen Acetylierungen zu. Es entsteht bei der oxydativen Decarboxylierung der Brenztraubensäure und kann auch aus freiem Acetat durch Reaktion mit Coenzym A in Anwesenheit von ATP synthetisiert werden. Acetyliertes Coenzym A ist ein Endprodukt der Fettsäureoxydation und unter anderen Bedingungen auch das Ausgangsmaterial für die Fettsäuresynthese. Es liegen, worauf bereits hingewiesen wurde und im Band X dieses Handbuches ausführlich eingegangen wird, Beweise dafür vor, daß bei vielen Organismen vom Acetyl-Coenzym A auch die Synthese der Carotinoide und Sterine ausgeht. Mit der Synthese des Kautschuks, bestimmter ätherischer Öle und anderer Stoffe mit geraden und verzweigten Ketten hat die Pflanze weitere Spielarten der Synthese mit Hilfe von acetyliertem Coenzym A „erfunden". Im Band XII dieses Handbuches wird das Acetylproblem im Rahmen der partiellen Oxydation von Kohlenhydraten und Kohlenhydratspaltprodukten behandelt[1].

Das Coenzym Pyridoxalphosphat (Formel s. S. 219) ist die physiologische Wirkungsform des Vitamins B$_6$. Pyridoxin, Pyridoxal, Pyridoxamin und ihre Phosphate besitzen Vitamincharakter und werden von den meisten zur Synthese dieses Vitamins nicht befähigten Organismen in Pyridoxalphosphat umgewandelt. Es wurde als das Coenzym von Bakteriendecarboxylasen, von Transaminasen, von Tryptophanase und einer Reihe anderer Enzyme erkannt. Bei der Transaminierung wirkt das Coenzym vermutlich als Acceptor und Donator der zu übertragenden Aminogruppe. Die Coenzymfunktion des Pyridoxalphosphats wurde erst vor reichlich einem Jahrzehnt von GUNSALUS, BELLAMY und UMBREIT (1944) entdeckt und näher untersucht.

Am P-Stoffwechsel beteiligte Enzyme (III F). Vor 50 Jahren wurden von SUZUKI, YOSHIMURA und TAKAISHI (1906) die ersten am P-Stoffwechsel beteiligten Enzyme, die Phosphatasen, entdeckt. Diese Enzyme zerlegen Phosphorsäureester in eine alkoholische Komponente und anorganische Phosphorsäure. Seither sind die Phosphatasen in vielen Pflanzen aufgefunden worden, doch haben die meisten Arbeiten, die ihnen gewidmet sind, nur begrenzten Wert, da häufig unreine Präparate, also Enzymgemische untersucht wurden und die Frage der Spezifität oft unbeachtet blieb. Trotz der Fülle der Untersuchungen ist auch ihre Funktion keineswegs klar. Auf eine ausführliche Darstellung der Phosphatasen wird in diesem Band wegen ihrer undurchsichtigen stoffwechselphysiologischen Rolle und der Mängel, die vielen Untersuchungen anhaften, verzichtet. Ihre physikalisch-chemischen Eigenschaften, ihre verschiedenen Arten und ihr Vorkommen werden z. B. von ROCHE (1950) ausführlich dargestellt.

Wesentlich besser sind wir über die übertragenden Enzyme informiert, jene Enzyme, welche die Phosphorylgruppe aus einer organischen Verbindung auf einen anderen organischen Acceptor übertragen:

$$RO \cdot \textcircled{P} + R'{-}OH \rightleftharpoons R{-}OH + R'O \cdot \textcircled{P}.$$

Die Zerlegung eines Esters durch eine Phosphatase kann als Sonderfall einer solchen Übertragung aufgefaßt werden; statt von einer alkoholischen organischen Verbindung wird hier die Phosphorylgruppe von Wasser übernommen:

$$RO \cdot \textcircled{P} + H{-}OH \rightleftharpoons R{-}OH + HO \cdot \textcircled{P}.$$

---

[1] DECKER, K., und F. LYNEN: Das Acetylproblem.

Da beiden Reaktionen die Übertragung der Phosphorylgruppe gemeinsam ist, können auch ihre Katalysatoren, wie es im Abschnitt III F geschieht, als übertragende Enzyme, als Transphosphorylasen oder Phosphotransferasen, bezeichnet werden. Daß es sich um eine Übertragung der Phosphorylgruppe handelt, ist für eine Reihe von Enzymen tierischer Herkunft und für einige Enzyme der Hefe nachgewiesen worden. Die Spaltung eines Kohlenhydratesters z. B. erfolgt zwischen dem P-Atom und dem die Bindung zum organischen Rest herstellenden Sauerstoffatom, und der Anlagerung der P-haltigen Gruppe geht die Abspaltung einer Hydroxylgruppe aus der Phosphorsäure voraus.

AXELROD (1947) hat gezeigt, daß eine Reihe bisher nur für Hydrolasen gehaltener Phosphatasen auch Übertragungen im engeren Sinn katalysieren können. Im Saft der Navel-Apfelsinen kommen Phosphatasen dieses Typs vor, und sie finden sich auch in Saft und Schalen anderer Citrusfrüchte, im Apfel, in der Luzerne und im Tabak. Nach AXELROD sind Phosphatasen mit der Fähigkeit zur Hydrolyse und Übertragung der Phosphorylgruppe auf einen organischen Acceptor allgemein im Pflanzenreich verbreitet. Eine Untersuchung der meisten beschriebenen Phosphatasen unter diesem neuen Gesichtspunkt steht jedoch noch aus.

Übertragungen der Phosphorylgruppe zwischen zwei organischen Verbindungen auf energetisch hohem Niveau werden durch die Transphosphatasen im engeren Sinn katalysiert. Nach der Hexokinase, dem schon lange bekannten Vertreter dieser Enzymgruppe, werden sie auch als Phosphokinasen bezeichnet. Es muß angenommen werden, daß sie in jeder Zelle vorkommen. In allen von ihnen katalysierten Übertragungsreaktionen ist das ATP/ADP-System als Coenzym beteiligt. Die Mehrzahl der Phosphotransferasen überträgt eine Phosphorylgruppe des ATP, das dabei selbst in ADP übergeht, auf organische Acceptoren. Beispiele sind die 3-Phosphorylkinase, die 3-Phosphoglycerinsäure reversibel in 1,3-Diphosphoglycerinsäure überführt, und die Hexokinase, deren Funktion in der Phosphorylierung der Glucose und einiger anderer Hexosen mit Hilfe von ATP besteht. Die Adenylkinase lagert zwei Moleküle ADP in je ein Molekül ATP und AMP um, eine Reaktion, die, wie die meisten Übertragungen dieser Art, reversibel ist. Andere übertragende Enzyme sind die Mutasen, z. B. die Phosphoglucomutase, die Glucose-1-phosphat in Glucose-6-phosphat umwandelt, wobei Glucose-6-diphosphat die Rolle eines Coenzyms spielt.

Auf energetisch niedrigem Niveau vollziehen sich die ohne ATP/ADP ablaufenden Übertragungen der Phosphorylgruppe (low energy transfers). Sie werden von Enzymen vom Typ der Citrusphosphatase katalysiert, die Hydrolasen und Transferasen gleichzeitig sind.

Zu den übertragenden Enzymen gehören auch die in Band VI eingehend behandelten Phosphorylasen[1]. Sie katalysieren die Aufspaltung und Synthese komplexer Kohlenhydrate durch Aufnahme bzw. Abspaltung von Phosphorsäure und ähneln damit formal den Hydrolasen.

$$R\text{---}R' + HO \cdot \textcircled{P} \underset{\text{Dephosphorolyse}}{\overset{\text{Phosphorolyse}}{\rightleftarrows}} RO \cdot \textcircled{P} + HR'$$

$$R\text{---}R' + HOH \underset{\text{Kondensation}}{\overset{\text{Hydrolyse}}{\rightleftarrows}} ROH + HR'.$$

---

[1] GOTTSCHALK, A.: The enzymes controlling hydrolytic, phosphorolytic and transfer reactions of the oligosaccharides, S. 86—124. — HASSID, W. Z.: The synthesis and transformations of the oligosaccharides in plants, including their hydrolysis, S. 125—136. — WHELAN, W. I.: Starch and similar polysaccharides, S. 154—240. — GOERDELER, I.: Glykogenasen, S. 241—251.

Gut bekannt ist die Rohrzuckerphosphorylase, welche bei bestimmten Bakterien die Synthese dieses wichtigen Disaccharids aus Glucose-1-phosphat und freier Fructose katalysiert und auch seine Spaltung vermittelt (HASSID und DOUDOROFF 1950). Die Energie für die Synthese, die Glykosidierung, liegt in der Phosphatbindung des Glucose-1-phosphats und stammt letzten Endes aus dem ATP. Durch Vermittlung der Hexokinase wird vom ATP ein Phosphorylrest auf Glucose übertragen und das zunächst entstandene Glucose-6-phosphat lagert sich unter dem Einfluß von Phosphoglucomutase zu Glucose-1-phosphat um. Rohrzuckerphosphorylase ist absolut spezifisch bezüglich des Glucoserestes, aber nur gruppenspezifisch bezüglich des Acceptors. Die Fructose kann durch andere Ketohexosen, durch Ketopentosen und sogar eine Aldopentose vertreten werden.

Die Saccharosephosphorylase erweist sich bei der Neuknüpfung glykosidischer Bindungen, wie bei der Saccharosesynthese, als echte Phosphorylase, sie tritt aber außerdem als Transglucosidase auf und vermittelt ohne die Beteiligung von Glucose-1-phosphat oder anorganischem Phosphat die Übertragung des Glucoseanteils von einem Glucosid auf verschiedene Acceptoren. Die Bildung von Glucosesorbid aus Saccharose und Sorbose ist ein Beispiel dafür:

$$\text{D-Glucose-1-fructosid} + \text{L-Sorbose} \xrightleftharpoons{\text{Transglucosidase}} \text{D-Glucose-1-sorbosid} + \text{Fructose}.$$

Diese Übertragbarkeit des Glucoserestes, für welche die energiereiche Glykosidbindung die Voraussetzung ist, macht die Saccharose zu einer Schlüsselsubstanz des gesamten Stoffwechsels.

In der höheren Pflanze wird die Rohrzuckersynthese nicht wie bei den genannten Bakterien durch eine einfache Phosphorylase vermittelt, sondern durch ein Enzym, als dessen prosthetische Gruppe Uridindiphosphat-Glucose (UDPG) erkannt wurde (LELOIR und CARDINI 1953). UDPG ist ähnlich wie ATP gebaut, nur findet sich an Stelle des Adenosins das uracilhaltige Ribosid Uridin und an Stelle des terminalen Phosphatrestes Glucose (Formel s. S. 164). In Weizen-, Mais- und Bohnenkeimlingen sowie Kartoffelsprossen ließ sich ein Enzym feststellen, das aus UDPG und freier Fructose Rohrzucker synthetisiert. Es handelt sich im Prinzip um den gleichen Prozeß, wie er in Bakterien nachgewiesen wurde, nur ist im ersten Fall Glucose-1-phosphat und im zweiten UDPG der Glucosedonator:

$$\text{UDPG} + \text{Fructose} \rightleftharpoons \text{Saccharose} + \text{UDP}.$$

Wie mit Hilfe von ATP Glucose in Glucose-1-phosphat überführt wird, ist ATP auch für die Synthese und Resynthese von UDPG aus UDP erforderlich.

Mit Hilfe von Phosphorylasen synthetisiert und spaltet die Pflanze auch Polysaccharide, z. B. die Stärke. Bahnbrechend für die diesbezüglichen Untersuchungen an der Pflanze war der Befund von CORI, COLOWICK und CORI (1937), daß die tierische Zelle Glykogen in Gegenwart von anorganischem Phosphat unter Bildung von Glucose-1-phosphat (Cori-Ester) abbaut. Eine zum Stärkeabbau befähigte Phosphorylase wurde von dem kanadischen Biochemiker HANES 1940 in Erbsensamen und Kartoffelknollen festgestellt. Dieses seither in vielen Pflanzen aufgefundene Enzym vermag nicht nur Stärke in Glucose-1-phosphat zu spalten, sondern umgekehrt auch Stärke aus diesem Ester aufzubauen:

$$\text{Stärke (Amylose)} + \text{H}_3\text{PO}_4 \xrightleftharpoons{\text{Phosphorylase}} \text{Glucose-1-phosphat}.$$

Nach welcher Richtung diese reversible Reaktion verläuft, hängt von der Konzentration der einzelnen Reaktionspartner ab. Der Anfall von Glucose-1-phosphat und niedrige Konzentration an Phosphorsäure sind Voraussetzung für die Stärke-

9*

synthese, während Verarmung des Gewebes an Glucose-1-phosphat, hoher Stärke-
gehalt und die Anwesenheit von anorganischem Phosphat die Spaltung durch
Phosphorolyse begünstigen. Die reversible Reaktion der Stärkesynthese entspricht damit im Wesen der Synthese des Rohrzuckers aus Glucose-1-phosphat
und freier Fructose. Die für die Knüpfung der Glykosidbindungen notwendige
Energie wird von der Phosphatbindung im Glucose-1-phosphat geliefert. In
diese Form wird sie bei der Phosphorolyse auch wieder transformiert.

   Die Rolle der Phosphate bei der Energieübertragung (III G). Schon bald
nach der Entdeckung der Hexosediphosphorsäure als Gärungsprodukt durch
Harden und Young wurde erkannt, daß Phosphorylierungen für den Kohlenhydratstoffwechsel ganz allgemein von großer Bedeutung sind. Jedoch erst
mehr als 30 Jahre später wurde man darauf aufmerksam, daß auch die Energieübertragung und Energiespeicherung mit Phosphorylierungs- und Dephosphory-
lierungsvorgängen verknüpft sind. Die grundlegenden Erkenntnisse wurden
wieder an tierischen Objekten, insbesondere beim Studium der Muskelkontraktion
erarbeitet. Sie gipfeln in der Feststellung, daß die Energie für die Muskelkontraktion durch die Spaltung von Kreatinphosphat in Kreatin und Phosphorsäure gewonnen wird, daß Kreatinphosphat im Prozeß der Glykolyse regeneriert
und die im Glykolyseprozeß freiwerdende Energie gespeichert werden kann.
Mit der Erkenntnis, daß Kreatinphosphat mit Hilfe von Adenosintriphosphat
(ATP) gebildet wird (Lohmann-Reaktion), war ein weiterer wichtiger Schritt
bei der Erforschung des Energieübertragungsmechanismus getan. In Form von
ATP kann Energie auch gespeichert werden. Dieses Nucleotid spielt, wie wir
heute wissen, eine fundamentale Rolle im Energiehaushalt wahrscheinlich aller
Gewebe, indem es an zahlreichen Beladungsreaktionen teilnimmt. Bei diesen
Beladungsreaktionen wird der endständige der drei Phosphatreste des ATP
abgespalten und an einen organischen Körper angeheftet, wodurch eine Verbindung mit höherem Energieinhalt entsteht. Das ATP ist eine Verbindung
mit hohem Gruppenpotential, d. h. bei der Abspaltung des dritten und auch
noch des zweiten Phosphatrestes wird viel Energie frei. Bewerkstelligen spezifische
übertragende Enzyme die Abspaltung (Hydrolyse), kommt es nicht nur zu einer
Übertragung des Phosphatrestes, sondern auch der in der Pyrophosphatbindung
festgelegten Energie. Im Gegensatz zum Adenosindi- und -triphosphat ist das
Adenosinmonophosphat wie viele andere einfache Phosphorsäureester eine Verbindung mit niedrigem Gruppenpotential; denn die nur mit energischen Mitteln
durchführbare Hydrolyse liefert wenig freie Energie.

   Die grundlegenden Arbeiten wurden von Meyerhof, Parnas, Embden, Cori,
Lohmann und ihren Mitarbeitern durchgeführt. Von wesentlicher Bedeutung
und richtungweisend für den Fortgang der Untersuchungen waren die Berichte
von Lipmann und Kalckar, beide 1941 erschienen, die sich mit der Rolle der
energiereichen Phosphate im Stoffwechsel befassen. Die Thermodynamik und
den Mechanismus der Phosphatbindung behandelt eine referierende Arbeit von
N. O. Kaplan (1951). In diesem Handbuch wird über die energiereichen Phos-
phatester im Band XII von R. S. Bandurski[1] ausführlich berichtet.

   Eine typische und eingehend untersuchte Beladungsreaktion ist die Überführung von Glucose in Glucose-6-phosphat, die in Gegenwart von ATP durch
die Hexokinase katalysiert wird:

$$\text{Glucose} + \text{ATP} \rightarrow \text{Glucose-6-phosphat} + \text{ADP}.$$

Die Phosphorylierung der Glucose ist eine endergone, eine energieverbrauchende
Reaktion; sie wird ermöglicht durch Koppelung mit der gleichzeitig ablaufenden

---

[1] Bandurski, R. S.: The group potentials and the rôle of phosphates in energy transfer.

exergonen, energieliefernden Abspaltung eines Phosphatrestes des ATP. Die Hexokinasereaktion ist der erste Schritt der Saccharose-, der Stärke- und der Glykogensynthese. Die freie Energie des Glykosemoleküls wird erhöht, um die Ausbildung glucosidischer Bindungen zu ermöglichen.

Spaltet man ATP im Reagensglas enzymatisch in ADP und *freie* Phosphorsäure auf, wird Wärme frei. Geht die enzymatische Spaltung im Muskel vor sich, tritt die freie Energie hauptsächlich als mechanische Arbeit auf. In den elektrischen Organen bestimmter Fische wird sie in elektrische Energie, in den biolumineszierenden Organen einiger Insekten in Lichtenergie transformiert.

ATP ist in den meisten, besonders den pflanzlichen Geweben nur in geringer Konzentration vorhanden, so daß sich der Nachweis schwierig gestaltet und bei der höheren Pflanze erst in wenigen Fällen gelungen ist. Trotzdem darf angenommen werden, daß es hier dieselbe zentrale Rolle im Energiestoffwechsel wie bei den tierischen Organismen spielt.

Adenosindi- und Adenosintriphosphat sind zweifellos die wichtigsten, jedoch nicht die einzigen an der Energieübertragung beteiligten Nucleotide. Die ebenfalls dazu befähigten Adenosin-, Guanosin- und Inosinphosphate enthalten den Purinring, die Cytidin- und Uridinphosphate den Pyrimidinring als basische Komponente. Kohlenhydratkomponente ist in allen Fällen D-Ribose, die in 5′-Stellung mit dem Phosphatrest verknüpft ist.

Zu den synthetischen Prozessen, die durch phosphatgebundene Energie in Gang gesetzt und gehalten werden, gehören neben der Saccharose-, Stärke- und Glykogensynthese die Verknüpfung von Aminosäuren zu Peptiden, die Amidbildung, die β-Carboxylierung, die Fettsäuresynthese und zweifellos viele andere. Es kann aber nicht jede energiereiche Phosphatbindung direkt für diese endergonen Vorgänge nutzbar gemacht werden, sondern, soweit wir wissen, eignet sich dafür nur das ATP/ADP-System. Der laufende Verbrauch und der geringe Gehalt der Gewebe an ATP machen seine dauernde Neubildung im Stoffwechsel erforderlich. ATP entsteht unmittelbar aus ADP durch Anlagerung eines Phosphatrestes. Kopplung an eine exergone Reaktion, wie sie die Spaltung eines der beim Kohlenhydratabbau entstehenden energiereichen Phosphorsäureesters darstellt, ermöglicht diese endergone Reaktion. Zusammen mit dem Phosphatrest wird die in der Bindung gelegene freie Energie mit auf das ADP übertragen. Ein Beispiel für das „Einfangen" freier Energie durch ADP ist die Reaktion dieses Nucleotids mit Diphosphoglycerinsäure:

$$
\begin{array}{ccc}
\mathrm{CH_2O{-}\textcircled{P}} & & \mathrm{CH_2O{-}\textcircled{P}} \\
\mathrm{\overset{|}{C}HOH} \quad + \mathrm{ADP} \rightleftharpoons & & \mathrm{\overset{|}{C}HOH} \quad + \mathrm{ATP.} \\
\mathrm{\overset{|}{C}\!\!\diagup^{\!O}_{\!O\sim\textcircled{P}}} & & \mathrm{\overset{|}{C}\!\!\diagup^{\!O}_{\!OH}}
\end{array}
$$

Es ist der Sinn des katabolischen Stoffwechsels, neben Bauelementen für die Synthesen von Zellsubstanzen auch die dafür erforderliche Energie bereitzustellen. Diese „einzufangen", zu speichern und ihre Weitergabe an energiebedürftige Reaktionen zu vermitteln, ist im wesentlichen die Funktion der Adenosinphosphate. Die Ausbeute an ATP ist bei den verschiedenen Formen des Kohlenhydratabbaues sehr unterschiedlich. Während bei vollständiger Verbrennung (Veratmung) je Mol Hexose schätzungsweise 36 Moleküle ATP gebildet werden, entstehen bei der alkoholischen Gärung nur etwa 2 Moleküle ATP.

Neben exergonen und endergonen Reaktionen wird die Gruppe der isergonen Reaktionen unterschieden. Es handelt sich um Phosphatübertragungen von einem Nucleotid auf ein anderes, wobei es zu keinen oder nur unwesentlichen Änderungen der freien Energie kommt. Die reversible Umwandlung von 2 Molekülen ADP in je ein Molekül ATP und AMP durch die Adenylkinase ist ein Beispiel dafür. Durch die Nucleotidkinasen kann die phosphatgebundene Energie des ATP auf andere Nucleotide übertragen werden.

Reserven an energiereichem Phosphat in Gestalt des Kreatin- oder Argininphosphates sind bei Pflanzen nicht festgestellt worden, möglicherweise übernehmen hier anorganische Polyphosphate, die schon vor mehr als 60 Jahren in höheren Pflanzen und in der Hefe nachgewiesen wurden, diese Rolle. Ein phosphatübertragendes Enzym, das Phosphatreste aus Polyphosphaten auf ADP überführt, konnten Hoffmann-Ostenhof und Mitarbeiter (1954) aus Hefe gewinnen.

Literatur.

Albaum, H. G., and W. W. Umbreit: Phosphorus transformations during development of the oat embryo. Amer. J. Bot. **30**, 553—558 (1943). — Arnon, D. I., M. B. Allen and F. T. Whatley: Photosynthesis by isolated chloroplasts. Nature (Lond.) **174**, 394—396 (1954). — Axelrod, B.: Citrus fruit phosphatase. J. of Biol. Chem. **167**, 57—72 (1947).

Beevers, H., and M. Gibbs: The direct oxidation pathway in plant respiration. Plant Physiol. **29**, 322—324 (1954).

Calvin, M.: Der Photosynthese-Cyclus. Angew. Chem. **68**, 253—264 (1956). — Cori, C. F., S. P. Colowick and G. T. Cori: The isolation and synthesis of glucose-1-phosphoric acid. J. of Biol. Chem. **121**, 465—477 (1937). — Czapek, Fr.: Biochemie der Pflanzen, Bd. III. Jena: Gustav Fischer 1925.

Emerson, R. L., I. F. Stauffer and W. W. Umbreit: Relationships between phosphorylation and photosynthesis in *Chlorella*. Amer. J. Bot. **31**, 107—120 (1944).

Gregory, J. D., G. D. Novelli and F. Lipmann: The composition of coenzym A. J. Amer. Chem. Soc. **74**, 854 (1952). — Gunsalus, I. C., W. D. Bellamy and W. W. Umbreit: A phosphorylated derivative of pyridoxal as the coenzyme of tyrosine decarboxylase. J. of Biol. Chem. **155**, 685—686 (1944).

Hanes, C. S.: Breakdown and synthesis of starch by an enzyme system from pea seeds. Proc. Roy. Soc. Lond., Ser. B **128**, 421—450 (1940). — The reversible formation of starch from glucose-1-phosphate catalysed by potato phosphorylase. Proc. Roy. Soc. Lond., Ser. B **129**, 174—208 (1940). — Harden, A., and W. J. Young: The alcoholic fermentation of yeast-juice. II. The coferment of yeast-juice. Proc. Roy. Soc. Lond., Ser. B **78**, 369—375 (1906). — Hassid, W. Z., and M. Doudoroff: Synthesis of disaccharides with bacterial enzymes. Adv. Enzymol. **10**, 123—143 (1950). — Hoffmann-Ostenhof, O., I. Kenedy, K. Keck, O. Gabriel u. H. W. Schönfellinger: Ein neues phosphat-übertragendes Ferment aus Hefe. Biochim. et Biophysica Acta **14**, 285 (1954). — Hoffmann-Ostenhof, O., u. W. Weigert: Über die mögliche Funktion des polymeren Metaphosphats als Speicher energiereichen Phosphats in der Hefe. Naturwiss. **39**, 303—304 (1952).

Kalckar, H. M.: The nature of energetic coupling in biological synthesis. Chem. Rev. **28**, 71—78 (1941). — Kandler, O.: Über die Beziehungen zwischen Phosphathaushalt und Photosynthese. I. Phosphatspiegelschwankungen bei *Chlorella pyrenoidosa* als Folge des Licht-Dunkel-Wechsels. Z. Naturforsch. **5b**, 423—437 (1950). — Kaplan, N. O.: Thermodynamics and mechanism of the phosphate bond. In Sumner-Myrbäck, The Enzymes, vol. II, part 1. New York: Academic Press 1951. — Karrer, P., K. Schöpp u. F. Benz: Synthesen von Flavinen. IV. Helvet. chim. Acta **18**, 426—429 (1935). — Kostytschew, S.: Lehrbuch der Pflanzenphysiologie. Chemische Physiologie. I. Berlin: Springer 1926. — Kuhn, R., K. Reinemund, H. Kaltschmitt, R. Ströbele u. H. Frischmann: Synthetisches 6,7-Dimethyl-9-d-riboflavin. Naturwiss. **23**, 260 (1935).

Leloir, L. F., and C. E. Cardini: The biosynthesis of sucrose. J. Amer. Chem. Soc. **75**, 6084 (1953). — Lipmann, F.: Metabolic generation and utilization of phosphate bond energy. Adv. Enzymol. **1**, 99—162 (1941). — Lynen, F., u. E. Reichert: Zur chemischen Struktur der „aktivierten Essigsäure". Z. angew. Chem. **63**, 47—48 (1951).

Pfeffer, W.: Pflanzenphysiologie, Bd. I. Leipzig: Wilhelm Engelmann 1897.

Roche, J.: Phosphatases. In Sumner-Myrbäck, The Enzymes, vol. I, part 1, p. 473—510 New York: Academic Press 1950.

SCHLENK, F., u. H. v. EULER: Cozymase. Naturwiss. **24**, 794—795 (1936). — SIMONIS, W.: Untersuchungen zur lichtabhängigen Phosphorylierung. II. Z. Naturforsch. **11b**, 354—363 (1956). — SIMONIS, W., u. M. ÉHRENBERG: Untersuchungen zur lichtabhängigen Phosphorylierung. IV. Die Wirkung des Lichtes auf die ^{32}P-Einlagerung bei chlorophyllfreien Pflanzenzellen und -geweben. Z. Naturforsch. **12b**, 156—163 (1957). — SIMONIS, W., u. K. H. GRUBE: Weitere Untersuchungen über Phosphathaushalt und Photosynthese. Z. Naturforsch. **8b**, 312—317 (1953). — SIMONIS, W., u. H. KATING: Untersuchungen zur lichtabhängigen Phosphorylierung. I. Die Beeinflussung der lichtabhängigen Phosphorylierung von Algen durch Glucosegaben. Z. Naturforsch. **11b**, 165—172 (1956). — Untersuchungen zur lichtabhängigen Phosphorylierung. Z. Naturforsch. **11b**, 704—708 (1956). — STICH, H.: Der Nachweis von Metaphosphaten in normalen, verdunkelten und Trypaflavin-behandelten Acetabularien. Z. Naturforsch. **8b**, 36—44 (1953). — STREHLER, B. L.: Firefly luminescence in the study of energy transfer mechanism. II. Adenosine triphosphate and photosynthesis. Arch. of Biochem. a. Biophysics **43**, 67—79 (1953). — SUZUKI, U., Y. YOSHI-MURA and M. TAKAISHI: On the occurrence of an enzyme which decomposes anhydrooxy-methylene phosphoric acid. Tokyo Chem. Soc. **27**, 1330—1342 (1906).

THEORELL, H., and A. P. NYGAARD: The combination of flavin mononucleotide and ribo-flavin with the protein of the old yellow enzyme. Acta chem. scand. (Copenh.) **8**, 1104—1105 (1954). — THOMAS, I. B., and A. M. I. HAANS: Photosynthetic activity of fragments of *Spirogyra* chloroplasts. Biochim. et Biophysica Acta **18**, 287—288 (1955).

VOGLER, K. G.: The nature of the chemosynthetic process. J. Gen. Physiol. **26**, 103—117 (1942). — VOGLER, K. G., and W. W. UMBREIT: The nature of the energy storage material active in the chemosynthetic process. J. Gen. Physiol. **26**, 157—167 (1942).

WARBURG, O., u. W. CHRISTIAN: Ein zweites sauerstoffübertragendes Ferment und sein Absorptionsspektrum. Naturwiss. **20**, 688 (1932). — Über das Oxydationsferment. Naturwiss. **20**, 980—981 (1932). — Pyridin, der wasserstoffübertragende Bestandteil von Gärungsfermenten (Pyridinnucleotide). Biochem. Z. **287**, 291—333 (1936). — Isolierung der prosthetischen Gruppe der D-Aminosäure-Oxydase. Biochem. Z. **298**, 158—168 (1938).

Accumulation de l'acide phosphorique
(Phytine, Polyphosphates).

Par

J. M. Wiame.

I. Les polyphosphates[1].

1. Introduction et chimie.

Les acides métaphosphoriques sont des anhydrides de l'acide phosphorique dont la composition correspond à la formule $(HPO_3)_n$. Cette composition peut s'expliquer si on admet une structure cyclique donnée par les formules 1 et 2,

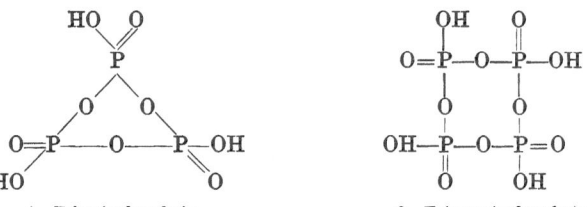

1. Trimetaphosphate.   2. Tetrametaphosphate.

qui sont celles d'un *trimétaphosphate* et d'un *tétramétaphosphate*. Dans ces composés cycliques, il n'existe qu'un seul type de fonction acide, qui est une fonction acide forte. On appelle également métaphosphate le sel de Graham, qui nous intéresse particulièrement ici. Le sel de Graham est obtenu par la déshydratation et fusion de phosphate disodique à 620°, suivie d'un refroidissement brusque donnant une substance vitreuse dont la composition est très proche de $NaPO_3$. On avait attribué à ce sel la formule d'un hexamétaphosphate. Des recherches récentes ont montré cependant que ce composé devait être considéré comme un homologue à poids moléculaire élevé de la série des polyphosphates (LAMM et MALMGREN 1940, LAMM 1944, SAMUELSON 1944, VAN WAZER et HOLST 1950), c'est-à-dire de polymères linéaires d'acide phosphorique.

Les premiers termes de cette série de substances sont le pyrophosphate (formule 3), le triphosphate (formule 4) et le tétraphosphate (formule 5). Lorsque

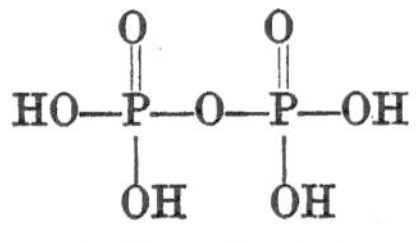

3. Pyrophosphate. 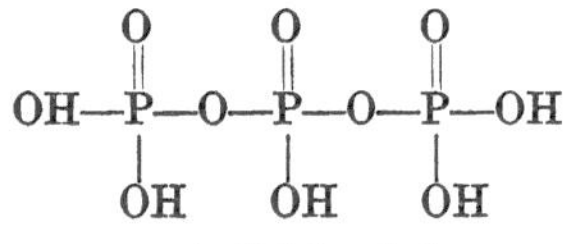   4. Triphosphate.

le poids moléculaire des polyphosphates augmente, leurs compositions se rapprochent de celle donné par la formule $(NaPO_3)_n$. Toutefois, aux extrémités de la chaîne se trouvent des groupes ayant deux fonctions acides, dont l'une est beaucoup plus faible, la proportion de ces fonctions acides plus faibles, qui peut être

---

[1] Une revue d'ensemble de cette question par G. SCHMIDT (1951) peut être consultée avec intérêt en complément à cette étude.

déterminée par électrotitration, permet de déterminer la longueur de la chaîne. Le sel de Graham peut donc être représenté par la formule 6.

$$
\begin{array}{cccc}
& \text{O} & \text{O} & \text{O} & \text{O} \\
& \| & \| & \| & \| \\
\text{HO—P—O—P—O—P—O—P—OH} \\
& | & | & | & | \\
& \text{OH} & \text{OH} & \text{OH} & \text{OH}
\end{array}
\qquad
\begin{array}{ccc}
\text{O} & \left[\ \text{O}\ \right] & \text{O} \\
\| & \| & \| \\
\text{HO—P—} & \text{O—P—} & \text{—P—OH} \\
| & | & | \\
\text{OH} & \text{OH}\ \ _n & \text{OH}
\end{array}
$$

5. Tetraphosphate.   6. Sel de Graham.

Parmi les propriétés importantes de ces dérivés, il faut mentionner que les tri- et tétramétaphosphates ne donnent pas de précipité avec les sels de barium à aucun p_H, tandis que les polyphosphates donnent un précipité. Le sel de Graham en particulier donne un sel qui est insoluble à p_H 2.5, ce qui le distingue aisément des autres polyphosphates qui ne précipitent qu'à des p_H plus élevés (JONES 1941). Les polyphosphates, comme le pyrophosphate, font partie des dérivés acidolabiles, c.à.d. hydrolysables à 100° en solution normale d'acide fort, en un temps ne dépassant pas 15 minutes; dans ces conditions, la transformation en orthophosphate est complète.

Une propriété intéressante, à la fois pour la détection chimique et cytochimique, est la possibilité pour les polyphosphates à haut poids moléculaire, de donner la *réaction métachromatique* (WIAME 1947). On entend par réaction métachromatique la propriété qu'ont certains polymères anioniques comme les esters polysulfuriques (LISON 1936), les acides polyglutamiques (WIAME, non publié, HANBY et RYDON 1946), de faire passer du bleu au rouge violacé certains colorants tels que le bleu de toluidine. Cette propriété correspond vraisemblablement à une polymérisation du colorant effectué au niveau du polyanion (MICHAELIS et GRANICK 1945, WIAME 1947). Alors que le bleu de toluidine à l'état monomère possède un maximum d'absorption pour la lumière de 630 mμ, le complexe formé avec les polyphosphates a un maximum d'absorption très marqué à 530 mμ. On peut définir la métachromasie quantitativement par le rapport d'absorption à ces deux longueurs d'onde. Parmi les polyanions phosphoriques, les acides nucléiques donnent une réaction métachromatique (MASSART *et al.* 1951). La réaction métachromatique des acides nucléiques est cependant beaucoup plus faible, étant donné que le maximum d'absorption dans ce cas est beaucoup moins marqué et se situe vers 580 mμ. Le sel de Graham est précipité par les colorants basiques tels que le bleu de toluidine.

Les acides polyphosphoriques précipitent les protéines en milieu légèrement acide. Ils peuvent former des composés cristallins avec celles-ci (PERLMAN 1938). L'acide trichloracétique, un autre précipitant des protéines, élimine l'acide polyphosphorique de ses combinaisons protéiques (WIAME 1949).

Lorsqu'une solution de polyphosphates (sel de Graham) est additionnée d'un sel de barium ou de calcium, il ne se forme pas un précipité immédiatement: une partie des ions Ba^{++} ou Ca^{++} se complexe au polyphosphate, en donnant un composé soluble, et c'est par l'ajoute d'un excès d'ions divalents que le polyphosphate précipite.

On peut enlever le barium des polyphosphates par échange sur résine telle que Amberlite IR-I00 H (EBEL 1952, DAMLE et KRISHNAN 1954). Jusqu'à présent, les divers essais pour effectuer une séparation chromatographique des polyphosphates à haut poids moléculaire ont échoué. Le sel de Graham ne migre pas dans le cas des solvants utilisés jusqu'à présent; ceci le distingue des tri- et tétramétaphosphates, ainsi que des polyphosphates de faible poids moléculaire (EBEL 1952, CROWTHER 1954). Nous avons obtenu une bonne séparation du métaphosphate en mélange avec d'autres substances polyanioniques (ac. nucléique, ac. polyglutamique) en effectuant une électrophorèse sur papier à 0⁰ en milieu très acide, et en révélant les polyanions par le bleu de toluidine.

Dans ce qui va suivre, nous adopterons le terme de *polyphosphate* et non celui de métaphosphate, pour les substances dont le comportement est proche du sel de Graham.

2. Les polyphosphates et la métachromasie des cellules.

La présence de polyphosphates fut établie depuis longtemps chez la levure par Lieberman (1888) et dans les graines de cotonier par Hardin (1892). Alors que sa présence dans les plantes supérieures ne semble pas avoir été retrouvée jusqu'à présent, dans le cas de la levure et des champignons inférieurs, cette substance a été étudiée par Kossel (1893), Ascoli (1899) qui la retrouvèrent simultanément à l'acide ribonucléique. Ils donnaient à ce mélange de polyphosphate et d'acide nucléique le nom d'*acide plasmique*. Ce n'est toutefois qu'en 1936 que Mac Farlane le signale comme un composé important de la levure, et en 1944, Mann le retrouve chez *Aspergillus niger* (1944a et b).

Indépendamment des travaux concernant l'analyse chimique des constituants cellulaires, les cytologistes appelaient *granulations métachromatiques* ou *volutine* des corpuscules capables de fixer des colorants basiques avec, dans certaines conditions, l'apparition du virage métachromatique (v. p. 139). Ces corpuscules furent décrits tout d'abord par Ernst (1888), Neisser (1888), Babes (1889) chez le bacille de la diphtérie; ils furent dénommés volutine par Meyer (1904) qui les observa chez la levure, et remarqua que la coloration métachromatique était beaucoup plus acido-résistante que la coloration normale (orthochromatique) que prennent les composés nucléiques. Les corpuscules métachromatiques furent souvent mentionnées chez les champignons, les levures et les algues (Grimme 1902, Guillermond 1906). D'une façon générale, la mise en évidence des corpuscules métachromatiques se fait en colorant les frottis au bleu de toluidine, lavage à l'alcool et différenciation pendant $^1/_2$ à 1 minute, dans HCl N/20. On peut également immerger les levures dans un mélange de bleu de toluidine dissous dans du formol acidifié par l'acide acétique. L'examen se fait directement dans ce mélange (Lindegren 1949).

Pendant longtemps, on a attribué aux corpuscules métachromatiques une nature nucléique. Brandt (1942) montra cependant que ces corpuscules n'absorbaient pas la lumière ultra-violette et il en conclut qu'ils pourraient être constitués d'esters sulfuriques, puisqu'à cette époque, les esters sulfuriques étaient les seules substances biologiques métachromatiques connues (Lison 1936).

La reprise des travaux à la fois chimiques et cytochimiques sur les polyphosphates résulte d'un travail de Jeener et Brachet (1943a et b) concernant les variations de la basophilie des cellules de levures. Ces auteurs ont préparé une levure très peu basophile en aérant une suspension de levures dans un milieu exempt de phosphate. Dans ces conditions, il se produit une croissance limitée et les cellules obtenues se colorent faiblement par le bleu de toluidine. Si on aère la suspension de levure dans un milieu qui, tout en étant exempt de phosphate, contient cependant des facteurs de croissance nécessaires à la croissance de *Saccharomyces cerevisiae* (Wiame 1947, 1949), la réduction de la basophile est encore plus marquée et seul se colore dans chaque cellule, un corpuscule contenant l'acide désoxyribonucléique. Dans le cas d'une levure n'exigeant pas de vitamines, l'épuisement est encore plus aisé à réaliser; c'est le cas de *Torulopsis utilis (Candida utilis)*.

Lorsque des levures ainsi appauvries en phosphates sont placées en présence à la fois de phosphate et d'un substrat carboné pouvant être métabolisé en aérobiose (alcool ou glucose) ou en anaérobiose (glucose), il se produit une synthèse très importante d'une substance basophile. On peut mesurer cette synthèse par la quantité de colorant retenu par les cellules (JEENER et BRACHET 1943, WIAME 1945, 1946, 1947). Les substances basophiles ne se forment que si les cellules sont métaboliquement actives (respiration ou fermentation), les poisons du métabolisme catabolique, l'acide monoïdacétique, l'ion fluor et l'acide cyanhydrique, dans le cas d'un catabolisme aérobique (avec l'éthanol comme substrat, p. ex.) inhibent la synthèse de la substance basophile. De plus, la quantité de substance basophile formée est proportionnelle à la quantité de substrat utilisé, montrant ainsi la liaison étroite des deux phénomènes (WIAME et LEFEBVRE 1946, WIAME 1947).

Sur la base des connaissances établies au préalable sur de nombreux types de cellules, JEENER et BRACHET (1943) avaient pensé que la substance basophile ainsi formée était de l'acide ribonucléique. En étudiant la composition des levures en phosphore, purine et pentose et surtout en extrayant la substance basophile formée, il apparut que cette substance était anormalement riche en phosphore; WIAME émit alors l'hypothèse que la substance était un composé nucléique polyphosphoré (1945, 1946). Toutefois, des traveaux ultérieurs (WIAME 1946—1948) ont montré que le mode de préparation n'excluait pas une coprécipitation d'acide nucléique et de polyphosphate.

En modifiant le procédé d'extraction (v. p. 140), des polyphosphates exempts d'acide nucléique ont pu être préparés.

La teinte métachromatique peut être observée directement ou au moyen d'un spectrophotomètre, sur des frottis denses de levures riches en métaphosphates, colorés au bleu de toluidine; elle contraste avec la teinte bleue (orthochromatique) des levures dont la basophile est dûe en majeure partie à l'acide ribonucléique. Les cellules de la levure ayant crû dans les conditions habituelles, contiennent normalement quelques granulations métachromatiques; les cellules riches en polyphosphates obtenues par la méthode décrite précédemment sont entièrement remplies de la substance, dont la combinaison avec les colorants basiques est acido-résistante, comme les granulations métachromatiques habituelles. D'une façon très générale, dans les cellules de champignons, d'algues et de bactéries riches en granulations métachromatiques, on a pu mettre en évidence des polyphosphates.

L'ensemble des données chimiques et cytochimiques ne permet pas de douter de l'identité de la volutine (corpuscule métachromatiques) avec les polyphosphates chez les organismes. Il faut rappeler que les esters sulfuriques d'ose, dans les tissus animaux (LISON 1936), certaines substances bactériennes capsulaires comme l'acide polyglutamique et les parois cellulaires d'algues marines (agar-agar) sont intensément métachromatiques, polyanioniques, mais non phosphorées.

3. Détection et détermination quantitative des polyphosphates cellulaires.

Les propriétés qui permettent l'identification et la détermination des polyphosphates cellulaires sont l'insolubilité des sels de barium en milieu acide, la labilité en solution acide (normale) à 100°C et la réaction métachromatique. Les deux premières propriétés sont essentielles pour les déterminations quantitatives, la métachromasie est un test très sensible, mais surtout utile à la détection qualitative. De toute façon, une identification certaine doit se faire par l'ensemble des propriétés auxquelles on peut encore ajouter des tests qui, quoique moins sensibles, constituent un bon contrôle: ce sont la précipitabilité des protéines, la formation de complexes avec le calcium et le barium, l'immobilité dans les solvants chromatographiques.

La méthode d'extraction généralement adoptée est celle qui fut décrite par Wiame (1949). MacFarlane (1936) et Mann (1944) ont extrait les cellules de levure ou de moisissures au moyen d'acide trichloracétique (TCA). Comme l'avait montré McFarlane, l'extraction avec TCA 5% laisse un résidu important de polyphosphate non extrait, et l'utilisation de TCA 10% augmente l'extraction (Wiame 1949).

On peut procéder à 2 ou 3 extractions à 0⁰ pendant 1 heure. Les polyphosphates sont précipités de cet extrait par un excès de barium en tampon acétate p_H 4.5. Quoique le sel de Graham précipite quantitativement à p_H 2.5, la présence de polyphosphates cellulaires différents du sel de Graham, oblige à effectuer la précipitation à p_H 4.5 (Wiame 1949). Ebel a confirmé ce fait et a montré que la fraction précipitée entre 2.5 et 4.5 était constituée de polyphosphate à faible poids moléculaire (1952d). Il est encore difficile actuellement de décider si ceux-ci sont préformés ou résultent d'une dégradation de polyphosphate macro-moléculaire. La précipitation se fait en agitant à la chambre froide l'extrait avec le sel de barium. Dans ces conditions, pour la levure, on a vérifié que le précipité ne contient pas d'orthophosphate et est entièrement constitué de phosphate acido-labile.

Lorsqu'on désire effectuer un *test de métachromasie* sur un tel extrait, il faut éloigner la majeure partie des sels (qui inhibe la réaction métachromatique); dans ce cas, on précipite les polyphosphates avec le nitrate de plomb, après avoir amené l'extrait à p_H 4. Le précipité redissous avec le minimum d'acide est débarrassé du plomb par H_2S et la solution neutralisée, tout en restant légèrement acide, est additionnée goutte à goutte de bleu de toluidine (30mg/litre). On peut détecter ainsi très aisément le polyphosphate à la concentration 10^{-4} M (Wiame 1949). Damle et Krishnan (1954) ont repris l'étude des facteurs qui influencent la réaction métachromatique, et standardisé la méthode.

Le polyphosphate contenu dans l'extrait précédent a été appelé »soluble dans l'acide trichloracétique«, en abrégeant: *acido-soluble*. Une difficulté résulte du fait qu'une seconde fraction, dont la signification physiologique paraît différente (Juni *et al* 1947, Wiame 1949) reste non extraite par le traitement précédent. Cette fraction a été appelée »*acido-insoluble*«; elle est liée à des constituants cellulaires dont la nature n'est pas encore précisée. Lorsque la quantité de polyphosphate acido-insoluble n'est pas faible par rapport aux autres composants phosphorés acido-insolubles, tels que les acides nucléiques, sa présence se révèle aisément par une forte proportion d'orthophosphates dans un extrait trichloracétique à 95⁰C, et surtout par un rapport du phosphore total à l'absorption à la lumière ultra-violette de 260 mμ, qui est plus élevé que pour les acides nucléiques. L'extraction à chaud dans l'acide trichlor-acétique hydrolyse les polyphosphates et leur identification à l'état non dégradé présente des difficultés. La nature polyphosphorique de ces phosphates labiles a cependant pu être mise en évidence par le test métachromatique, après une extraction alcaline (entre p_H 8 et 9), qui extrait simultanément des acides nucléiques et des protéines (non métachromasique). Wiame (1949) et Ebel (1952c) ont déterminé cette fraction par une hydrolyse en milieu acide à 100⁰, en effectuant une correction pour le phosphore libéré dans ces conditions par les acides nucléiques. Ceci ne constitue qu'une approximation qui, si elle vaut pour la levure, riche en polyphosphate acido-insoluble, peut manquer de sélectivité dans d'autres cas. Ultérieure-ment, de Deken (non publié) a montré que dans une mélange d'acide nucléique et de poly-phosphate précipités par le Ba^{++}, on pouvait éliminer l'acide nucléique par un lavage avec NaOH 0,1 N. La méthode d'électrophorèse déjà citée serait probablement très utile.

D'autres méthodes d'extraction, visant surtout à isoler les polyphosphates à l'état natif, ont été préconisées. Schmidt *et al.* (1946) extraient à l'eau après un traitement de la levure à l'alcool à froid, et à l'éther. Ingelman (1948) et Ingelman et Malmgren (1950) ont extrait les mycelium d'*Aspergillus*, au moyen d'alcalis légers, après les avoir hachés; de plus, ils ont introduit une dialyse dans la préparation. Dans ces conditions, ils ont obtenu un polyphos-phate dont le poids moléculaire, déterminé par sédimentation, était de 6.000 à 7.000. Ebel (1952d) a trouvé par analyse potentiométrique des groupes terminaux, pour un polyphosphate non dialysé, un poids moléculaire de 1300, soit 13 unités phosphoriques. Yoshida (comm. pers.) a comparé les poids moléculaires de la fraction acido-soluble et insoluble. Le premier a un p.m. d'environ 1000 et l'autre est beaucoup plus élevé.

La quantité de métaphosphates accumulée dans la levure peut atteindre 1% de P pour la fraction soluble et 0.5% P pour la fraction insoluble. L'ensemble des métaphosphates (calculé en $NaPO_3$) peut donc représenter 4.5% du poids sec des cellules. Le détail des compositions dans diverses conditions a été donné par WIAME (1949) et CHAYEN *et al.* (1955).

Les métaphosphates ont été identifiés non seulement chez la levure et l'*Aspergillus niger*, mais aussi dans un grand nombre d'organismes inférieurs, ce sont *Neurospora grassa* (HOULAHAN 1948), la cyanophycée *Phormidium ambiguum* et *Corynebacterium diphteriae* (EBEL 1952c), *Polytomella ceca* (WIAME), *Euglena* (ALBAUM *et al.* 1950), l'algue unicellulaire *Acetabularia mediterranea* (NEUGNOT, comm. pers.), *Mycobacterium smegmatis* (1954), *Aerobacter aerogenes* dans des conditions acides de culture (DUGUID *et al.* 1954). Dans ce dernier cas, les granulations de volutine sont visibles au microscope a contraste de phase et au microscope électronique.

S. NIEMIERKO et W. NIEMIERKO (1950a et b) ont signalé la présence de polyphosphate dans les excreta de larves de *Galleria mellonella* et *Achroea grisella*, deux insectes capables de digérer la cire; d'autres excreta d'insectes (*Dixippus, Ephestia* et *Bombyx mori*) ne contiennent pas de polyphosphate. La présence de polyphosphate semble exceptionnelle dans les organes ou tissus animaux.

4. La signification biologique des polyphosphates.

La structure polyanhydride des polyphosphates, ainsi que la similitude de sa vitesse d'hydrolyse en milieu HCl à 100° avec celle des groupes polyphosphoryles de l'acide adenosine-triphosphorique (ATP), avaient conduit WIAME (1946) à supposer que ce composé pourrait représenter une réserve énergétique pour les cellules. La structure polymérique aurait l'intérêt de concentrer cette énergie en un système ayant le minimum d'activité osmotique. La chaleur d'hydrolyse des liaisons anhydrides du trimétaphosphate a été mésurée par MEYERHOF *et al.* (1953), qui ont trouvé une valeur de 19.000 calgr. par mole, soit 6.300 par liaison P—O—P. D'autre part, YOSHIDA (1955) a obtenu une valeur d'environ 10.000 cal. par liaison P—O—P pour la fraction acido-insoluble, ainsi que pour le sel de Graham.

Ceci est en accord avec les observations déjà signalées p. 139 concernant la dépendance étroite entre un catabolisme capable de produire des groupes phosphoryle riches et la synthèse de polyphosphates. Un catabolisme intact, s'il est nécessaire, n'est toutefois pas suffisant, la dissociation des deux phénomènes peut être réalisée par l'azoture de sodium, qui, à la concentration de 10^{-4} M, inhibe la formation de polyphosphate en laissant inchangée l'intensité de la fermentation anaérobique (WIAME 1947c, MEYERHOF *et al.* 1952). Des résultats analogues ont été obtenus avec les dinitrophénols. La synthèse des polyphosphates est donc comparable à cet égard aux autres synthèses cellulaires en général (CLIFTON 1946, DE DEKEN 1955, NICKERSON 1949, YOSHIDA 1953a et b).

La distinction des deux fractions de polyphosphates n'est pas seulement relative à l'extraction par l'acide trichloracétique à froid, mais concerne également l'activité physiologique des deux fractions (JUNI *et al.* 1947, WIAME 1949). C'est la fraction liée, non extractible par le TCA (acido-insoluble) qui est métaboliquement la plus active. Elle échange plus rapidement ses groupes phosphorés avec de l'orthophosphate marqué, placé dans le milieu extérieur. Elle se transforme en orthophosphate dans une levure où le catabolisme est arrêté (p. ex. par le froid); le passage inverse d'orthophosphate à métaphosphate se réalise dès que la cellule métabolise du glucose. On a un système rapide de passage de l'ortho- au métaphosphate insoluble. Si une levure riche en métaphosphate est cultivée dans un milieu sans phosphate, c'est la fraction acido-insoluble qui est utilisée en premier lieu.

La structure des polyphosphates suggère une relation directe avec l'acide adenosine-triphosphorique du type suivant:

$$ATP + (NaPO_3)_n \rightleftharpoons ADP + (NaPO_3)_{n+1}$$

Hoffmann-Ostenhof et Weigert (1952).

Ce n'est que récemment qu'une telle réaction a été démontrée. Yoshida et Yamataka (1953a et b) ont pu préparer un extrait de levure qui pouvait oxyder différents substrats (alcool, glucose, etc.) et simultanément transformait l'orthophosphate en polyphosphate. En l'absence de substrat oxydable, l'ATP seul permet également la synthèse de polyphosphates. Les mêmes auteurs (comm. pers.: Symp. of Enzym. Chem., 1953, p. 86, Japon) ont également montré que le métaphosphate additionné d'ADP donne lieu à une formation d'ATP. Le même résultat a été obtenu par Hoffmann-Ostenhof et al. (1954).

D'autre part, Winder et Denneny (1955) ont réalisé avec des extraits de *Mycobacterium smegmatis* un système de phosphorylation utilisant le métaphosphate chimique et fonctionnant par la somme des deux réactions:

$$
\begin{array}{ll}
\text{polyphosphate} + \text{ADP} & \longrightarrow \quad \text{ATP} \\
\text{ATP} + \text{glycérol} & \longrightarrow \quad \text{glycérol-phosphate} + \text{ADP} \\
\hline
\text{polyphosphate} + \text{glycérol} & \xrightarrow{\text{trace}}{\text{ADP}} \quad \text{glycérol-phosphate}
\end{array}
$$

L'ensemble de ces travaux ne laisse aucun doute sur le rôle du métaphosphate comme donneur de groupes phosphoryles riches en énergie. Les structures macromoléculaires des polyphosphates pourraient peut-être les impliquer dans des réactions plus complexes. La signification des différentes fractions retrouvées dans les cellules est encore inconnue.

Chayen et al. (1955) ont étudié l'évolution des polyphosphates insolubles et de l'acide nucléique de *Torulopsis utilis* au cours des différentes phases d'une croissance. La quantité de polyphosphate atteint une valeur minimale au moment où les synthèses de protéine sont les plus actives et où la quantité d'acide nucléique est maximum. Les polyphosphates pourraient donc intervenir comme source de phosphoryles riches non seulement pour former de l'ATP, mais peut-être directement dans des synthèses de macromolécules.

Une liaison éventuelle des polyphosphates avec les acides nucléiques a été récemment soutenue par Belozerski (1955).

Des phosphatases ayant la propriété d'hydrolyser les fonctions anhydrides des polyphosphates ont été décrites depuis longtemps et sont extrêmement répandues; leur présence n'est pas limitée aux tissus et cellules contenant les polyphosphates. Une revue concernant ces enzymes, plus détaillée que nous ne pouvons le donner ici, a été présentée par Schmidt (1951) et par Ingelman (1950).

Seule la phosphatase hydrolysant le pyrophosphate, le composé le plus simple de la série, a été obtenue jusqu'à présent à l'état cristallin par Kunitz (1951). Cet enzyme est spécifique. D'autres phosphatases qui seraient moins spécifiques ont été étudiées, dans des préparations plus ou moins purifiées. Des hydrolyses enzymatiques, triphosphatasiques, ont également été observées dans de nombreux cas depuis leur découverte par Neuberg et Fischer en 1937. Dans le laboratoire de Neuberg, Kitasato (1928) montra la présence de phosphatase des polymères supérieurs du type du sel de Graham. Les polyphosphatases diffèrent des pyrophosphatases, par une sensibilité plus grande à la chaleur. Alors que les enzymes précités agissent en libérant de l'orthophosphate, une polyphosphate dépolymérase a été trouvée dans des extraits de levures, de moisissures et de bactéries — absente dans les tissus animaux — qui agit sur des polyphosphates de très haut poids moléculaire (P.M. 10^6, sel de Currol); elle scinde la molécule sans libérer d'ortho-phosphate (INGELMAN et MALMGREN 1947, 1948, 1949).

L'action de l'ensemble de ces phosphatases, contrairement aux phosphotrans-férases, produit la perte de l'énergie accumulée dans les liaisons anhydrides des polyphosphates et, faute d'autres données, on ne peut leur assumer qu'un rôle de mobilisation de phosphate ou de destruction d'un excès de polyphosphate dont l'action inhibitrice fut signalée entre autres par VISHNIAC (1950) dans le cas du triphosphate.

Selon LINDEGREN, les polyphosphates participeraient directement à la division mitotique chez la levure (1947, 1949, 1951). Etant donné la complexité et la diffi-culté d'interprétation des figures cytologiques chez la levure, il semble encore difficile de donner une opinion définitive sur les interprétations de LINDEGREN.

KORNBERG, KORNBERG et SIMMS [Biochim. et Biophysica Acta 20, 215 (1956)], ont purifié un enzyme extrait de *Escherichia coli* réalisant la synthèse de poly-phosphate suivant l'équation:

$$x\,ATP+ (PO_3^-)_n \rightarrow x\,ADP + (PO_3^-)_{n+x}$$
initiateur

Seul le phosphate terminal de l'ATP est mis en jeu dans la synthèse, l'ADP est un inhibiteur de la réaction. La polymère formé est entièrement sous la forme non dialysable, sans production apparente de molécules intermédiaires. La nécessité et la nature de l'initiateur dont les auteurs font l'hypothèse n'ont pas pu être précisées. La réversibilité de la réaction n'est pas apparue dans les conditions où les auteurs ont opéré.

Il faut noter que la présence de métaphosphate sous forme d'éléments méta-chromatiques n'a pas été trouvée jusqu'à présent chez *E. coli*.

II. L'acide phytique (phytine)[1].

1. Composition et structure.

L'acide phytique est une substance phosphorée très répandue chez les plantes. Elle est également le composé phosphoré qui y est présent le plus souvent à la plus forte concentration. Il s'y trouve sous forme de sel calco-magnésien. Dans de nombreuses graines, il représente 60 à 90% du phosphore total.

L'acide phytique est un ester hexaphosphorique de l'inositol; il fut découvert par PFEFFER (1872) et PALLADIN (1895) dans les graines. Sa nature chimique fut précisée par WINTERSTEIN (1897) et sa synthèse fut réalisée à partir d'inositol et d'anhydride phosphorique par POSTERNAK (1921). Quoique l'on puisse préciser qu'il est un dérivé hexaphosphoré de l'inositol, le détail de sa structure n'est pas encore définitivement établi. L'inositol est un polyalcool cylique à six fonctions alcool, la formule la plus simple de l'acide phytique pourrait donc être représentée par la formule I qui fut proposée par ANDERSON (à qui l'on doit de nombreux travaux sur ce composé; 1914—1920).

La formule I rend compte de la plupart des propriétés de l'acide phytique, qui aurait donc la formule $C_6H_{18}O_{24}P_6$. Cependant, on connait des sels bien cristallisés tels que le phytate de calcium et de sodium $C_6H_{12}O_{27}P_6Ca_2Na_8$ et le phytate de sodium $C_6H_{12}O_{27}P_6Na_{12}$, qui correspondent à un acide phytique de formule $C_6H_{24}O_{27}P_6$, c.à.d. $C_6H_{18}O_{24}P_6 \cdot 3\,H_2O$ (POSTERNAK 1921). Les 3 molécules d'eau se retrouvent dans certains des sels d'alcaloïdes; dans un sel de barium et, dans

---

[1] Une revue d'ensemble détaillée, traitant en particulier de la structure de ce composé, ainsi que de divers dérivés de moindre importance, a été donnée par COURTOIS (1951) (voir aussi DANGSHAT 1955).

ce cas, elles sont éliminées par le chauffage vers 110—120⁰. Dans le cas des sels de sodium et calcium ainsi que du sel de sodium, il est impossible d'éliminer les 3 molécules d'eau.

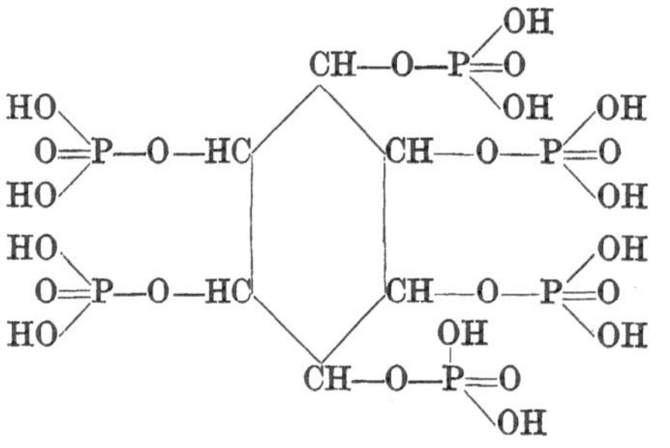

I
acide phytique (formule d'Anderson)

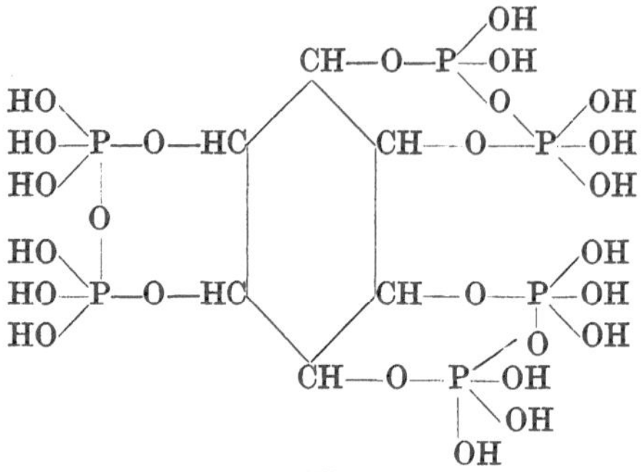

II
acide phytique (formule de Neuberg)

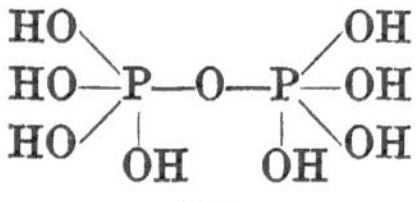

III
acide hydroxyphosphorique

Il n'est pas exclu que ces 3 molécules d'eau soient constitutives et que la formule de l'acide phytique doive être écrite par exemple sous la forme II, qui fut proposée par Neuberg (1907, 1908) et qui ferait que l'acide phytique serait un ester de l'acide hydroxyphosphorique (III) qui est présent dans les solutions d'acide orthophosphorique (Sanfourche 1933, 1938).

2. Propriétés chimiques; détermination et préparation.

L'hydrolyse de l'acide phytique est très lente: elle n'est complète qu'après 6 heures, dans H_2SO_4 10% à l'autoclave à 150—160⁰. L'hydrolyse complète a été obtenue en 2 heures à 160—170⁰ au moyen d'acide formique à 20% (Linden-feld 1934).

La structure de l'acide phytique permet de prévoir un certain nombre de propriétés, telles que la possibilité de former de nombreux sels mixtes, par exemple le sel calco-sodique, que nous avons signalé et le sel calco-magnésien qui est la forme sous laquelle il se trouve chez les végétaux. L'acide phytique forme avec les métaux des sels insolubles, même à des p_H acides, cette propriété est très utilisée pour la détermination quantitative. Les sels alcalins sont très solubles. Avec les ions alcalino-terreux, on obtient comme pour les polyphosphates des complexes. Le phytate de sodium ne donne de précipité avec l'ion Ca^{++} qu'après l'ajoute d'une certaine quantité de cet ion. Le calcium commence par se complexer et le complexe est très stable. Le calcium n'y est plus précipitable par les oxalates (Yang 1940).

Dans certaines conditions de p_H, les sels phytiques forment également des complexes ou des précipités avec les protéines. Lors de l'extraction des protéines des graines, on peut obtenir des substances simulant des dérivés phosphorés appelés «caséines végétales» et qui sont en réalité des combinaisons de protéines avec l'acide phytique (Barré 1951). Cependant, Bourdillon (1951) a obtenu à partir de graines de haricots un complexe protéine-acide phytique cristallisable à p_H 4.0 (dodécahèdre anisotrope). Après dialyse à p_H 8.0, la protéine est séparée de l'acide phytique, celle-ci cristallise (p_H 5.0) sous une autre forme (bisphénoïdes isotropes). A partir de la protéine exempte d'acide phytique, la forme initiale ne peut être réobtenue qu'en présence d'un grand excès d'acide phytique, d'où l'auteur conclut que le complexe préexiste dans la graine.

Que ce soit pour la séparation ou la détermination quantitative, la phytine est extraite à plusieurs reprises par HCl 0.1 à 0.2 N. Pour la préparation, elle est reprécipitée par neutralisation, et elle est purifiée par la méthode de POSTERNAK (1921).

Diverses méthodes de dosage quantitatif ont été décrites (DANGSHAT 1955); elles sont généralement basées sur l'insolubilité du sel ferrique en milieu acide. On peut mesurer le fer utilisé en présence de sulfocyanure de fer comme indicateur (HEUBNER et STADLER 1914), ou doser sur le précipité soit le phosphore (MAC CANCE et PRINGLE 1945), soit le fer (BECKER 1950) ou encore l'inositol par l'acide periodique après hydrolyse acide (HEGGEN, REITH 1950). SMITH (1952) ont séparé différents dérivés phosphorés de l'inositol par chromatographie sur colonne. ALBAUM et UMBREIT (1943) ont employé une phytase tirée de germes de glands de chênes pour la détermination quantitative.

Une étude critique des données existant sur la répartition de la phytine du grain de blé, a été présentée par BIGWOOD (1951). Il ressort de cette étude que les méthodes actuelles de dosage de la phytine ne sont pas encore entièrement satisfaisantes.

3. Distribution et rôle biologique.

A quelques exceptions près, l'acide phytique n'est présent que dans les végétaux. Il s'y trouve sous forme de sel calco-magnésien. C'est dans les graines qu'il atteint la plus forte concentration: 2% dans le grain de blé et 8% dans le riz. La richesse ne semble pas liée à la filiation botanique (COURTOIS et PEREZ 1948). Dans les céréales c'est principalement dans les cellules à aleurone et le scutellum que la phytine est concentrée. Les farines à fort taux d'extraction seront donc les plus riches (GUILLEMET 1946). On la rencontre également dans les divers organes de réserves (tubercules, rhizomes) et, éventuellement dans les feuilles, les tiges, *etc.* La phytine se retrouve dans les sols, apportée par les débris végétaux. Ceci est compréhensible, étant donné la stabilité de cette substance à l'hydrolyse chimique et enzymatique.

En dehors du règne végétal, la phytine est peu répandue, quoique l'inositol lui-même soit connu depuis longtemps comme une substance indispensable à la croissance de nombreux microorganismes et qu'il est un constituant normal des tissus animaux. De petites quantités d'inositol phosphoré ont été rétrouvées dans le lait et l'urine (STARKENSTEIN 1910) et dans les hématies nuclées (RAPAPORT 1940). Un inositol ayant deux groupes phosphoryles et formant un phosphatide a été identifié par FOLCH (1949) dans les cerveau. Parmi les substances proches de l'inositol phosphoré, on a identifié qu'une souche de *E. coli* ayant un bloc génétique sur la voie des synthèses aromatiques, excrétait un acide phospho-shikimique (DAVIS, B. C., et MINGIOLI 1953).

La rôle biologique de la phytine est mal connu. La seule réaction métabolique où ce composé intervient est son hydrolyse par la *phytase*, un enzyme qui, lui-même, a été peu étudié. C'est un enzyme que l'on extrait généralement des grains de blé ou de seigle; il est concentré dans les cellules à aleurone, comme la phytine. Toutefois, il n'y a de pas relation entre la teneur en phytase et en phytine (COURTOIS et PEREZ 1948); au cours de la germination, l'activité de la phytase augmente et la phytine disparait. Ceci suggère que la phytine pourrait être une simple réserve phosphorée (ALBAUM et UMBREIT 1943, COURTOIS et PEREZ 1948).

Une description des différents types de phytase est donnée par Courtois (1951).

Le rôle de donneur de groupe phosphoryle riche est pratiquement exclu d'après les diverses données qui ont été décrites précédemment. Aucune activité transphosphatasique mettant en jeu la phytine, n'a été signalée jusqu'à présent.

La phytine est un mauvais donneur de phosphore pour les animaux qui l'excrètent presqu'entièrement. D'autre part, par son rôle complexant sur les métaux, elle peut provoquer des déficiences calciques (Bruce et Callow 1934). Cette déficience se manifeste particulièrement avec les pains préparés à partir de farine à taux de blutage élevé. Pendant la guerre de 1940, des ajoutes de calcium (CaCO₃) ont été préconisées par MacCance et Widdowson (1944). Ce procédé a été effectivement utilisé à cette époque contre le rachitisme infantile avec de bons résultats.

Bibliographie.

1. Polyphosphates.

Albaum, H. G., A. Schatz, S. H. Hutner and A. Hirschfeld: Phosphorylated compounds in *Euglena*. Arch. of Biochem. **29**, 210 (1950). — Ascoli, A.: Über die Plasminsäure. Z. physiol. Chem. **28**, 426 (1899).

Babes, V.: Z. Hyg. **5**, 173 (1889). — Belozerski, A. N.: Complexe métaphosphate-acide nucléique des levures et nature chimique de la volutine. Comm. au 3. Congr. de Biochimie, Bruxelles, 1955. — Brandt, K. M.: Physiologische Chemie und Cytologie der Preßhefe. Protoplasma **36**, 77 (1941).

Chayen, R., S. Chayen and E. R. Roberts: Observations on nucleic acid and polyphosphate in *Torulopsis utilis*. Biochim. et Biophysica Acta **16**, 117 (1955. — Clifton, C. E.: Microbial assimilation. Adv. Enzymol. **6**, 269 (1946). — Crowther, J. P.: Filter paper chromatographic analysis of phosphate mixture. Nature (Lond.) **173**, 486 (1954).

Damle, S. P., and P. S. Krishnan: Studies on the role of metaphosphate in molds. I. Quantitative studies on the metachromatic effect of metaphosphate. Arch. of Biochem. a. Biophysics **49**, 58 (1954). — Deken, R. de: Relations entre la structure moléculaire et le pouvoir inhibiteur de nitro- et halophénols. Biochim. et Biophysica Acta **17**, 457 (1955). — Duguid, J. P., I. W. Smith and J. F. Wilkinson: Volutin production in *Bacterium aerogenes* due to the development of an acid reaction. J. of Path. **67**, 289 (1954).

Ebel, J. P.: Sur le dosage des métaphosphates dans les microorganismes par hydrolyse différentielle; technique et application aux levures. C. r. Acad. Sci. Paris **226**, 2184 (1948). — Recherches sur les polyphosphates contenus dans diverses cellules vivantes. I. Mise au point d'une méthode d'extraction. Bull. Soc. Chim. biol. Paris **34**, 321 (1952a). — II. Etude chromatographique et potentiométrique des polyphosphates de levure. Bull. Soc. Chim. biol. Paris **34**, 330 (1952b). — III. Recherche et dosage des polyphosphates dans les cellules de divers organismes et animaux supérieurs. Bull. Soc. Chim. biol. Paris **34**, 491 (1952c). — IV. Localisation cytologique et rôle physiologique des polyphosphates dans la cellule vivante. Bull. Soc. Chim. biol. Paris **34**, 498 (1952d). — Ernst, P.: Z. Hyg. **4**, 25 (1888).

Grimme, A.: Z. Bakter. 1, **32**, 161 (1902). — Guillermond, A.: Bull. Inst. Pasteur **4**, 145 (1906).

Hanby, W. E., and H. N. Rydon: The capsular substance of *Bacillus anthracis*. Biochemic. J. **40**, 297 (1946). — Hardin, M. B.: The presence of metaphosphoric acid in cottonseed meal. Carolina Agricult. Exper. Stat. Bull., New ser. **10** (1892). — Hoffmann-Ostenhof, O., J. Kenedy, K. Keck, O. Gabriel u. H. W. Schönfellinger: Ein neues phosphatübertragendes Ferment aus Hefe. Biochim. et Biophysica Acta **14**, 285 (1954). — Hoffmann-Ostenhof, O., u. W. Weigert: Über die mögliche Funktion des polymeren Metaphosphats als Speicher energiereichen Phosphats in der Hefe. Naturwiss. **39**, 303 (1952). — Houlahan, M. B., and H. K. Mitchell: The accumulation of acidlabile, inorganic phosphate by mutant of *Neurospora*. Arch. of Biochem. **19**, 257 (1948).

Ingelman, B.: Metaphosphate and its enzymatic breakdown. Dans: Sumner-Myrbäck, The enzymes, vol. I, part 1, p. 511—516. New York: Academic Press 1950. — Isolation of polymetaphosphate of high molecular weight from *Aspergillus niger*. Svensk. kem. Tidskr. **60**, 222 (1948). — Ingelman, B., and H. Malmgren: Enzymatic breakdown of polymetaphosphate. Acta chem. scand. (Copenh.) **1**, 422—432 (1947); **2**, 365—380 (1948); **3**, 157—162 (1949). — Investigations of high molecular weight isolated from *Aspergillus niger*. Acta chem. scand. (Copenh.) **4**, 478—486 (1950a).

JEENER, R., et J. BRACHET: Recherches sur la synthèse de l'acide pentosenucléique par les levures. Relations avec la fermentation et la respiration. Bull. Cl. Sci. Acad. roy. Belg. **29**, 476 (1943). — JONES, L. T.: Estimation of ortho-, pyro-, meta- and polyphosphates in the presence of one another. Industr. Engin. Chem., Anal. Ed. **14**, 536 (1942). — JUNI, E., M. D. KAMEN, S. SPIEGELMAN and J. M. WIAME: Physiological heterogenety of metaphosphate in yeast. Nature (Lond.) **160**, 717 (1947).

KITASATO, T.: Über Meta-phosphatase. Biochem. Z. **197**, 257 (1928). — KOSSEL, A.: Über die Nucleinsäure. Arch. f. Physiol. **160**, (1893). — KUNITZ, M.: Isolation of cristallin pyrophosphatase from baker's yeast. J. Amer. Chem. Soc. **73**, 1387 (1951).

LAMM, O.: Notiz über die Ladungseffekte bei Sedimentations- und Diffusionsmessung und die Molekulargewichtsbestimmung an hochmolekularen Metaphosphaten. Ark. Kemi (Stockh.) A 18, Nr 8 (1944). — LAMM, O., u. H. MALMGREN: Z. anorg. Chem. **245**, 103 (1940). — Messung und Berechnung von Sedimentations-Gleichgewichten an hochmolekularen Metaphosphaten. Ark. Kemi (Stockh). A 17, Nr 26 (1944). — LIEBERMANN, O.: Über Nuclein. Pflügers Arch. **43**. 97 et 1890 (1888). — LINDEGREN, C. C.: Function of volutin (metaphosphate) in mitosis. Nature (Lond.) **159**, 63 (1947). — The yeast cell. Saint-Louis: Educational Publisher 1949. — The relation of metaphosphate formation to cell division in yeast. Exper. Cell. Res. **2**, 275 (1951). — LISON, L.: Histochimie animale. Paris: Masson & Cie. 1936.

MACFARLANE: Phosphorylation in living yeast. Biochemic. J. **30**, 1369—1379 (1936). — MALMGREN, H.: A contribution to the physical chemistry of colloïdal metaphosphate. Acta chem. scand. (Copenh.) **2**, 147—165 (1948). — Enzymatic breakdown of polymetaphosphate. IV. The activation and the inhibition of the enzyme. Acta chem. scand. (Copenh.) **3**, 1331—1342 (1949). — MALMGREN, H., u. O. LAMM: Dispersitätsmessungen an hochmolekularen Kalium-metaphosphaten. Z. anorg. Chem. **252**, 255 (1944). — MANN, Y.: Studies on the metabolism of mould fungi. Biochemic. J. **38**, 339, 345 (1944). — MASSART, L., R. CONSSENS et M. SILVER: Métachromasie et structure des acides nucléiques. Bull. Soc. Chim. biol. Paris **33**, 514 (1951). — MEYER, A.: Bot. Z. **62**, 113 (1904). — MEYERHOF, O., and P. OHLMEYER: Purification of adenosinetriphosphatase of yeast. J. of Biol. Chem. **195**, II (1952). — MEYERHOF, O., R. STATAS and A. KAPLAN: Heat of hydrolysis of trimetaphosphate. Biochim. et Biophysica Acta **12**, 121 (1953). — MICHAELIS, L., and S. GRANICK: Metachromasy of basic dyestuffs. J. Amer. Chem. Soc. **67**, 1212 (1945).

NEISSER, A.: Z. Hyg. **4**, 165 (1888). — NEUBERG, C., u. H. A. FISCHER: Über die enzymatische Spaltung von Triphosphorsäure. III. Hydrolyse durch tierische Fermente. Enzymologia (Den Haag) **2**, 241 (1937—1938). — NEUGNOT: Com. pers. 1950. — NICKERSON, W. J.: Dependance in yeast of phosphate uptake and polymerization upon the occurence of glucose polymerization. Experientia (Basel) **5**, 202 (1949). — NIEMIERKO, S., and W. NIEMIERKO: Metaphosphate in the excreta of the wax-moth, *Galleria mellonella*. Nature (Lond.) **166**, 268 (1950). — Studies in the biochemistry of waxmoth *(Galleria mellonella)*, 6, Metaphosphate in the excreta of *Galleria mellonella*. Acta Biol. exper. (Pol.) **15**, 111 (1950).

PERLMAN, G.: On the preparation of cristallized egg albumine metaphosphate. Biochemic. J. **32**, 931 (1938).

SAMUELSON, O.: Cité par EBEL 1952a. Svensk. kem. Tidskr. **56**, 343 (1944). — SCHMIDT, G.: The biochemistry of inorganic pyrophosphates and metaphosphates. Dans: McELROY, W. D., and BENTLEY GLASS, Phosphorous metabolism. Vol. I. Baltimore: John Hopkins Press 1951. — SCHMIDT, G., L. HECHT and S. J. THANNHAUSER: The enzymatic formation and the accumulation of large amounts of metaphosphate in bakers yeast under certain conditions. J. of Biol. Chem. **166**, 775 (1946). — SMITH, I. W., J. F. WILKINSON and J. P. DUGUID: Volutin production in *Aerobacter aerogenes* due to nutrient imbalence. J. Bacter. **68**, 450 (1954).

VISHNIAC, W.: Antagonism between sodium tripolyphosphate and adenosine-triphosphate in yeast. Arch. of Biochem. **26**, 167 (1950).

WAZER, J. R., VAN, and K. A. HOLST: Structure and properties of the condensed phosphates. J. Amer. Chem. Soc. **72**, 639, 906 (1950). — WIAME, J. M.: Sur l'existence d'un nouveau composé phosphoré dans la levure. C. r. Soc. Biol. Paris **139**, 784 (1945). — Remarques sur la métachromasie des cellules de levure. C. r. Soc. Biol. Paris **140**. 897 (1946). — Sur la présence d'un dérivé nucléique polyphosporé dans la levure. C. r. Soc. Biol. Paris **140**, 825 (1946). — Etude d'une substance polyphosphorée, basophile et métachromatique chez les levures. Biochim. et Biophysica Acta **1**, 234 (1947a). — The metachromatic reaction of hexametaphosphate. J. Amer. Chem. Soc. **69**, 3146 (1947b). — Yeast metaphosphate. Federat. Proc. **6**. No I (1947c). — Métaphosphate et corpuscules métachromatiques chez la levure. Rev. Ferm. et Ind. Alim. **3**, 83 (1948). — The occurrence and physiological behavior of two metaphosphate fractions in yeast. J. of Biol. Chem. **178**. 919 (1949). — WIAME, J. M., et P. H. LEFEBVRE: Condition de formation dans la levure, d'un composé nucléique polyphosphoré. C. r. Soc. Biol. Paris **140**, 921 (1946). — WINDER, F.. and J. M. DENNENY:

Metaphosphate in mycobacterial metabolism. Nature (Lond.) **174**, 353 (1954). — Utilization of metaphosphate for phosphorylation by cell-free extracts of *Mycobacterium smegmatis*. Nature (Lond.) **175**, 636 (1955).

YOSHIDA, A.: Biochemical studies on metaphosphate. Sci. Pap. Coll. Gen. Educ. Univ. Tokyo **3**, 151 (1953b). — Studies on metaphosphate. II. Heat of hydrolysis of metaphosphate extracted from yeast cells. J. of Biochem. (Japon) **42**, 163 (1955). — YOSHIDA, A., and A. YAMATAKA: On the metaphosphate of yeast. I. J. of Biochem. **40**, 85 (1953a).

2. Phytine.

ALBAUM, H. G., and W. W. UMBREIT: Phosphorus transformations during the development of the oat embryon. Amer. J. Bot. **30**, 553 (1943). — ANDERSON, R. J.: J. of Biol. Chem. **17**, 141, 151, 165, 171 (1914); **18**, 441 (1914); **20**, 441 (1915); **43**, 469 (1920); **44**, 429 (1920).

BARRÉ, E.: Préparation et purification des protéines des amandes. Bull. Soc. Chim. biol. Paris **33**, 1473 (1951). — BECKER, M.: Cité par DANGSCHAT 1955. — BIGWOOD, E. J.: Observations concernant l'acide phytique du graine de froment. Bull. Soc. chim. biol. Paris **33**, 1261 (1951). — BOURDILLON, J.: A cristalline bean seed proteine in combination with phytic acid. J. of Biol. Chem. **189**, 65 (1951). — BRUCE, H. M., and R. K. CALLOW: Cereals and rickets. The role of inositolhexaphosphoric acid. Biochemic. J. **28**, 517 (1934).

COURTOIS, J. E.: Les esters phosphoriques de l'inositol. Bull. Soc. Chim. bil. **33**, 1075 (1951). — COURTOIS, J. E., et CH. PEREZ: Recherches sur la phytase. VIII. Teneur en inosito-phosphates et activité phytasique de diverses graines. Bull. Soc. Chim. biol. Paris **30**, 195 (1948a).

DANGSHAT, G.: Inosite und verwandte Naturstoffe. Dans: PAECH-TRACEY, Moderne Methoden der Pflanzenanalyse. Heidelberg: Springer 1955. — DAVIS, B. C., and E. S. MINGIOLI: Aromatic biosynthesis. VII. Accumulation of two derivatives of shikimic acid by bacterial mutants. J. Bacter. **66**, 129 (1953).

FOLCH, J.: Brain diphosphoinositide, a new phosphatide having inositol metadiphosphate as a constituent. J. of Biol. Chem. **177**, 505 (1949).

GUILLEMET, R.: Cité par COURTOIS 1951. — Biol. Med. **35**, 88 (1946).

HEGGEN, M., u. J. REITH: Acta pharmac. internat. **1**, 133 (1950). — HEUBNER, W., u. M. STADLER: Biochem. Z. **64**, 422 (1914).

LINDENFELD, K.: Über die Gewinnung von Inositol aus inosit-phosphorsauren Salzen. Biochem. Z. **272**, 284 (1934).

MACCANCE, R., and W. J. S. PRINGLE: Biochem. J. **39**, 3, 123, (1945). — MACCANCE, R., and E. WIDDOWSON: Biochemic. J. **29**, 2694 (1935). — Activity of the phytase in different cereals and its resistance to dry heat. Nature (Lond.) **153**, 650 (1944).

NEUBERG, C.: Cité par COURTOIS 1951. — Biochem. Z. **5**, 443 (1907); **9**, 551, 557 (1908). — PALLADIN, W.: Cité par COURTOIS 1951. — Z. Biol. **13**, 191 (1895). — PFEFFER, E.: Cité par COURTOIS 1951. — Pringsheim Jb. wiss. Bot. 8, 429 et 475 (1872). — POSTERNAK, S.: Sur la synthèse de l'acide inosito-hexaphosphorique. Helvet. chim. Acta **4**, 150 (1921). — POSTERNAK, S., et TH. POSTERNAK: Sur la configuration de l'inosite inactive. Helvet. chim. Acta **12**, 1165 (1929).

RAPAPORT, S.: J. of Biol. Chem. **135**, 403 (1940).

SANFOURCHE, A.: Recherches sur l'acide phosphorique et les phosphates. I. La formation des phosphates alcalino-terreux basiques. Bull. Soc. chim. France **53**, 951 (1933). — Préparation et propriétés du phosphate neutre de lithium. C. r. Acad. Sci. Paris **206**, 1820 (1938). — SMITH, D. H.: Chromatographic separation of soil org. compounds. Iowa State Coll. J. Sci. **26**, 287 (1952). — STARKENSTEIN, E.: Biochem. Z. **30**, 56 (1910).

WINTERSTEIN: Cité par COURTOIS 1951. Ber. dtsch. chem. Ges. **30**, 2299 (1897).

YANG, E. K.: Ionization of calcium phytate. Nature (Lond.) **145**, 745 (1940).

The phosphatides.

By

J. A. Lovern.

The phosphatides of plant tissues are discussed in detail in Vol. 7, VIII, where full references are also given. In the present condensed account the same general subdivision will be made into microorganisms, seeds and vegetative tissues of higher plants. The fatty acids of the phosphatides will not be discussed here.

Composition.

Among the microorganisms it is noteworthy that whereas fungi possess phosphatides consisting essentially of lecithin (phosphatidyl choline) and the classical "cephalin" (now generally named phosphatidyl ethanolamine), bacteria commonly contain highly complex and unusual phosphatides. Some bacteria do, indeed, contain lecithin and phosphatidyl ethanolamine as their predominant phosphatides, e.g. Phytomonas tumefaciens (GEIGER and ANDERSON), but the general picture, with the organisms so far studied, is quite different from this. In the Mycobacteria the phosphatide fraction is almost nitrogen-free, acidic and of high molecular weight (ANDERSON; BARBIER and LEDERER). It consists mainly of fatty acid esters of phosphorylated carbohydrates of two types, which may, so far as we know, be combined in a larger unit. One phosphorylated carbohydrate is a neutral substance containing mannose and inositol, and the other is a mannose-glycerol-diphosphoric acid (ANDERSON), or an inositol-glycerol-diphosphoric acid in certain cases (DE SÜTÖ-NAGY and ANDERSON). The diphtheria bacillus is particularly confusing since each worker seems to find a different range of hydrolysis products from the phosphatide. Thus an unidentified aldohexose, dihydroxyacetone, a galactose-mannose-phosphoric acid and choline and inositol have all been reported as hydrolytic fragments in four completely contradictory studies (CHARGAFF; GUBAREV, LUBENETS and GALAEV; PUDLES; TAKAHASHI). Several species of Salmonella contain acidic phosphatides, which are not based on choline, ethanolamine, inositol or carbohydrate, but apparently on a series of amino acids, including a large proportion of an unknown amino acid (CMELIK). In the phosphatides of Lactobacillus acidophilus, in spite of the presence of a little choline, most of the nitrogen is unidentified and there is a phosphorylated polysaccharide, formed from galactose, glucose and fructose (CROWDER and ANDERSON).

The phosphatides of seeds and similar tissues generally contain large proportions of lecithin and phosphatidyl ethanolamine. When salts of phosphatidic acids have been found, as in wheat germ (CHANNON and FOSTER), it is now known that these were probably autolytic artifacts (ACKER and ERNST). The "cephalin" fraction is highly complex in many cases, e.g. soyabean and groundnut. Thus soybean cephalin contains at least two types of inositol-containing phosphatide (SCHOLFIELD, DUTTON, TANNER and COWAN), one of which also contains carbohydrate units: galactose, mannose and arabinose (SCHOLFIELD, DUTTON and DIMLER; HAWTHORNE and CHARGAFF). From groundnut cephalin a fraction has been isolated which seems to be an N-glycosyl derivative of the ethanolamine ester of phosphatidyl inositol phosphate, the carbohydrate components being arabinose and galactose (MALKIN and POOLE).

Vegetative tissues, such as leaves and roots, have often been reported to contain salts of phosphatidic acids, but these are now known to be most probably the products of enzymic action during drying and extraction of the lipids (HANA-HAN and CHAIKOFF). The phosphatides of living tissues almost certainly consist of lecithin and phosphatidyl ethanolamine, although metal phosphatidates have been found as one component of the undegraded phosphatides of rubber latex (SMITH). Neither the "lecithin" nor the phosphatidic acid of rubber latex, incidentally, were the simple classical compounds, both containing sugar (mainly galactose) and the phosphatidic acid also containing inositol.

Metabolism.

It is uncertain to what extent phosphatides are mobilised or synthesised during various stages of plant growth and development, probably owing to the difficulty of their quantitative determination in tissues. It is likely, however, that a net synthesis of phosphatides occurs during the germination of seeds (DUCET; HOUGET).

Studies of phosphatide turnover in plants are only now beginning (MAZELIS and STUMPF), although it is well known that highly-active phosphatide-splitting enzymes are of widespread occurrence. The best known, in higher plants, splits the choline from lecithin to give phosphatidic acid (ACKER and ERNST; HANAHAN and CHAIKOFF; SMITH). In leaves this enzyme is concentrated entirely in the chloroplasts (KATES), and in cereal grains mainly in the embryo (ACKER and ERNST), in the latter case its activity increasing greatly on germination. Enzymes have also been found in leaf chloroplasts which can further degrade phosphatidic acid, ultimately to fatty acids, glycerol and inorganic phosphate (KATES).

Bacteria exhibit some interesting sequences of phosphatide degradation, e.g. in Serratia plymuthicum the first stage of breakdown is apparently the loss of one fatty acid, to give a lysophosphatide, and then loss of the remaining fatty acid to give (from lecithin) glycerylphosphorylcholine (HAYAISHI and KORNBERG). This can be degraded by a third enzyme to give choline and glycero-phosphoric acid. Clostridium welchii contains an enzyme which splits lecithin into phosphorylcholine and a diglyceride (MACFARLANE and KNIGHT). Thus phosphatide metabolism possibly follows a different course in bacteria and higher plants.

Literature.

ACKER, L., u. G. ERNST: Über das Vorkommen eines phosphatidspaltenden Ferments in Cerealien. Biochem. Z. **325**, 253–257 (1954). — ANDERSON, R. J.: The chemistry of the lipids f the tubercle bacillus and certain other microorganisms. Fortschr. Chem. organ. Naturstoffe **3**, 145–202 (1939).

BARBIER, M., and E. LEDERER: Sur un acide aminé du phosphatide de *Mycobacterium phlei*. Biochim. et Biophysica Acta **8**, 590–591 (1952).

CHANNON, H. J., and C. A. M. FOSTER: The phosphatides of wheat germ. Biochemic. J. **28**, 853–864 (1934). — CHARGAFF, E.: Über das Fett und das Phosphatid der Diphtherie-bakterien. Hoppe-Seylers Z. **218**, 223–240 (1933). — CMELIK, S.: Über Bakterienlipoide. III. Untersuchung verschiedener Lipoidfraktionen von *Salmonella paratyphi* C. Hoppe-Seylers Z. **296**, 67–73 (1954). — CROWDER, J. A., and R. J. ANDERSON: A contribution to the chemistry of *Lactobacillus acidophilus*. III. The composition of the phosphatide fraction. J. of Biol. Chem. **104**, 487–495 (1934).

DUCET, G.: La choline des vegetaux. Internat. Congr. Biochem. Abstr. of Communications. 1. Congr. Cambridge, Engl. 1949, p. 495–496.

GEIGER jr., W. B., and R. J. ANDERSON: The chemistry of *Phytomonas tumefaciens*. I. The lipids of *Phytomonas tumefaciens*. The composition of the phosphatide. J. of Biol. Chem. **129**, 519–529 (1939). — GUBAREV, E. M., E. K. LUBENETS and Y. V. GALAEV: Chemical composition of some fractions of lipids of diphtheria bacteria. Biochimija 18, 37–46 (1953).

HANAHAN, D. J., and I. L. CHAIKOFF: On the nature of the phosphorus-containing lipides of cabbage leaves and their relation to a phospholipide-splitting enzyme contained in these leaves. J. of Biol. Chem. **172**, 191–198 (1948). — HAWTHORNE, J. N., and E. CHARGAFF: A study of inositol-containing lipides. J. of Biol. Chem. **206**, 27–37 (1954). — HAYAISHI, O., and A. KORNBERG: Metabolism of phospholipides by bacterial enzymes. J. of Biol. Chem. **206**, 647–663 (1954). — HOUGET, J.: Sur le mécanisme de la transformation des lipides en glucides au cours de la germination du ricin. C. r. Acad. Sci. Paris 216, 821–822 (1943).

KATES, M.: Lecithinase activity of chloroplasts. Nature (Lond.) **172**, 814–815 (1953).

MACFARLANE, M. G., and B. C. J. G. KNIGHT: The biochemistry of bacterial toxins. I. The lecithinase activity of *Cl. welchii* toxins. Biochemic. J. **35**, 884–902 (1941). — MALKIN, T., and A. G. POOLE: The structure of the glyceroinositophosphatide of ground nut. J. Chem. Soc. **1953**, 3470–3478. — MAZELIS, M., and P. K. STUMPF: Fat metabolism in higher plants. VI. Incorporation of P^{32} into peanut mitochondrial phospholipids. Plant Physiol. **30**, 237–243 (1955).

PUDLES, J.: Étude chimique des lipides du bacille diphtherique. Thesis, Faculty of Sciences, University of Paris, for the degree of Docteur de l'Université 1953.

SCHOLFIELD, C. R., H. J. DUTTON and R. J. DIMLER: Carbohydrate constituents of soybean "lecithin". J. Amer. Oil Chem. Soc. **29**, 293–298 (1952). — SCHOLFIELD, C. R., H. J. DUTTON, F. W. TANNER jr. and J. C. COWAN: Components of soybean "lecithin". J. Amer. Oil Chem. Soc. **25**, 368–372 (1948). — SMITH, R. H.: The phosphatides of the latex of *Hevea brasiliensis*. 3. Carbohydrate and polyhydroxy constituents. Biochemic. J. **57**, 140–144 (1954). — SÜTÖ-NAGY, G. I. DE, and R. J. ANDERSON: The chemistry of the lipides of tubercle bacilli. LXXVI. Concerning inositol glycerol diphosphoric acid, a component of the phosphatide of human tubercle bacilli. J. of Biol. Chem. **171**, 761–765 (1947).

TAKAHASHI, H.: Bacterial components of *Corynebacterium diphtheriae*. V. Phospholipids. II. Structure of CHARGAFF's corynin. J. Pharmaceut. Soc. Jap. 68, 292–296 (1948.)

The phosphoric acid esters
in carbohydrate metabolism: Lower plants.

By

W. W. Umbreit.

It is a curious, but not necessarily unexpected, fact that studies on phosphorylation in microorganisms are generally far more detailed and far more advanced than similar studies on higher plants. This has been true from the very start, and knowledge of the phosphate esters in yeast frequently preceded the comparable knowledge in muscle, and both far preceded this knowledge in plants. The reasons for this lag in knowledge of plant tissue appear to be not only the relatively fewer number of workers but also the more difficult methodology that is required in plant tissues. To review the entire mass of data on phosphorylation and phosphate esters in carbohydrate metabolism in microorganisms is both too detailed for comprehension and would divert too much space from the main purpose of this volume. We shall, therefore, summarize the present position relying on review summaries for references to further work. A particularly good review is GUNSALUS, HORECKER and WOOD (1955).

There are at least two major pathways between glucose and pyruvate. One of these, the EMBDEN-MEYERHOF pathway, has as its unique character (that is, a reaction series not part of other pathways) the reaction series fructose-6-phosphate to fructose-1,6-diphosphate to triose phosphate. All of the other reactions could be present as part of other known pathways. This in itself is of considerable interest since it points out that many reactions are part of more than one reaction pathway and are common to several routes. To specifically identify the EMBDEN-MEYERHOF system when others are likely to be present thus depends today upon the specific presence of *all* of these enzymes. This pathway can also be identified by way of isotopic tracers, but only in the sense that tracer data would say it could occur, but not that it *does* occur. Specific enzymatic data is generally lacking but tracer studies show results compatable with the EMBDEN-MEYERHOF pathway in several lactic acid bacteria, and in some aerobic and some anaerobic spore formers. Data which are inconsistent with this pathway have been found for a variety of other microorganisms which indicates that while the EMBDEN-MEYERHOF system may be present, there are other pathways present as well.

A second pathway, termed the hexosemonophosphate pathway, has been established in many microorganisms in recent years. This consists of the following complex of reactions. Glucose-6-phosphate is oxidized to 6-phospho-gluconic acid (actually the glucopyranose-6-phosphate is converted to the 6-phosphogluconolactone). The latter substance is converted along either of two pathways. In a minor one (in the sense that it is known in only a few organisms) [*Pseudomonas* (ENTNER and DOUDOROFF 1952, MAC GEE and DOUDOROFF 1954, KOVACHEVICH and WOOD 1955), *Acetobacter* (KOVACHEVICH and WOOD 1955) and *Azotobacter* (MORTENSON and WILSON 1954)], by way of an enzyme termed 6-phosphogluconate dehydrase, 6-phosphogluconate is converted to 2-keto-3-desoxy-6-phosphogluconate and this substance is split to yield pyruvate and glyceraldehyde-

3-phosphate. The major pathway (in the sense that it has been identified in many more microorganisms, and is present in higher plants as well as animals) involves an oxydation to 3-ketophosphogluconate, which is decarboxylated to yield ribulose-5-phosphate. This substance is also a key compound whose subsequent pathways may be varied considerably. These are discussed separately below:

One: Ribulose phosphate is in equilibrium with ribose-5-phosphate. Other pentoses (arabinose, xylose) enter and leave these main metabolic paths by way of their 5-phosphates and isomerization with ribulose-5-phosphate.

Two: Ribulose-5-phosphate, activated by an enzyme termed transketolase, is split into glyceraldehyde-3-phosphate and an "active glycol aldehyde" which must be removed before the reaction can proceed. One way of removing it is by condensation with ribose-5-phosphate (from the first reaction series above) to form heptulose-7-phosphate. The glyceraldehyde-3-phosphate can be converted to pyruvate but it may also be used in a further reaction series mentioned in three. The "active glycol aldehyde" is also used in several reactions mentioned in five.

Three: The heptulose phosphate, found in two, plus glyceraldehyde-3-phosphate, forms fructose-6-phosphate (which may break down to pyruvate *via* the EMBDEN-MEYERHOF system if present, or may be isomerized to glucose-6-phosphate to start the hexosemonophosphate system over) and a tetrose phosphate, which appears to be L-erythrulose phosphate. These reaction series, outlined in two and three, are particularly pertinent for higher plants where there is distinct evidence that they occur and may, indeed, be related to the photosynthetic process.

Four: There appears to be a "side reaction", in the sense that it does not seem to lead anywhere, between the tetrose-phosphate formed in three and the glyceraldehyde-3-phosphate formed in two, to yield a heptulose-diphosphate. Its importance cannot yet be assessed, except that it may be a precursor of shikimic acid and thus of aromatic rings (KALAN and SRINIVASAN 1955).

Five: The tetrose phosphate, formed in three, and the active glycol aldehyde formed in two, condense to yield glucose-6-phosphate, which will be recognized as the starting material. GUNSALUS, HORECKER and WOOD (1955) provide references and discussion of each of these steps and cite the evidence for their occurence in microorganisms (bacteria, yeasts, molds, and algae) as well as their existence in higher plants and animals so that detailed references will not be given here.

However, one further point is of interest. It appears that ribulose-5-phosphate may react with adenosine-triphosphate to form ribulose-diphosphate. This substance can react with CO_2 to form a diphosphohexonic acid, which yields, on cleavage, phosphoglyceric acids. It is supposed that this may be one of the major routes for the incorporation of CO_2 in photosynthesis. So far this system has been found only in green plants and algae (WEISSBACH, SMYRNIOTIS and HORECKER 1954).

There thus exists, in our present knowledge of metabolism of phosphate esters in microorganisms, a wide variety of indications of pathways so far largely unexplored in many plants. Among these paths are many which are known to exist in one plant or another and these should serve as the background to further studies of higher plants.

There are certain phosphate esters, of which acetyl phosphate is the prototype, which are produced by exchange with CoA-acyl groups. The occurrence of these esters seems to be confined to the bacteria. There are other substances, for example, the 2-phospho-4-hydroxy-4-carboxyadipic acid (RAPOPORT and WAGNER 1951, UMBREIT 1953), involved in carbohydrate metabolism which are widely distributed in various organisms but whose discussion has been omitted, along with discussions of phosphate compounds in nucleic acid, amino acid, and lipid metabolism.

Literature.

ENTNER, N., and M. DOUDOROFF: Glucose and gluconic acid oxidation of *Pseudomonas saccharophila*. J. of Biol. Chem. **196**, 853–862 (1952).

GUNSALUS, I. C., B. L. HORECKER and W. A. WOOD: Pathways of carbohydrate metabolism in microorganisms. Bacter. Rev. **19**, 79–128 (1955).

KALAN, E. B., and P. B. SRINIVASAN: Synthesis of 5-dehydro-shikimic acid from carbohydrates in a cell-free extract. Amino Acid Metabolism, p. 826–830. Edit. by W. D. MCELROY and B. GLASS. Baltimore, Md.: Johns Hopkins University Press 1955. — KOVACHEVICH, R., and W. A. WOOD: [1] Carbohydrate metabolism of *Pseudomonas fluorescens*. III. Purification and properties of a 6-phosphogluconate dehydrase. J. of Biol. Chem. **213**, 745–756 (1955). — [2] Carbohydrate metabolism of *Pseudomonas fluorescens*. IV. Purification and properties of 2-keto-3-deoxy-6-phosphogluconate aldolase. J. of Biol. Chem. **213**, 757–767 (1955).

MACGEE, J., and M. DOUDOROFF: A new phosphorylated intermediate in glucose oxidation. J. of Biol. Chem. **210**, 617–626 (1954). — MORTENSEN, L. E., and P. W. WILSON: Initial steps in breakdown of glucose by the azotobacter. Bacter. Proc. **1954**, 108.

RAPOPORT, S., and R. H. WAGNER: A phosphate ester of a tricarboxylic acid in liver. Nature (Lond.) **168**, 295–296 (1951).

UMBREIT, W. W.: The action of streptomycin. VI. A new metabolic intermediate. J. Bacter. **66**, 74–81 (1953).

WEISSBACH, A., P. Z. SMYRNIOTIS and B. L. HORECKER: The enzymatic formation of ribulose diphosphate. J. Amer. Chem. Soc. **76**, 5572–5573 (1954).

The phosphoric acid esters
in carbohydrate metabolism: Higher plants.

By
H. G. Albaum.

I. Introduction.

Phosphorylated compounds are involved in many aspects of carbohydrate metabolism. This has been conclusively shown for animal tissues and many microorganisms. Evidence obtained during recent years shows that similar compounds are involved in the carbohydrate metabolism of plant tissue. Phosphorylated esters are intermediates in the conversion of glucose energy into the high energy bonds of compounds like adenosine triphosphate (ATP). They are essential in the synthesis of starch, just as they are in the synthesis of glycogen in animal tissues and yeast, and many bacterial polysaccharides. Phosphorylated esters have also been implicated in the synthesis and utilization of fatty acids and glycerol, and are essential in the formation of pentoses. The latter are integral parts of nucleic acids and of many of the coenzymes of certain enzyme systems. Green plants in contrast to animals are able to carry on photosynthesis. An accumulating body of evidence has shown that during this process, phosphorylated compounds are produced. It would appear that plants, like all living things, possess a number of common pathways for the breakdown of sugar and related substances; and in most cases, these pathways require phosphorylated compounds.

II. Methods.

The identification of the esters involved in the above reactions has not always been clear-cut and direct. In only a few cases, have these compounds been isolated. The earliest investigations on phosphorylated compounds in plants simply distinguished between phosphorus readily extracted on treatment of the tissue with trichloracetic acid (TCA) and phosphorus which was bound to or associated with the TCA insoluble material. In these early studies, the acid soluble fraction was analyzed for inorganic orthophosphate and organic phosphate. With the discovery of adenosine triphosphate (ATP) in animal tissues and the demonstration that two-thirds of the phosphorus of this compound is labile (hydrolyzable in 7 min. at 100^0 C. in N hydrochloric acid), labile phosphorus determinations were carried out on acid soluble extracts prepared from plants (ARNEY 1939, JAMES and ARNEY 1939). The assumption was made in this work that labile phosphorus was probably synonymous with high energy or ATP phosphorus.

An extension of this method, based on rate of phosphorus hydrolysis was applied to studies on barley (HEARD 1945). In this study, barley grains were germinated for seven days. The plant was removed above the grain and extracted with trichloracetic acid. Hydrolysis of extracts, and of authentic compounds, was carried out in N hydrochloric or N sulfuric acid at 100^0 for 7 minutes, for

60 minutes, and for 180 minutes. It was assumed that any phosphatehydrolyzed in 7 minutes, after correction was made for triosephosphate phosphorus (which was hydrolyzed to the extent of 44 per cent in that interval) must be due to ADP or ATP. The 60 minute hydrolysis value was used as an index, after appropriate correction was made, of phosphopyruvic acid. The phosphorus liberated at 180 minutes, again after appropriate correction, was used to determine hexose diphosphate and hexose-6-phosphate. On the basis of this type of approach, it was found that about 70 per cent of the acid soluble phosphorus in either etiolated seedlings or in green leaves was present as inorganic orthophosphate. Approximately 6 per cent of the phosphorus was ascribed to adenyl pyrophosphate. No triosephosphate was measurable. The remainder of the phosphorus was believed to be largely hexose-6-phosphate with little or no hexosediphosphate being present. It was apparent that this approach gave presumptive evidence for the above compounds, but did not identify them with any surety.

The above procedure was extended and made more reliable by LE PAGE and UMBREIT (1943) in their studies on *Thiobacillus thiooxidans*. They were interested in the phosphorus compounds associated with glycolysis. By using pure compounds and fractionating with barium, these authors were able to show that the following compounds form insoluble barium salts at p_H 7: inorganic orthophosphate, adenosinetriphosphate (ATP), adenosinediphosphate (ADP), phosphoglyceric acid (PGA), and hexosediphosphate (HDP). The following compounds remain soluble as the barium salts at the same p_H: glucose-1-phosphate, glucose-6-phosphate, fructose-6-phosphate, 5-adenylic acid, and diphosphopyridine nucleotide (DPN). They did not imply in their initial publication that these compounds cannot cross-contaminate the different fractions, but they did show that when the amounts of tissue, as well as the volume of reagents were carefully controlled, barium soluble and insoluble compounds did not cross-contaminate to any measurable extent. Briefly, the procedure as in earlier work, consisted of deproteinizing with trichloracetic acid and treating the acid soluble fraction with barium at neutral p_H. The barium insoluble precipitate was removed and both the precipitate and the supernatant analyzed for the compounds present. Inorganic orthophosphate was measured directly. Phosphoglyceric acid was identified both from the stability of its phosphate during acid hydrolysis and by what appeared to be a specific chemical reaction of RAPOPORT (1937). Hexose diphosphate was calculated from hydrolysis data and from fructose (ROE 1934). The ratio of ATP to ADP and the specific quantities present were measured by determining total nitrogen as an index of purine present, pentose, and the ratio of labile phosphorus to total organic phosphorus present, after deducting that phosphorus due to the other compounds detected. The labile phosphorus value had to be corrected for the phosphorus hydrolyzed from hexose diphosphate, which was considerable (26.5 per cent in 7 minutes).

For the estimation and identification of the compounds in the barium soluble fraction, the following techniques were employed: Fructose-6-phosphate was again calculated from fructose; glucose-1-phosphate by hydrolysis of the phosphorus which was completely split off in 7 minutes, at 100° C. in N hydrochloric acid, as well as from its reducing value after removal of phosphorus; adenylic acid from the nitrogen and ribose estimation after correcting for the levels of these present in DPN, which in turn, was estimated from the concentration of nicotinamide. The phosphorus remaining was assumed to result from glucose-6-phosphate; this could be checked independently from its reducing value, making correction, of course, for the reducing values of the other compounds. Results obtained with this procedure in *Thiobacillus* were excellent. Well over

90 per cent of the total phosphorus could be accounted for in terms of the compounds listed above. With slight modification, this technique was readily applicable to animal tissues and has since been used extensively in a wide variety of studies.

In the procedure described above, a complete identification of the different phosphorylated compounds depends on all the compounds being known. Since the procedure worked so well in animal tissues and bacteria, it did not seem unreasonable to assume that it might work equally well for plants. The first application of this technique to plant tissues was carried out by ALBAUM and UMBREIT (1943) on developing oat seedlings. Contrary to expectation, the fractionation procedure turned out to be only partially successful. A few compounds only could be identified with certainty. There was a considerable amount of labile phosphorus but it was already apparent that one could not justifiably equate lability with the presence of high energy phosphorus. As shown above, not only do ATP and ADP have phosphorus which is readily hydrolyzable, but the phosphorus of glucose-1-phosphate is completely labile while the phosphorus of hexose diphosphate is partially so. Furthermore, on the assumption that the corrected labile phosphorus can be used as an index of ATP and ADP content, the apparent amount of pentose was always in great excess. This could be explained, in part, by the observation that sugar formed from the hydrolysis of starch reacts with the pentose reagent giving increased density with non-specific absorption. In addition, when large amounts of starch were present, as frequently occurred, it was found that the phosphoglyceric acid reaction instead of yielding the expected blue color produced one which was almost black due to charring.

When all the possible corrections for known compounds were made, it was apparent that there was still a large amount of phosphorus which could not be accounted for. This led to a search for other types of compounds which might be present. Indeed, the earlier literature had revealed that there are present in cereal plants, large quantities of phytic acid in the form of phytin, and part of the phosphorus which could not be accounted for turned out to be due to this compound. Correcting for phytate still did not give a complete balance indicating the presence of still other phosphorus-containing compounds, not yet identified. The only compounds, therefore, that could be accounted for with certainty in oat seedlings were: inorganic orthophosphate, fructose-6-phosphate, hexose diphosphate, phytic acid, and labile phosphorus which might be associated with ATP. The same kinds of difficulties were encountered by EMERSON, STAUFFER and UMBREIT (1944) in their attempts to fractionate the phosphorus-containing compounds in *Chlorella*. In this study the only compounds which could be identified with certainty were inorganic phosphate, and hexose diphosphate. Again, evidence was obtained for the presence of ATP, but the expected ratios for nitrogen, pentose, and phosphorus were not realized.

Additional difficulties in the fractionation procedure described above, when applied to green tissues, were demonstrated in some experiments carried out with *Euglena* (ALBAUM, SCHATZ, HUTNER and HIERSHFELD 1950). Labile phosphorus in the tissue extracts was present in great excess of that known to be associated with the nucleotides (in these experiments, adenylic acid, ADP, and ATP were enzymatically determined (ALBAUM and LIPSHITZ 1950). If one relied merely on lability of phosphorus it would appear that *Euglena* contained exceedingly large amounts of an ATP-like material. Further analysis using enzymes, however, revealed that a portion of this highly labile phosphorus resulted from inorganic pyrophosphate. The remainder appeared to be inorganic metaphosphate. A perusal of the early literature revealed that inorganic metaphosphate also occurs in the tissues of higher plants (HARDIN 1892). Using a modified fractionation procedure together with specific enzymes, it was therefore possible to demonstrate the presence of the following compounds in *Euglena*: inorganic orthophosphate, pyrophosphate, metaphosphate, adenylic acid, ATP, ADP, DPN, glucose-1-phosphate, fructose-6-phosphate, hexosediphosphate, phosphoglyceric acid, and riboflavin phosphate.

It is important to point out that in the isolation and in the fractionations described above, relatively large amounts of tissue were used. In the case of *Euglena*, cells were harvested from 200 liters of medium; this was equivalent to 24.3 gm. of dried cells. In physiological studies it is difficult to obtain such large amounts of material and consequently difficult to demonstrate phosphate-containing compounds with any great degree of reliability. One method of approach to this problem was suggested by BENSON, BASSHAM, CALVIN, GOODALE, HAAS and STEPKA (1950) on photosynthesis. This work will be referred to in more detail later. From the viewpoint of methodology, however, they were able to demonstrate the presence of various phosphorylated compounds by a combination of chromatographic and radioautographic techniques. We have applied this method to mung bean seedlings using approximately 3 gm. of tissue (ALBAUM and SCHER, unpublished experiments). The seedlings were soaked for short periods of time in solutions containing radioactive orthophosphate. After washing their outer surfaces exhaustively, the seedlings were deproteinized with trichloracetic acid. Protein-free filtrates after neutralization were chromatographed by either upward or downward migration for periods of 12 to 18 hours in one or more of the following developers: (a) butanol-acetic acid-water, consisting of 74 ml. n-butanol, 19 ml. glacial acetic acid, and 50 ml. distilled water; (b) isoamyl alcohol-malonic acid, consisting of 200 ml. of 0.1 M malonic acid brought to p_H 6 plus 75 ml. of isoamyl alcohol; (c) ethanol-acetic acid-water, consisting of 80 ml. ethanol and 0.8 ml. acetic acid brought after mixing to p_H 3.5 with hydrochloric acid, and with distilled water added to give a final volume of 100 ml. After drying, the paper chromatograms were placed on x-ray film for approximately 48 hr. Using known compounds as reference standards and determining their position on the paper by enzyme cleavage with alkaline phosphatase and chemical identification of the position of the phosphorus released, the following compounds were demonstrated in mung beans: glucose-1-phosphate, glucose-6-phosphate, fructose-6-phosphate, inorganic orthophosphate, phosphoglyceric acid, phytic acid, and phosphopyruvic acid. The R_f values obtained for these compounds are listed in Table 1.

The same technique combined with barium fractionation was used by AXELROD, BANDURSKI and SALTMAN (1951) in pea meal extracts where they demonstrated the following phosphorylated compounds: glucose-1-phosphate, phosphoenolpyruvate, hexosediphosphate, adenosine-5-phosphate, a compound which they believed to be fructose-6-phosphate, and a substance which possessed mobilities identical to adenosine triphosphate. No 3-phosphoglyceric acid could

Table 1. *R_f-values for various phosphorylated compounds chromatographed by downward migration in an ethanolacetic acid developer for 12 hr. in a chromatocab[1].*

Compound	R_f value
Hexosediphosphate . . .	0.17
Phosphoglycerate	0.21
Glucose-1-phosphate. . .	0.30
Glucose-6-phosphate. . .	0.40
Inorganic orthophosphate	0.44
Phosphopyruvate	0.62

[1] University Apparatus Company, Berkeley, California.

be detected. The chromatographic procedure was a two-dimensional one described by BANDURSKI and AXELROD (1951). In this procedure Schleicher and Schuell No 589 filter paper, 28 centimeters square was used. They found that the chromatography of phosphorus esters, in general, was best accomplished with free acids. The salts of the esters were readily converted to the corresponding acids and the metallic ions removed from the extracts by shaking the solutions or suspensions of the salts with a cation exchanger such as Dowex No 50. The chromatographic runs were done in cylindrical jars. Development was carried out first in an acid solvent for $6^1/_2$ hours at 2^0 at which time the solvent front reached the upper edge of the paper. After removal from the solvent, the paper was air-dried and developed at right angles to the first development, in an alkaline solvent. The second development in the alkaline solvent required from 12–15 hours. The acid solvent consists of 80 volumes of methanol, 15 volumes of formic acid (88 per cent by weight) and 5 volumes of water. The alkaline one consists of 60 volumes of methanol, 10 volumes of ammonium hydroxide (specific gravity 0.9015) and 30 volumes of water.

The recent experiments of the BENSON *et al.* (1950) group have also provided evidence for the presence of many of the phosphorylated intermediates described above. While their interest was mainly in the pathways of photosynthesis, they nevertheless have not only verified the presence of the intermediates described above, but have presented evidence for new intermediates. They worked with a variety of green plants including *Chlorella, Scenedesmus, Euglena, Sugar Beets*, etc. They were interested in the photosynthetic dark reactions and were able to show that exposure to light for short periods of time leads to the reduction of labeled carbon dioxide to yield phosphorylated compounds. Radioactivity in hexosephosphates, triosephosphates, phosphopyruvic acid, and in phosphoglyceric acid was readily demonstrated. In effect they have detected the synthesis of all of the phosphorylated intermediates of the glycolytic cycle during photosynthesis. In one of the more recent publications from a member of this group (BUCHANAN 1953) glucose-6-phosphate, fructose-6-phosphate, mannose and septulose phosphates were identified. Also demonstrated was sucrose phosphate, which on acid hydrolysis yielded fructose-1-phosphate. The implication of this last observation on the sucrose synthesis problem will be discussed later.

One other procedure for identifying phosphorylated compounds in plant tissues should be mentioned. This involves the use of column chromatography. It has been used by the author in following changes in concentration of ATP during the growth of mung bean seedlings (ALBAUM 1952). Recently ZILL and KRALL (1954) reported the separation of phosphorylated compounds in plants using similar procedures. The details of their column procedure were not available for inclusion into this manuscript at the time of writing.

III. Individual esters and enzymes of their metabolism.

It is clear from the above that a large number of phosphorylated compounds have been identified in different plant tissues. Most of these compounds appear to be identical with those found in other living things. Evidence for the participation of these in essential reactions in the plant also comes from studies in which the enzymes involved in the synthesis and breakdown of these phosphorus esters have been carried out. In the remainder of the article, the individual compounds will be described together with information concerning the enzymes which bring about their transformations.

The main storage carbohydrate in plants is, of course, starch. Within the plant, starch is transformed into glucose-1-phosphate, largely through the action of a phosphorylase. The details of this transformation are the subject of a separate chapter in these volumes (vol. VI). This conversion of starch to glucose-1-phosphate requires a source of inorganic phosphorus. The young seedling in the oat has relatively little inorganic phosphate until about the 48th hour after germination (ALBAUM and UMBREIT 1944). Inorganic phosphorus does not become available in this case to any appreciable extent until *phytic acid*, which is present in high concentration, breaks down under the influence of the enzyme, phytase. The structure of phytic acid is shown below:

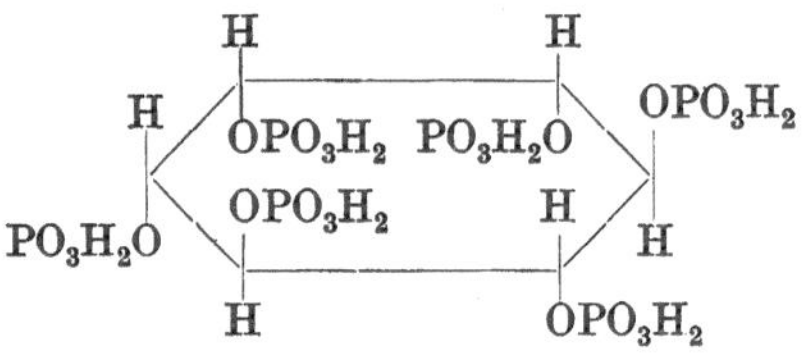

$$
\begin{array}{c}
\text{H} \qquad\qquad \text{H} \\
\text{H} \diagup\; \text{OPO}_3\text{H}_2 \;\; \text{PO}_3\text{H}_2\text{O} \quad \diagdown\; \text{OPO}_3\text{H}_2 \\
\text{OPO}_3\text{H}_2 \qquad \text{H} \\
\text{PO}_3\text{H}_2\text{O} \diagdown \qquad\qquad \diagup\; \text{H} \\
\text{H} \qquad\qquad \text{OPO}_3\text{H}_2
\end{array}
$$

phytic acid

It is found in all cereal plants. Its presence has also been demonstrated in mung beans (ALBAUM 1952). It would appear that its main role in plant metabolism is to act as a storehouse for inorganic phosphorus. In the plant, it exists as the calcium-magnesium salt (phytin). It forms a barium salt, which is completely precipitated at p_H 4. It is very stable to acid hydrolysis, probably more so than in any of the other phosphorus-containing esters.

Glucose-1-phosphate, whose structure is shown below, can readily be isolated from plant tissue.

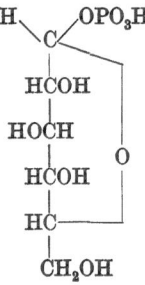

α-D-glucose-1-phosphate

In fact, when it is required as a substrate, it can be prepared enzymatically from starch either by the method of HANES (1939) or by a modification of this procedure by SUMNER and SOMERS (1944). The procedure is outlined below:

8 gms. of soluble starch were boiled with about 100 cc. of water; cooled and added 12 gms. of Na_2HPO_4 (anhyd.) and 5 g. of KH_2PO_4 (anhyd.) dissolved in about 300 cc. of water. Added 100 cc. of potato-cyanide extract. Diluted to 100 cc. Added toluene and mixed. Kept at 20–25⁰ C. for 24 hours. At the end of this time, the potato enzymes were inactivated by the addition of iodine solution. The latter was removed by the addition of sodium thiosulfate. The preparation was then treated with a two per cent pancreatin solution and allowed to stand at room temperature from 3–4 hours. At this point, barium acetate and ammonia were added until the solution was alkaline to phenol red. The barium phosphate precipitate was discarded and the supernate solution was treated with two volumes of 95 per cent alcohol. This precipitated the barium

salt of glucose-1-phosphate. The barium salt was converted into the di-potassium salt, which was crystallized from alcohol. The yield was about 3.5 gms. The glucose-1-phosphate prepared in this way had a purity of about 85 per cent and contained some potassium sulfate. It was obtained practically pure by recrystallizing once according to the procedure described by HANES (1940). The pure material had the empirical formula $C_6H_{11}O_9PK_2 . 2H_2O$ and had a specific rotation with sodium light of about $+79^0$ and a phosphorus content of 8.33 per cent.

Glucose-1-phosphate as indicated above forms a barium salt soluble in water, but is precipitable with two volumes of alcohol. Its phosphorus is split off in 7 minutes when heated in N HCl at 100^0 C. Prior to hydrolysis, it has no reducing value; afterward, however, it has a value equivalent to that of glucose.

The enzyme source for the above preparation was a cyanide extract of potato and was prepared by disintegrating about 325 g. of potato (sliced a few seconds previously) in 100 cc. of 0.01 N hydrochloric acid (KCN neutralized with hydrochloric acid) in a "Powermaster" blender. The disintegrated mass was pressed through cheesecloth, and the juice, amounting to about 175 cc., was centrifuged to free it from starch and cellular matter.

Glucose-1-phosphate is converted into *glucose-6-phosphate* shown below; through the action of a *phosphoglucomutase*:

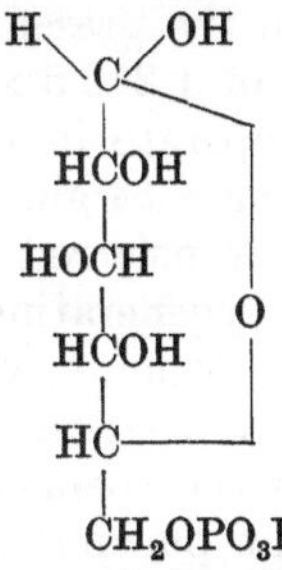

$$
\begin{array}{c}
H\diagdown C\diagup OH \\
HCOH \\
HOCH \\
HCOH \\
HC \\
CH_2OPO_3H_2
\end{array}
$$

D-glucose-6-phosphate

This enzyme has been purified from animal sources. It was first shown to be present in pea extracts by HANES (1939), and more recently it has been partially purified from the green gram, *Phaseolus radiatus* (RAMASARMA, RAM and GIRI 1954). The procedure for isolation is outlined below:

The resting seeds of green gram were finely powdered and the powder was extracted with five times its own weight of water under a layer of toluene for 12 hours at 0^0. After centrifugation, the supernatant was dialyzed against running distilled water overnight and clarified once more by centrifugation. This crude extract was adjusted to p_H 5 with acetic acid and was heated to 45^0 with stirring. It was kept at this temperature for 10 minutes and was rapidly cooled to 0^0; the insoluble material was removed by centrifugation. 30 gm. of ammonium sulfate was added to each 100 cc. of the supernatant. The precipitate formed was discarded. 15 additional gm. of solid ammonium-sulfate was added to the supernatant and the precipitate this time was collected. It was dissolved in a convenient volume of water and dialyzed as before to free it of ammonium sulfate which interferes with the estimation of enzyme activity. By this procedure a fourfold purification was achieved. Attempts at further fractionation resulted in much loss of the activity. The preparation was devoid of phosphorylase activity. It was activated by Mg^{++}, Mn^{++}, and Co^{++}. Sodium sulfite inhibited the enzyme in the presence of veronal buffer but activated in its absence. The enzyme was inhibited by Na^+, Hg^{++}, Ag^+, Cu^{++}, KCN, and Ca^{++}. Metal-binding agents such as cysteine did not activate.

The optimum p_H of the enzyme was around p_H 7.5. Temperature optimum was 45^0. The enzymatic conversion of glucose-1-phosphate to glucose-6-phosphate attained equilibrium when about 95 per cent of the substrate was converted.

Enzyme activity was determined by estimating the unconverted glucose-1-phosphate. This was done by measuring the inorganic phosphorus liberated on hydrolysis in N sulfuric acid in a boiling waterbath for 7 minutes. Glucose-6-phosphate is not hydrolyzed under these conditions. The decrease in inorganic phosphorus after hydrolysis is an index of the glucose-6-phosphate formed.

The work of the LELOIR group in Argentina has shown that in addition to a metallic activator, phosphoglucomutase requires *glucose-1-6-diphosphate* as a coenzyme. CARDINI (1951) has studied the requirements for this coenzyme by extracts from seeds of the legume, *Canivalia ensiformis*. For these experiments, he soaked seeds for 24 hours, skinned them, homogenized in a Waring blender in one half their weight of cold water. This material was strained through cheese-cloth and the extract centrifuged. The supernatant was dialyzed overnight against distilled water and used as an enzyme source. In the absence of added glucose-1-6-diphosphate, there was practically no activity. Maximum activation was obtained at a concentration of 0.2×10^{-9} moles of glucose-1-6-diphosphate. CARDINI found no effect with added cysteine on the reaction when no glucose-1-6-diphosphate was added, but in the presence of the added coenzyme, activation occurred at a concentration of $1.8-2.5 \times 10^{-2}$ M.

The coenzyme is believed to operate in the following way:

In going from glucose-1-phosphate to glucose-6-phosphate, phosphorus from the 1 position of the coenzyme is transferred to the 6 position of the substrate, leaving glucose-6-phosphate and regenerating the coenzyme. In going from glucose-6-phosphate to glucose-1-phosphate, phosphorus is transferred from the 6 position of the coenzyme to the 1 position of the substrate, leaving glucose-1-phosphate and again regenerating the coenzyme. It is interesting to point out that the equilibrium for the plant phosphoglucomutase appears to be similar to that found in muscle and in yeast.

Two mechanisms are known for the formation of the coenzyme. Extracts of muscle and yeast convert glucose-1-phosphate to glucose-1-6-diphosphate in the presence of ATP. In *Escherichia coli* two moles of glucose-1-phosphate interact to form one mole of glucose-1-6-diphosphate and one mole of glucose. The extracts in CARDINI's experiment form glucose-1-6-diphosphate by the first method.

Glucose-6-phosphate may also be produced by another pathway, the phosphorylation of glucose in the presence of ATP. The enzyme involved is a *hexokinase*. In yeast this enzyme can phosphorylate fructose and mannose as well. Presumptive evidence for the presence of such an enzyme has been available for some time. In 1953, however, SALTMAN (1953) partially purified this enzyme from wheat germ and showed that it was present in a wide variety of other plant tissues. Whereas in animal tissues and in yeast, the enzyme appears to be soluble, in plant tissues it appears to be partially soluble and partially bound to the mitochondrial fraction of the cell. The insoluble or mitochondrial fraction was prepared as follows: 15 gm. of wheat germ and 60 ml. of cold 0.5 M mannitol were homogenized with sand in a cold mortar. The mixture was centrifuged at $500 \times g$ for 10 minutes to remove starch and cell debris. The supernatant fluid was removed and recentrifuged at $18,000 \times g$ for 15 minutes, and the supernatant fluid from this centrifugation was removed by suction. The residue was resuspended in 20 ml. of 0.5 M mannitol and recentrifuged at $18,000 \times g$ for 10 minutes, and the supernatant fluid was again removed, and the residue suspended in 5.0 ml. of cold distilled water. This suspension was used as the enzyme.

It may be stored at 2^0 for 10 days with the loss of about ten per cent of the original activity.

The soluble fraction was obtained by extracting the wheat germ with 4 times its weight of cold water for 30 minutes at 2^0 with agitation. The mixture was centrifuged at $18,000 \times g$ for 20 minutes, and the cloudy supernatant fluid was decanted, adjusted to p_H 5.5 with N acetic acid, and immediately centrifuged. The precipitate was discarded and the clear supernatant fluid was brought to p_H 7.0 with N KOH. A saturated solution of ammonium sulfate, p_H 7.0, was added to bring the supernatant solution to 55 per cent saturation with respect to ammonium sulfate and the resulting precipitate discarded. Additional saturated ammonium sulfate was added to bring the enzyme solution to 65 per cent saturation, and the precipitate was collected by centrifugation, dissolved in distilled water, and dialyzed overnight against distilled water. By this procedure a 5-fold increase in the specific activity was obtained.

The distribution of the enzymes of several plant tissues is shown in table 2, which is taken from SALTMAN's paper.

Table 2. *Distribution of hexokinase in soluble and insoluble fraction of several tissues.*

Tissue	Insoluble units[1] per fraction	Per cent activity	Soluble units per fraction	Per cent activity
Wheat germ (commercial)	46.8	33	93.6	67
Wheat germ (dissected)	100.8	68	46.8	32
Potato	16.2	100	0.0	0
Potato (blendor)	11.7	30	27.9	70
Pea seed (var. Alaska)	6.1	30	13.8	70
Mung bean seed	16.4	28	42.0	72
Mung bean hypocotyl	14.0	19	61.0	81
Avena seed	3.5	19	14.7	81
Spinach leaf	1.3	28	3.4	72

[1] 1 unit = 1 mole of glucose phosphorylated per 10 minutes in 1.5 ml. of standard reaction mixture.

The enzyme appears to have a broad p_H optimum ranging from about p_H 8 to p_H 11. The Km (glucose) is 4.4×10^{-4} M and the Km (ATP), 8.7×10^{-4} M. As in the case of the yeast enzyme, Mg^{++} ions are essential for activity, the optimum concentration being 0.01 M. At this concentration Mn^{++} ions are only 80 per cent as effective. Co^{++} is completely ineffective while $CuSO_4$, $ZnCl_2$, and $HgCl_2$ act as potent inhibitors. The temperature optimum for the insoluble hexokinase was found to be 37^0. The energy of activation based on the low temperature portion of the plot of log activity versus reciprocal of absolute temperature yielded a value of 11,670 calories per mole. The enzyme is capable of phosphorylating glucose, fructose, mannose, and glucosamine. The relative rates of phosphorylation at a substrate concentration of 0.0033 molar are 1.00, 0.62, 0.68, and 0.52 respectively. Substances which influence SH groups exert only a small effect on plant hexokinase. This indicates that the hexokinase of higher plants is not markedly dependent upon functional-SH groups for activity. In this respect plant hexokinase resembles the hexokinase of yeast.

STAFFORD, BARNETT, CONN and VENNESLAND (1952) have prepared ammonium sulfate fractions from water extracts of wheat germ, which in addition to phosphorylating glucose, fructose, and mannose, are also able to phosphorylate galactose in the presence of ATP. The partially purified hexokinase described above is unable to phosphorylate galactose. This suggests that wheat germ also contains a *galactokinase*, which in yeast phosphorylates galactose in the

1 position. The above authors also present evidence that glucose-6-phosphate is an end product of galactose phosphorylation. If this is true, it must be assumed

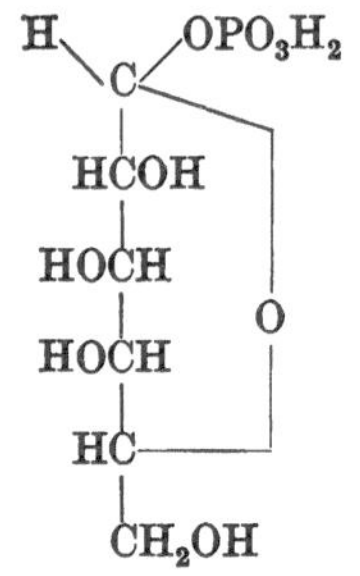

α-D-galactose-1-phosphate

that wheat germ contains a *galacto-waldinase*, which converts *galactose-1-phosphate* into glucose-1-phosphate, which in turn is converted into glucose-6-phosphate via the phosphoglucomutase reaction. The galacto-waldinase requires as a coenzyme, *uridine diphosphoglucose*, shown below:

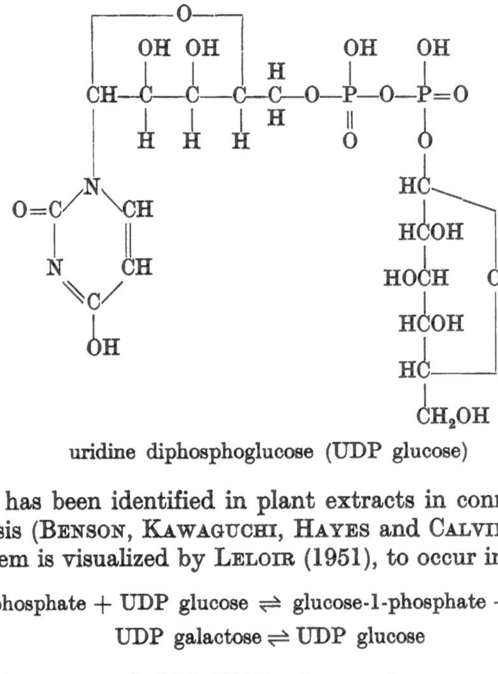

uridine diphosphoglucose (UDP glucose)

This compound has been identified in plant extracts in connection with studies on photosynthesis (BENSON, KAWAGUCHI, HAYES and CALVIN 1952). The operation of this system is visualized by LELOIR (1951), to occur in the following way:

galactose-1-phosphate + UDP glucose ⇌ glucose-1-phosphate + UDP galactose

UDP galactose ⇌ UDP glucose

The mechanism by means of which UDP galactose is reconverted to UDP glucose is not yet clear.

The kinase reactions described above require ATP. Although SALTMAN (1953) reports that inosine triphosphate can function as a phosphate donor in the hexokinase reaction, it is only 35 per cent as effective. Muscle tissues can utilize ADP as well as ATP as a phosphate donor. This is due to the presence in muscle of an enzyme, adenylate kinase, which effects the following transformation:

$$2\ ADP \rightleftharpoons ATP + AA$$

action of adenylate kinase

It is clear from the above reaction that ADP is not the phosphate donor, where it appears to function in the hexokinase reaction, but ATP, which is generated. Adenylate kinase has been partially purified from plant sources by ammonium sulfate fractionation (CAMPBELL and BANDURSKI 1952).

ATP with a purity of about 70 per cent has been isolated *from mung beans* (ALBAUM, OGUR and HIRSHFELD 1950). An outline of the method follows:

Mung bean sprouts were extracted with trichloracetic acid. The trichloracetic acid extracts were neutralized and treated with barium. The barium precipitates were suspended in water and the p_H adjusted to 2.0. The supernatant was discarded. The material remaining insoluble was freed of barium by stirring in an excess of M sodium sulfate. The nucleotide was precipitated as the silver salt. Silver was removed as the chloride and the nucleotide once more precipitated as the barium salt. Further purification was effected by treatment with amberlite, IR 100. This preparation was chemically identical with ATP isolated from animal sources, but behaved differently in some of the enzyme reactions, in which ATP is known to participate. After further treatment with mercury, normally used in isolation procedures for animal ATP, its behaviour was indistinguishable from the animal product.

The enzyme which converts glucose-6-phosphate into *fructose-6-phosphate*, *phosphohexosisomerase*, has not been isolated from plant tissue. Its presence, however, has been shown in the investigations of SOMERS and COSBY (1945). In these studies the *Thomas Laxton* variety of peas were used. These were ground to a fine powder in a porcelain mill. The ground meal was extracted with water. The extract was centrifuged and exhaustively dialyzed against distilled water. This was used as an enzyme source. This preparation was able to convert fructose-6-phosphate into an aldosemonophosphate which appeared to be indistinguishable from glucose-6-phosphate. About 70 per cent of the added fructose-6-phosphate was converted. The presence of fluoride and iodoacetate did not effect this equilibrium. The relative proportions of the ketose monophosphate and the aldomonophosphate when equilibrium had been reached agreed well with the relative proportions found at equilibrium in yeast and muscle extracts. This disappearance of fructose-6-phosphate was measured by three separate criteria, the rate of phosphorus hydrolysis, the disappearance of fructose, and the appearance of aldose. The structure of the ester is shown below:

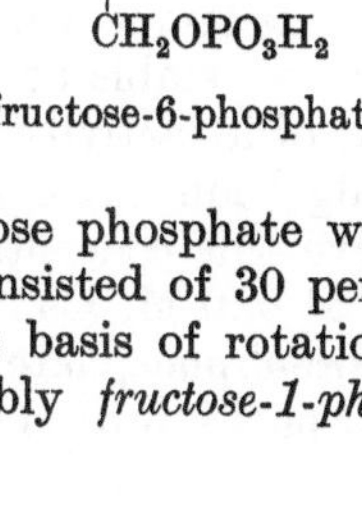

$$
\begin{array}{l}
\text{CH}_2\text{OH} \\
\quad | \quad\diagup\text{OH} \\
\quad \text{C} \\
\quad | \\
\text{HOCH} \qquad \\
\quad | \qquad\quad \text{O} \\
\text{HCOH} \\
\quad | \\
\text{HC}\!\!-\!\!-\!\!-\!\! \\
\quad | \\
\text{CH}_2\text{OPO}_3\text{H}_2
\end{array}
$$

D-fructose-6-phosphate

A mixture of fructose and glucose phosphate was isolated from pea leaves by HASSID (1938). The mixture consisted of 30 per cent glucose and 70 per cent fructosemonophosphate. On the basis of rotation, HASSID concluded that the fructose component was probably *fructose-1-phosphate*. This ester has been described in other tissues, but is not in the direct glycolytic pathway. It may, however, be involved in the synthesis of sucrose. This will be considered later.

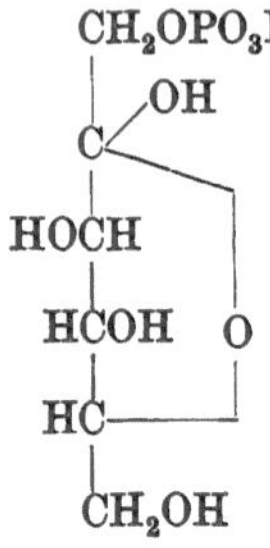

$$CH_2OPO_3H_2$$

D-fructose-1-phosphate

Fructose-6-phosphate in the presence of ATP is converted into *fructose-1-6-diphosphate*, with ATP acting as the phosphate donor.

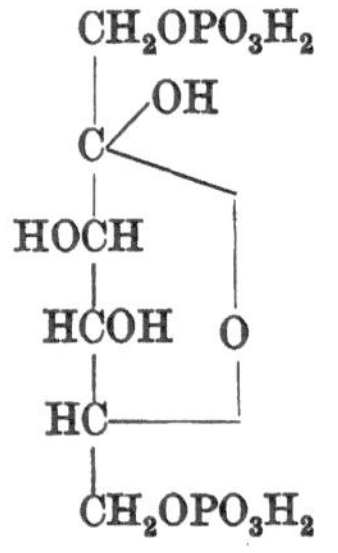

$$CH_2OPO_3H_2$$

D-fructose-1-6-diphosphate

The enzyme involved is *phosphohexokinase*. It has been partially purified by Axelrod, Saltman and Bandurski (1952) from pea meal. Pea seed meal (40 mesh), *Alaska* variety, was suspended in 2 to 4 times its weight of cold (2⁰) water for 30 minutes and agitated occasionally. This mixture was centrifuged (16,000 × g) for 20 minutes. The supernatant was poured off and adjusted to p_H 7.5 with 2 N NaOH and made 28 per cent saturated with respect to ammonium sulfate by the addition of the proper volume of saturated ammonium sulfate solution, p_H 7.5. After standing 30 minutes, the mixture was centrifuged 20 minutes at 16,000 × g and the supernatant made 38 per cent saturated with the required volume of saturated ammonium sulfate solution. The precipitate collected after 30 minutes standing and 20 minutes centrifugation at 16,000 × g was dissolved in a volume of water equal to approximately one-third of the original volume of extract and dialyzed with gentle agitation against three changes of 40 volumes of freshly prepared 10^{-4} M. cysteine. All preparative procedures were carried out at 2⁰.

The above fractionation resulted in a 6-fold increase in specific activity. The enzyme has a sharp p_H maximum at p_H 6 and another maximum, approximately 16 per cent lower, at p_H 9. Its temperature optimum is 38⁰, energy of activation, 3700 calories per mole. It was stable at –5⁰ centigrade for as long as two weeks, but lost over 50 per cent of its activity in 20 hours at 2⁰, and more than 90 per cent in 6 hours at 25⁰. Mg^{++} ions are essential for activity. The optimum concentration was 0.01 M. Above this concentration, inhibition occurred. Mn^{++} ions are about half as effective at 0.01 M. Cu^{++} acted as a potent inhibitor. The Michaelis constant for fructose-6-phosphate as determined graphically was 7.1×10^{-3} M. Adenosine diphosphate proved to be half as effective as ATP as a phosphate donor under standard test conditions. The apparent activity of ADP according to the authors may well have been due to the presence of adenylate kinase in the enzyme preparation. The enzyme apparently occurs rather widely in materials from higher plants. It has been demonstrated in oat seeds, in sunflower fruit, in tomato seeds, in pine pollen, and the seeds of radish, soybean, broadbean, and corn.

The conversion of fructose-1-diphosphate into *dihydroxyacetone phosphate* and *3-phosphoglyceraldehyde* is brought about by the enzyme, *aldolase*. This enzyme has been isolated by STUMPF (1948) from peas. The enzyme was readily extracted

$$\begin{array}{ll}
\mathrm{CH_2OPO_3H_2} & \mathrm{H}\diagdown\mathrm{C}\diagup^{\mathrm{O}} \\
| & | \\
\mathrm{C{=}O} & \mathrm{HCOH} \\
| & | \\
\mathrm{CH_2OH} & \mathrm{CH_2OPO_3H_2}
\end{array}$$

dihydroxyacetone phosphate   D-phosphoglyceraldehyde-3-phosphate

from soaked pea seeds with one-tenth per cent potassium carbonate as the extracting solvent. The extract then was subjected to (1) ammonium sulfate fractionation, (2) isoelectric precipitation with dilute acetic acid at p_H 5.5 by which procedure much inert protein was precipitated while the aldolase remained in solution and finally (3) by acetone fractionation followed by dialysis and a final isoelectric precipitation. This procedure resulted in a 92-fold purification. The enzyme is rather stable; solutions can be frozen, and stored indefinitely without loss of activity. It is unstable below p_H 5.5 and above 10, but stable within these ranges. The enzyme is completely inactivated by exposure to 60^0 for five minutes. Pea aldolase like animal aldolase exhibits the same high specificity to fructose-1-6-diphosphate. The monophosphates of glucose-1-phosphate and glucose-6-phosphate and the monophosphate of fructose-6-phosphate are inert in the enzyme system. The p_H optimum is approximately 8.5 in 0.1 M veronal buffer. The MICHAELIS-MENTEN constant is approximately 0.8×10^{-3} M, which suggests that the pea aldolase has an affinity some 10-fold greater than that of animal aldolase. In sharp contrast to the animal enzyme, pea aldolase is not very sensitive to heavy metal inhibition. Thus, in a final concentration of the 10^{-4} M, copper sulfate, mercuric acetate, phenylmercuric acetate, and silver nitrate do not inhibit the enzyme.

Attempts to resolve the enzyme into a protein moiety and a specific group were unsuccessful. After dialysis against 10^{-2} M acetate buffer at p_H 4.5, the enzyme was completely inactive; addition of either zinc, magnesium, manganese, or cobaltous ions to the dialyzed enzyme did not restore activity. The enzyme has also been shown to be present in 29 different species studied: among them, fungi, conifers, monocotyledons, and dicotyledons (TEWFIK and STUMPF 1949).

The enzyme which catalyzes the equilibrium between dihydroxyacetone phosphate, and 3-phosphoglyceraldehyde, *triosephosphate isomerase* has not been purified from plant sources, but its presence has been established (STUMPF 1950). In this work, acetone powders from pea meal were employed. These powders were suspended in five times their weight of distilled water, adjusted to p_H 6.5 with 0.1 M sodium bicarbonate. After 10 minutes they were centrifuged. The supernatant was dialyzed for 12 hours against 10^{-5} M thioglycolate at 4^0. Such preparations were able to convert fructose-diphosphate to phosphoglyceric acid. For every mole of fructose-diphosphate used, two moles of phosphoglyceric acid were formed. If the triosephosphate isomerase were not present, one would expect only one mole of phosphoglyceric acid formed from each mole of hexosediphosphate. If the formation of phosphoglyceric acid is blocked by the addition of iodoacetamide and if triosephosphate isomerase is not present, one would expect equimolar quantities of dihydroxyacetone phosphate and phosphoglyceraldehyde to be formed. In animal tissues where a triosephosphate isomerase is present, one gets through the action of aldolase on hexosediphosphate, a mixture consisting of 97 per cent dihydroxyacetone phosphate and 3 per cent phosphoglyceraldehyde. In the plant system, 96 per cent dihydroxyacetone phosphate and 4 per cent phosphoglyceraldehyde were found, thus establishing the presence of a triosephosphate isomerase.

When a yeast fermentation mixture is treated with an aldehyde fixative, one half of each hexose molecule is converted to acetaldehyde and the other half into glycerol. It has been determined that the mechanism of glycerol formation involves the reduction of dihydroxyacetone phosphate to glycerophosphate and the subsequent hydrolysis of the latter by phosphatase to yield glycerol.

$$
\begin{array}{ccccc}
CH_2OPO_3^= & & CH_2OPO_3^= & & CH_2OH \\
| & DPN^+ \quad DPNH & | & & | \\
C=O & \longleftrightarrow & HCOH & \longrightarrow & CHOH \\
| & & | & H_2O & | \\
CH_2OH & & CH_2OH & & CH_2OH \\
\text{dihydroxyacetone} & \alpha\text{-glycerophosphate} & \text{L-}\alpha\text{-glycerophosphate} & \text{phosphatase} & \text{glycerol} \\
\text{phosphate} & \text{dehydrogenase} & & &
\end{array}
$$

The only step in this sequence of reactions which is irreversible is the last one and it is generally agreed that the rephosphorylation of glycerol can only occur in the presence of a *glycerol kinase* and ATP. When extracts of the peanut (*Arachis hypogaea* L., var. *Virginia Jumbo*) were fractionated into a mitochondrial portion (MP) and a soluble protein fraction (SPF), a combination of both fractions led to the rapid formation of radioactive CO_2 from radioactive glycerol (STUMPF 1955). The SPF contains the glycolytic enzymes. The mitochondrial fraction contains the enzymes of the tricarboxylic acid cycle together with the cytochrome system. This system was completely inactive if either fraction, or ATP were left out. STUMPF has shown that plant mitochondria contain a glycerol kinase which apparently converts glycerol in the presence of ATP into *α-glycerophosphate*. Also attached to the mitochondria is an α-glycerophosphate dehydrogenase, which then oxidizes α-glycerophosphate to dihydroxyacetone phosphate, which in turn is converted into phosphoglyceraldehyde and enters the main carbohydrate pathway. It would appear from these observations that plant materials contain both a glycerol kinase and α-glycerophosphate dehydrogenase. It is also interesting to point out that as in muscle, the latter enzyme is associated with the mitochondria.

The transformation of 3-phosphoglyceraldehyde into *1-3-diphosphoglyceric acid* is mediated through a *triosephosphate dehydrogenase*.

$$
\begin{array}{c}
O\diagdown \\
\quad C\diagup OPO_3H_2 \\
| \\
HCOH \\
| \\
CH_2OPO_3H_2
\end{array}
$$

1-3-diphospho-D-glyceric acid

In the pea seed, this enzyme is DPN mediated. It has recently been purified (HAGEMAN and ARNON 1955). The procedure briefly consisted of extracting the acetone powder from pea seeds twice with potassium phosphate buffer at p_H 7.2 containing EDTA. The combined filtrates were heated to 55^0 and rapidly cooled.

The solution was then made 60 per cent saturated with respect to ammonium sulfate and the combined precipitate discarded. The supernatant solution was made 95 per cent saturated with respect to ammonium sulfate and the precipitate in this instance was collected and dissolved in 0.0015 M EDTA, p_H 7. Solid ammonium sulfate was again added to 40 per cent saturation and the precipitates discarded. The ammonium sulfate concentration was increased to 60 per cent and the p_H adjusted to 7.9. This precipitate contained the major portion of the enzyme activity and was stable over a period of five weeks at 3^0 in the 60 per cent saturated ammonium sulfate solution. The final purification was effected by dilution. Small (2 ml.) aliquots were taken from the protein suspension (60 per cent saturated ammonium sulfate), and the supernatant fluid was discarded. The precipitate was dissolved in 2 ml. of 0.0015 M EDTA and diluted with 10 ml. of cold (0^0 C.) water. The solution was allowed to stand for 30 minutes at 0^0 C. before the insoluble material was centrifuged and discarded. The original material extracted had a specific activity of 113. After dilution, the specific activity increased to 2930. At this final purification stage, the specific activity is of the same order of magnitude as the crystalline rabbit muscle triosephosphate dehydrogenase. The enzyme had no activity when TPN was substituted for DPN.

When the same method used for first demonstrating the enzyme in pea seeds (STUMPF 1950) was applied to pea leaves, it was not possible to detect the presence of the enzyme (TEWFIK and STUMPF 1951). This observation occasioned a good deal of interest since the accumulating evidence had indicated that photosynthesis occurs via a reversal of the same process which leads to the breakdown of sugar, namely glycolysis. If these reactions are to be reversed, a *triosephosphate dehydrogenase* must be present. This problem has recently been re-examined in acetone powders prepared from leaves of sugar beets, sunflower and tobacco (ARNON 1952). In such preparations, if proper precautions are taken, it is possible to show that a triosephosphate dehydrogenase is present, which can operate either with TPN or DPN. It operates much more effectively, however, with the former and does not show the same high sensitivity to iodoacetate exhibited by the DPN mediated enzyme from pea seeds.

ARNON was of the tentative opinion that only one enzyme is involved, which is TPN dependent "in vivo", but which can "in vitro" bring about the oxidation of phosphoglyceraldehyde with DPN as well. This assumption was also supported by the observation that TPN rather than DPN, is the dominant nucleotide in green leaves.

The ability to transfer phosphate from *1-3-diphosphoglyceric acid* to adenosine diphosphate has been shown to occur in a wide variety of plant extracts (AXELROD and BANDURSKI 1953). The enzyme involved is a *phosphoglyceryl kinase*. It has been shown to be present in the tomato stem, leaf, green and ripe fruit, in the Irish potato, in the sweet potato, in mung bean seeds, corn grains and wheat germ.

$$
\begin{array}{ccc}
\underset{\displaystyle \underset{\displaystyle \underset{\displaystyle \text{1-3-diphospho-D-glycerate}}{CH_2OPO_3^=}}{HCOH}}{O\diagdown C \diagup OPO_3^=} + ADP & \longleftrightarrow & \underset{\displaystyle \underset{\displaystyle \underset{\displaystyle \text{3-phospho-D-glycerate}}{CH_2OPO_3^=}}{HCOH}}{O \diagdown C \diagup O^-} + ATP
\end{array}
$$

The enzyme has been partially purified from pea seeds fractionated with ammonium sulfate. The preparation briefly consisted of grinding pea seeds in a WILEY mill and extracting for 10 minutes with water. The supernatant after centrifugation for 10 minutes was then fractionated with ammonium sulfate. Fractionations were made at p_H 7.5, at 5^0 C. Maximum activity was obtained in that fraction which precipitated between 65 and 70 per cent saturation. This material showed a 10-fold purification over the original extract. The enzyme has a broad p_H optimum, between 7 and 10. It is activated by Mg^{++}, the maximum effect occurring at about 0.07 M Mn^{++} and Co^{++} ions are equally as effective. Ca^{++} ions are 52 per cent as effective, and Fe^{+++} only about 30 per cent. The enzyme is inhibited by copper, zinc, and mercury. It is also inhibited by sodium fluoride at a concentration of approximately 0.01 M. The MICHAELIS-MENTEN constant was 7.6×10^{-3} M for phosphoglyceric acid and 4.1×10^{-3} M for ATP. ADP was five per cent as effective as ATP and inosine triphosphate 16.3 per cent as effective.

The transformations of *3-phosphoglyceric acid* into *2-phosphoglyceric acid* and the conversion of the latter to *phosphoenolpyruvic acid* are catalyzed by a *phosphoglyceromutase* and an *enolase* respectively.

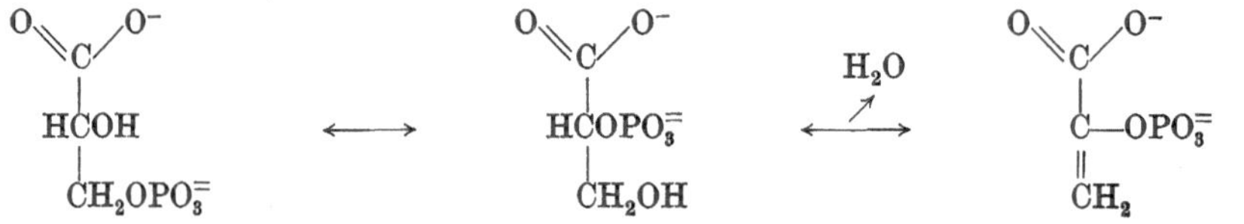

| 3-phospho-D-glycerate | 2-phospho-D-glycerate | phospho-enol-pyruvate |

These enzymes have not been isolated from plant sources although their presence has been demonstrated. Pea leaf acetone powder preparations or fresh homogenates when incubated with 3-phosphoglyceric acid formed measurable amounts of pyruvate (TEWFIK and STUMPF 1951). This points to the presence of these two enzymes. The addition of sodium fluoride, which is known to inhibit enolase completely inhibited the formation of pyruvate.

In muscle, the conversion of 3-phosphoglyceric acid into 2-phosphoglyceric acid involves the participation of a coenzyme, *2-3-diphosphoglyceric acid*, which is believed to operate in much the same fashion as glucose-1-6-phosphate in the phosphoglucomutase reaction.

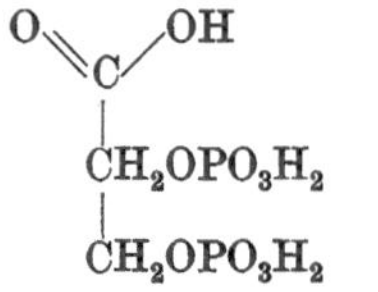

2-3-diphosphoglyceric acid

This coenzyme, thus far, has not been identified in plant materials.

The transfer of phosphate from phosphoenolpyruvic acid to adenylic acid has been shown to take place in pea seed extracts and extracts of pea leaves (TEWFIK and STUMPF 1951). Whether AMP is the direct phosphate acceptor or whether AMP is first converted to ADP before it accepts phosphate from phosphoenolpyruvic acid is not clear. More recently, BANDURSKI, GREINER and BONNER (1953) have shown that purified, dialyzed extracts of spinach leaves, catalyze the formation of oxalacetate from phosphoenolpyruvic acid and $C^{14}O_2$ with the liberation of inorganic phosphate.

It is clear from the material presented that not only do plants possess the inter-mediates in the classic glycolytic cycle, but they also contain enzymes which are able to bring about transformations similar to those occurring in muscle and in yeast.

One other sugar phosphate is worthy of special mention. The problem of sucrose synthesis has long intrigued the plant physiologist. CALVIN and BENSON (1949) has shown that it is the first free sugar formed during photosynthesis. When DOUDOROFF *et al.* (1943) demonstrated that sucrose in bacteria is synthesized from glucose-1-phosphate and fructose through the action of a sucrose phosphorylase, attempts were made to show that a similar enzyme operates in higher plants. Such an enzyme has not been demonstrated. The problem of sucrose synthesis, therefore, in higher plants still remains obscure. Recently, however, BUCHANAN (1953) showed that extracts from sugar beet leaves contained in the area generally designated on chromatographs as the hexosemonophosphate area, a new compound which could be separated from the usual hexosemonophosphates by rechromatographing in different solvents. This ester appeared to be a disaccharide. By a number of independent tests, BUCHANAN established that this new ester was a *sucrose phosphate* which on acid hydrolysis yielded glucose and fructose-1-phosphate. The structure proposed for this new ester is shown below:

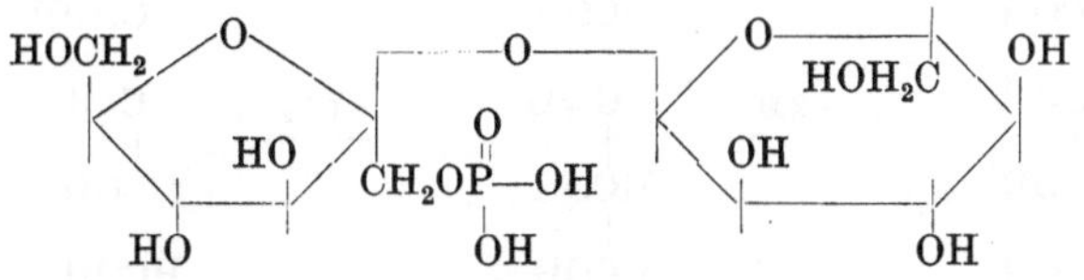

sucrose phosphate

BUCHANAN also suggested that this new compound is formed from uridine diphosphoglucose and fructose-1-phosphate. According to the author, this conclusion is partially supported by the observation that in kinetic experiments, using $C^{14}O_2$, the uridine diphosphoglucose becomes labeled shortly before sucrose (BENSON, KAWAGUCHI, HAYES and CALVIN 1952).

Until several years ago, no evidence was available for the occurrence in plants of a hexosemonophosphate *Shunt* pathway, as originally described by DICKENS (1938) for yeast. That such a pathway does function in higher plants was first suggested by CONN and VENNESLAND (1951). They were interested in a glutathione reductase present in wheat germ. The reaction required reduced TPN. They generated the latter by using glucose-6-phosphate as a substrate in the presence of glucose-6-phosphate dehydrogenase (Zwischenferment). They found that the reaction proceeded with or without the addition of the Zwischenferment. This meant that the enzyme was present in the wheat germ. This enzyme converts glucose-6-phosphate into *phosphogluconic acid*, and is TPN mediated.

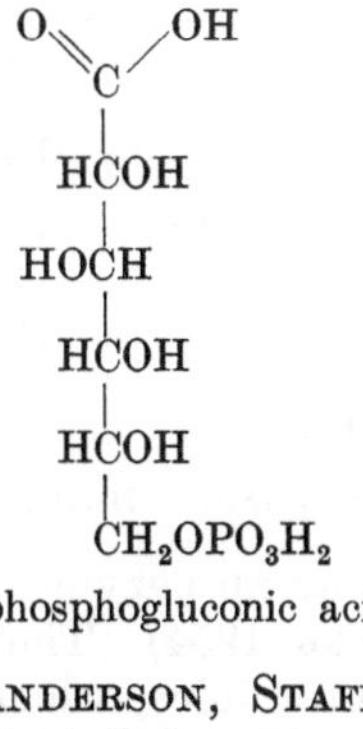

6-phosphogluconic acid

In a subsequent publication (ANDERSON, STAFFORD, CONN and VENNESLAND 1952) the enzyme was demonstrated in extracts of spinach leaves, cabbage leaves, potato tubers, avacado fruit and cantaloupe, as well as in parsley leaf and stem, etiolated pea seedlings and cucumber fruit. In general, the concentration of the enzyme was lower in roots.

AXELROD, BANDURSKI, GREINER and JANG (1953) showed that water extracts from acetone powders of spinach leaves contain a system that not only carries out the latter step, but also converts 6-phosphogluconic acid into a mixture of ribose and *ribulose phosphates*. The first step in the conversion leads to the transformation of 6-phosphogluconic acid into pentose phosphate and CO_2. This has been visualized as a two-step reaction:

The first which converts 6-phosphogluconic acid into a hypothetical *3-keto-6-phosphogluconate* and the second which leads to the decarboxylation of the latter to form *ribulose-5-phosphate*. These reactions are shown below:

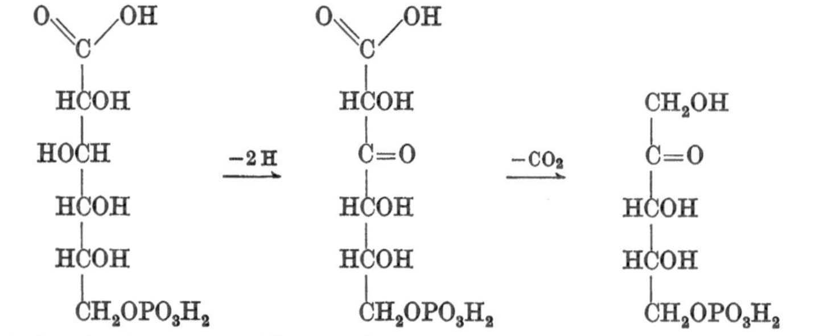

6-phosphogluconate   3-keto-6-phosphogluconate   D-ribulose-5-phosphate

The first of these reactions is catalyzed by a *phosphogluconic dehydrogenase* and is TPN mediated. Its presence has been demonstrated in wheat germ, in cantaloupe fruit, in parsley leaves, spinach leaves, parsnip root, cucumber fruit and turnip root (BARNETT, STAFFORD, CONN and VENNESLAND 1953). Of the tissues tested, the parsley leaf and cucumber fruit showed the highest activity. In general, the enzyme was prepared in the following way:

The plant was ground with sand, if necessary, but without the addition of water. Leaves were ground in the frozen state, fruits and roots were peeled and ground in the meat grinder. The coarse solids were strained through cheesecloth and the smaller particulate matter was removed by centrifugation at 22,000 × g. Extracts were dialyzed against .025 M phosphate buffer, p_H 7.4 for 24 hours at 37⁰ C.

Ribulose-5-phosphate is converted by the plant into *ribose-5-phosphate*.

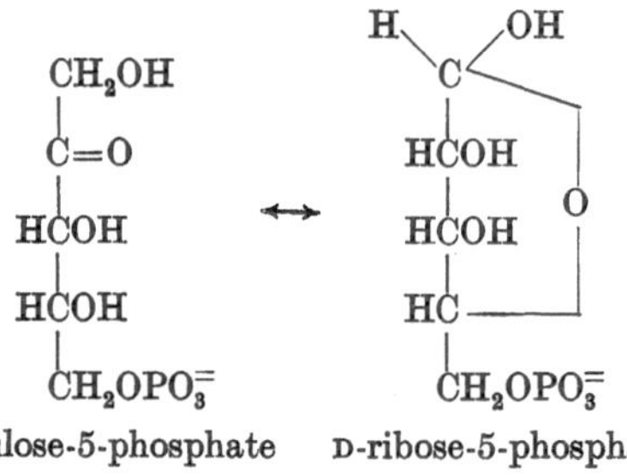

D-ribulose-5-phosphate   D-ribose-5-phosphate

This isomerization is catalyzed by an enzyme which has been partially purified from alfalfa (AXELROD and JANG 1954). The method of purification involves the preparation of an alfalfa press juice, which was briefly heated to 60⁰, and filtered. The filtrate was made 0.7 saturated with respect to solid ammonium sulfate. The precipitate was collected, suspended in water and dialyzed against running tap water for 15 hours. The remainder of the steps involved refractionation with ammonium sulfate resulting in a preparation which increased from a specific activity of 380 to a specific activity of 145,000. The enzyme was found to be without action on arabinose, D-arabinose-5-phosphate, D-ribose-3-phosphate, DL-glyceraldehyde phosphate, dihydroxyacetone phosphate and ribose. It was strongly inhibited by phosphoribonic acid. At equilibrium, approximately 25 per cent of the pentose was present as ribulose-5-phosphate and 75 per cent as ribose-5-phosphate. The enzyme shows a rather sharp p_H optimum at p_H 7 with only a slight falling off in activity on the alkaline side of the p_H range, but a marked decrease on the acid side. When the enzyme was stored at 25^0 for $2^1/_2$ hours at varying p_H's, it was quite stable over a wide p_H range between 4 and 9. The turnover number of the enzyme is approximately 240,000.

AXELROD, BANDURSKI, GREINER and JANG (1953), were also able to show that ribose-5-phosphate can be further degraded to yield *sedoheptulose phosphate* and triosephosphate. This conversion has also been demonstrated by HORECKER and SMYRNIOTIS (1952) with enzymes of animal origin. The reaction is pictured below:

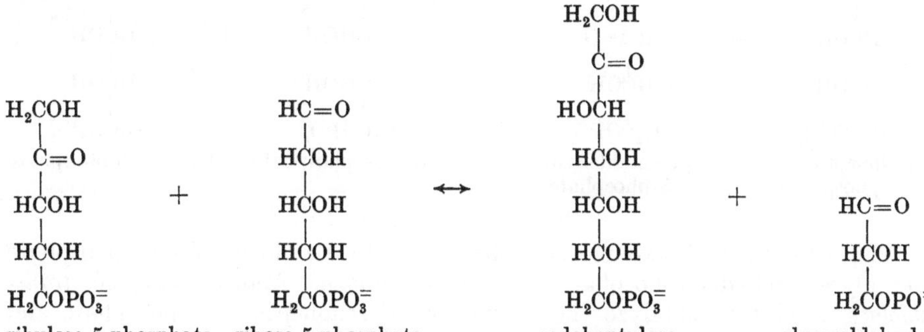

H_2COH	$HC=O$	H_2COH	
$C=O$	$HCOH$	$C=O$	
$HCOH$	$HCOH$	$HOCH$	
$HCOH$	$HCOH$	$HCOH$	$HC=O$
$H_2COPO_3^=$	$H_2COPO_3^=$	$HCOH$	$HCOH$
ribulose-5-phosphate	ribose-5-phosphate	$H_2COPO_3^=$	$H_2COPO_3^=$

ribulose-5-phosphate + ribose-5-phosphate ↔ sedoheptulose-7-phosphate + glyceraldehyde-3-phosphate

The formation of sedoheptulose phosphate has special interest in connection with photosynthesis. It has been described as one of the phosphorus esters found in the early products of photosynthesis (BENSON, BASSHAM and CALVIN 1951). The hexosemonophosphate fraction isolated from yeast also has been found to contain this ester (ROBISON, MACFARLANE and TAZCLAAR 1938). The enzyme which brings about this transformation, *transketolase*, has been purified from spinach leaves (HORECKER, SMYRNIOTIS and KLENOW 1953). The fresh spinach leaves were homogenized in a Waring Blender with cold 50 per cent saturated ammonium sulfate and adjusted to p_H 7.8. The homogenates were filtered and additional ammonium sulfate added. The precipitate was collected and adjusted to p_H 7.1. This step was followed by a second ammonium sulfate fractionation. The second step in the purification involved the adsorption of the enzyme on a calcium phosphorus gel followed by a third ammonium sulfate precipitation step. The next step in the procedure was one in which the enzyme was precipitated with acetone and finally once more precipitated with ammonium sulfate. The specific activity during this series of steps increased from .04 of a unit per ml. to 47. Transketolase preparations contained no aldolase activity. In the presence of excess substrate, activity of the purified spinach preparation with ribose-5-phosphate was about 40–50 per cent of that obtained with ribulose-5-phosphate. However, pentose phosphate isomerase was still present. The reaction does not go to completion and the pentose phosphate recovered from the reaction is an equilibrium mixture to ribose-5-phosphate and ribulose-5-phosphate. The sedoheptulose phosphate formed in the reaction was isolated by ion exchange chromatography and was presumed to be sedoheptulose-7-phosphate. The other product of the reaction is triosephosphate.

Evidence for the reversibility of the transketolase reaction has been obtained in the following way: In the presence of the purified spinach, pentose phosphate is formed from sedoheptulose phosphate and glyceraldehyde-3-phosphate.

Thiamine pyrophosphate is the coenzyme of transketolase. The spinach enzyme after acid precipitation showed an absolute requirement for this substance. Mg^{++} ions are also necessary for complete reactivation. Condensation reactions have also been observed with L-erythrulose and glyceraldehyde-3-phosphate catalyzed by spinach transketolase producing pentose-5-phosphate and heptulose.

$$
\begin{array}{c}
H_2COH \\
| \\
C=O \\
| \\
HOCH \\
| \\
HCOH \\
| \\
HCOH \\
| \\
HCOH \\
| \\
H_2COPO_3^= \\
\text{sedoheptulose-} \\
\text{7-phosphate}
\end{array}
\;+\;
\begin{array}{c}
\\ \\ \\ \\
HC=O \\
| \\
HCOH \\
| \\
H_2COPO_3^= \\
\text{glyceraldehyde-} \\
\text{3-phosphate}
\end{array}
\;\longrightarrow\;
\begin{array}{c}
\\ \\
HC=O \\
| \\
CHOH \\
| \\
CHOH \\
| \\
H_2COPO_3^= \\
\text{tetrose-4-phosphate}
\end{array}
\;+\;
\begin{array}{c}
H_2COH \\
| \\
C=O \\
| \\
HOCH \\
| \\
HCOH \\
| \\
HCOH \\
| \\
H_2COPO_3^= \\
\text{fructose-6-phosphate}
\end{array}
$$

transaldolase reaction

The conversion of two pentose phosphates to a sedoheptulose-7-phosphate and glyceraldehyde-3-phosphate is described above. Another enzyme, *transaldolase*, is able to catalyze the conversion of sedoheptulose-7-phosphate and glyceraldehyde-3-phosphate into fructose-6-phosphate and *tetrosephosphate*. The conversion of pentose phosphate to hexosemonophosphate is assumed to occur in this fashion. These reactions have been demonstrated in liver (HORECKER, GIBBS, KLENOW and SMYRNIOTIS 1954) and in spinach leaves (AXELROD, BANDURSKI, GREINER and JANG 1953). Using extracts from pea roots and labeled ribose-5-phosphate, GIBBS and HORECKER (1954) were able to show that this tissue also carries out these reactions. Pea leaf preparations carry out similar reactions, but from the labeling of the products obtained, GIBBS and HORECKER concluded that part of this conversion occurs via an additional pathway. This latter observation may be of importance in the rapid formation of uniformily labeled carbohydrates in photosynthesis.

The data presented above would make it appear that a complete Shunt mechanism is present in the tissues of higher plants. The early appearance of sedoheptulose and other intermediates in this pathway during photosynthesis has led to the growing conviction that these may be extremely important in the photosynthetic sequence. This will be discussed elsewhere in these volumes.

Literature.

ALBAUM, H. G.: The incorporation of radiophosphorus during growth. Symposium on phosphorus metabolism. Mich. State College Press, 55, East Lansing, Mich., 1952. — ALBAUM, H. G., and R. LIPSHITZ: Determination of adenosine triphosphate based on deamination rates. Arch. of Biochem. **27**, 102 (1950). — ALBAUM, H. G., M. OGUR and A. HIRSHFELD: The isolation of adenosine triphosphate from plant tissue. Arch. of Biochem. **27**, 130 (1950). — ALBAUM, H. G., A. SCHATZ, S. H. HUTNER and A. HIRSHFELD: Phosphorylated compounds in *Euglena*. Arch. of Biochem. **29**, 210 (1950). — ALBAUM, H. G., and R. SCHER: Paper chromatography of phosphorylated intermediates. (Unpublished Experiments). — ALBAUM, H. G., and W. W. UMBREIT: Phosphorus transformations during the development of the oat embryo. Amer. J. Bot. **30**, 553 (1943). — ANDERSON, D. G., H. A. STAFFORD, E. E. CONN and B. VENNESLAND: The distribution in higher plants of triphosphopyridine nucleotide-linked enzyme systems capable of reducing glutathione. Plant Physiol. **27**, 675 (1952). — ARNEY, S. E.: Phosphate fractions in barley seedlings. Biochemic. J. **33**, 1078 (1939. — ARNON, D. I.: The glycolytic cycle in the breakdown and synthesis of carbohydrates in green leaves. A symposium on phosphorus metabolism, Vol. 2, 67. Baltimore, Md. 1952. — AXELROD, B., and R. S. BANDURSKI: Phosphoglyceryl kinase in higher plants. J. of Biol. Chem. **204**, 939 (1953). — AXELROD, B., R. S. BANDURSKI, C. M. GREINER and R. JANG: The metabolism of hexose and pentose phosphates in higher plants. J. of Biol. Chem. **202**, 619 (1953).— AXELROD, B., R. S. BANDURSKI and P. SALTMAN: Phosphate uptake by pea meal extracts. Federat. Proc. **10**, 158 (1951). — AXELROD, B., and R. JANG: Purification and properties of phosphoriboisomerase from alfalfa. J. of Biol. Chem. **209**, 847 (1954). — AXELROD, B., P. SALTMAN, R. S. BANDURSKI and R. S. BAKER: Phosphohexokinase in higher plants. J. of Biol. Chem. **197**, 89 (1952).

BANDURSKI, R. S., and B. AXELROD: The chromatographic identification of some biologically important phosphate esters. J. of Biol. Chem. **193**, 405 (1951). — BANDURSKI, R. S., C. M. GREINER and J. BONNER: Enzymatic carboxylation of phosphoenolpyruvate to oxalacetate. Federat. Proc. **12**, 173 (1953). — BARNETT, R. C., H. A. STAFFORD, E. E. CONN, and B. VENNESLAND: Phosphogluconic dehydrogenase in higher plants. Plant Physiol. **28**, 115 (1953). — BENSON, A. A. *et al.*: Local citation 1950. — BENSON, A. A., J. A. BASSHAM and M. CALVIN: Sedoheptulose in photosynthesis by plants. J. Amer. Chem. Soc. **73**, 2970 (1951). — BENSON, A. A., J. A. BASSHAM, M. CALVIN, T. C. GOODALE, V. A. HAAS and W. STEPKA: The path of carbon in photosynthesis. V. Paper chromatography and radioautography of the products. J. Amer. Chem. Soc. **72**, 1710 (1950). — BENSON, A. A., S. KAWAGUCHI, P. HAYES and M. CALVIN: The path of carbon in photosynthesis. XVI. Kinetic relationships of th intermediates in steady state photosynthesis. J. Amer. Chem. Soc. **74**, 4477 (1952). — BUCHANAN, J. G.: The path of carbon in photosynthesis. XIX. The identification of sucrose phosphate in sugar beet leaves. Arch. of Biochem. a. Biophysics **44**, 140 (1953).

CALVIN, M., and A. A. BENSON: The path of carbon in photosynthesis. IV. The identity and sequence of the intermediates in sucrose synthesis. Science (Lancaster, Pa.) **109**, 140–142 (1949). — CAMPBELL, J. M., and R. S. BANDURSKI: Adenylate kinase in plant tissue. Amer. Soc. Plant Physiol. (AIBS), Sept. 1952 (Abstr.)— CARDINI, C. E.: Activation of plant phosphoglucomutase by Glucose 1.6 diphosphate. Enzymologia (Den Haag) **15**, 44 (1951). — CONN, E. E., and B. VENNESLAND: Glutathione reductase of wheat germ. J. of Biol. Chem. **192**, 17 (1951).

DICKENS, F.: Oxidation of phosphohexonate and pentose phosphoric acids by yeast enzymes. I. Oxidation of phosphohexonate. II. Oxidation of pentose phosphoric acids. Biochemic. J. **32**, 1626 (1938). — DOUDOROFF, M., N. KAPLAN and W. Z. HASSID: Phosphorolysis and synthesis of sucrose with a bacterial preparation. J. of Biol. Chem. **148**, 67–75 (1943).

EMERSON, R. L., J. F. STAUFFER and W. W. UMBREIT: Relationships between phosphorylation and photosynthesis in *Chlorella*. Amer. J. Bot. **31**, 107 (1944).

GIBBS, M., and B. L. HORECKER: The mechanism of pentose phosphate conversion to hexose monophosphate. II. With pea leaf and pea root preparations. J. of Biol. Chem. **208**, 813 (1954).

HAGEMAN, R. H., and D. I. ARNON: The isolation of triosephosphate dehydrogenase from pea seeds. Arch. of Biochem. a Biophysics **55**, 162 (1955). — HANES, C. S.: The breakdown and synthesis of starch by an enzyme system from pea seeds. Proc. Roy. Soc. Lond., Ser. B **128**, 421 (1939). — The reversible formation of starch from glucose-1-phosphate catalysed by potato phosphorylase. Proc. Roy. Soc. Lond., Ser. B **129**, 174 (1940). — HARDIN, M. B.: On the occurrence of meta-phosphoric acid and pyrophosphoric acid in cotton seed meal. S. Carolina Exper. Sta. Bull. **8**, 10 (1892). — HASSID, W. Z.: Isolation of a hexose-monophosphate from pea leaves. Plant Physiol. **13**, 641 (1938). — HEARD, C. R. C.: On phosphoric esters in barley. New Phytologist **44**, 184 (1945). — HORECKER, B. L., M. GIBBS, H. KLENOW and P. Z. SMYRNIOTIS: The mechanism of pentose phosphate conversion to hexose monophosphate. I. With a liver enzyme preparation. J. of Biol. Chem. **207**, 393 (1954). — HORECKER, B. L., and P. Z. SMYRNIOTIS: The enzymatic formation of sedoheptulose phosphate from pentose phosphate. J. Amer. Chem. Soc. **74**, 2123 (1952). — HORECKER, B. L., P. Z. SMYRNIOTIS and H. KLENOW: The formation of sedoheptulose phosphate from pentose phosphate. J. of Biol. Chem. **205**, 661 (1953).

JAMES, W. O., and S. E. ARNEY: Phosphorylation and respiration in barley. New Phytologist **38**, 340 (1939).

LELOIR, L. F.: The metabolism of hexosephosphates. A symposium on Phosphorus metabolism, Vol. I, 67. 1951. — LePAGE, G. A., and W. W. UMBREIT: Phosphorylated carbohydrate esters in autotrophic bacteria. J. of Biol. Chem. 147, 263 (1943).

RAMASARMA, T., SRI J. RAM and K. V. GIRI: Phosphoglucomutase of green gram *(Phaselolus radiatus)*. Arch. of Biochem. a. Biophysics 53, 167 (1954). — RAPOPORT, S.: Über die Bestimmung der Glycerinsäure in freier und veresterter Form. Biochem. Z. 289, 406 (1937). — ROBISON, R., M. S. MacFARLANE and A. TAZOLAAR: A new phosphoric ester isolated from the products of yeast juice fermentation. Nature (Lond.) 142, 114 (1938). — ROE, J. H.: A colorimetric method for the determination of fructose in blood and urine. J. of Biol. Chem. 107, 15 (1934).

SALTMAN, P.: Hexokinase in higher plants. J. of Biol. Chem. 200, 145 (1953). — SOMERS, G. F., and E. L. COSBY: The conversion of fructose-6-phosphate into glucose-6-phosphate in plant extracts. Arch. of Biochem. 6, 295 (1945). — STAFFORD, H., R. C. BARNETT, E. E. CONN and B. VENNESLAND: The oxidation of monosaccharides by TPN dependent enzymes. Amer. Soc. Plant Physiol. (AIBS) Sept. 1952 (Abstr.). — STUMPF, P. K.: Carbohydrate metabolism in higher plants. I. Pea aldolase. J. of Biol. Chem. 176, 233 (1948). — Carbohydrate metabolism in higher plants. III. Breakdown of fructose diphosphate by pea extracts. J. of Biol. Chem. 182, 261 (1950). — Fat metabolism in higher plants. III. Enzymic oxidation of glycerol. Plant Physiol. 30, 55 (1955). — SUMNER, J. B., and G. F. SOMERS: The preparation of glucose-1-phosphate. Arch. of Biochem. 4, 11 (1944).

TEWFIK, S., and P. K. STUMPF: Carbohydrate metabolism in higher plants. II. The distribution of aldolase in plants. Amer. J. Bot. 36, 567 (1949). — Carbohydrate metabolism in higher plants. J. of Biol. Chem. 192, 519 (1951).

ZILL, L. P., and A. R. KRALL: Phosphorylated compounds in plants: Separation in groups by ion-exchange chromatography. Amer. Soc. Plant Physiol. (AIBS), Gainesville, Sept. 1954 (Abstr.).

Pyridinnucleotide.

(Diphospho-pyridinnucleotid und Triphospho-pyridinnucleotid) *.

Von

K. Hasse.

Mit 2 Abbildungen.

Einleitung.

Diphospho-pyridinnucleotid und Triphospho-pyridinnucleotid sind die physiologischen Wirkungsformen des Vitamins Nicotinsäure. Beide Pyridinverbindungen wurden entdeckt auf Grund ihrer katalytischen Funktion als Komponenten von Fermentsystemen. HARDEN und YOUNG fanden 1906, daß Phosphat die Fähigkeit eines Hefeextraktes, Glucose zu vergären, steigert. Sie beobachteten aber auch einen anderen hitzestabilen Faktor, der sich durch Dialyse vom Hefesaft abtrennen ließ. Dieses Coenzym des komplexen Enzymsystems der alkoholischen Gärung — der „Zymase" — wurde später als Cozymase bezeichnet. EULER und seinem Arbeitskreis gelang es, eine wesentliche Anreicherung dieser Verbindung zu bewirken (Zusammenfassung bei MYRBÄCK 1933). Reine Cozymase wurde von SCHLENK und EULER 1936 erhalten.

Ein zweiter Faktor, der für die Oxydation von Glucose-6-phosphat zu 6-Phospho-gluconsäure benötigt wird, wurde von WARBURG und CHRISTIAN 1931 entdeckt. WARBURG und Mitarbeiter (1935) konnten diese Verbindung zuerst aus Pferdeerythrocyten isolieren. Beide Coenzyme sind einander chemisch sehr ähnlich. Diese Verbindungen sind Nicotinsäureamid-adenin-dinucleotide. Sie unterscheiden sich darin, daß das Cozymasemolekül 2, das WARBURGsche Coferment 3 Äquivalente Phosphorsäure enthält. Nach einem Vorschlag von WARBURG (WARBURG und CHRISTIAN 1936) bezeichnet man sie als „Diphospho-pyridinnucleotid" (DPN) und „Triphospho-pyridinnucleotid" (TPN). In Diphospho-pyridinnucleotid ist Nicotinsäureamid-ribotid mit Adenosin-5'-phosphat, im Triphospho-pyridinnucleotid mit Adenosin-2',5'-diphosphat verbunden (Zusammenfassung bei SCHLENK 1951).

Das Auffinden der biologischen Bedeutung der Nicotinsäure in den Pyridinnucleotiden führte dann bald zu der Entdeckung, daß Nicotinsäure und Nicotinsäureamid Wuchsstoffaktoren für verschiedene Mikroorganismen sind und Vitamincharakter für eine Reihe höherer Tiere besitzen (ELVEHJEM und Mitarbeiter 1938, FOUTS und Mitarbeiter 1937). Diphospho-pyridinnucleotid und Triphospho-pyridinnucleotid kommen allgemein verbreitet in tierischen und pflanzlichen Zellen vor. Sie sind Coenzyme zahlreicher Dehydrogenasen. Zur Zeit sind etwa 50 pyridinnucleotidabhängige Dehydrogenasen bekannt. Ihre Zahl wächst jedoch in dem Maße, wie unsere Kenntnisse über den intermediären Stoffwechsel

---

* Siehe auch die Beiträge von HASSE und MEEUSE in Band XII.
Folgende Abkürzungen werden verwendet: DPN Diphospho-pyridinnucleotid, DPNH reduziertes Diphospho-pyridinnucleotid, TPN Triphospho-pyridinnucleotid, TPNH reduziertes Triphospho-pyridinnucleotid, PN Pyridin-nucleotid (DPN oder TPN), NMN Nicotinsäureamid-mononucleotid, NR Nicotinsäureamid-ribosid, ATP Adenosin-triphosphat, ADP Adenosin-diphosphat, 5'-AMP Adenosin-5'-phosphat.

zunehmen. Die Dehydrogenasen können sich hinsichtlich ihrer Pyridinnucleotid-spezifität unterscheiden. Der chemische Wirkungsmechanismus dieser Coenzyme

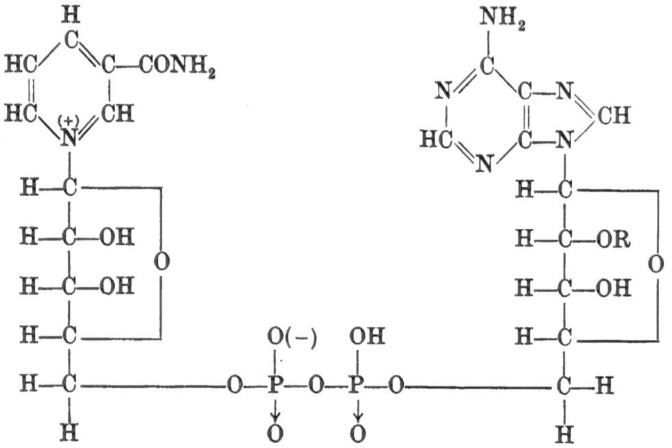

Diphospho-pyridinnucleotid (R = H).
Triphospho-pyridinnucleotid (R = $-PO_3H_2$).
Abb. 1.

bei dem Dehydrierungsvorgang besteht in einer reversiblen Reduktion des Pyridin-ringes (Zusammenfassung bei Warburg 1938):

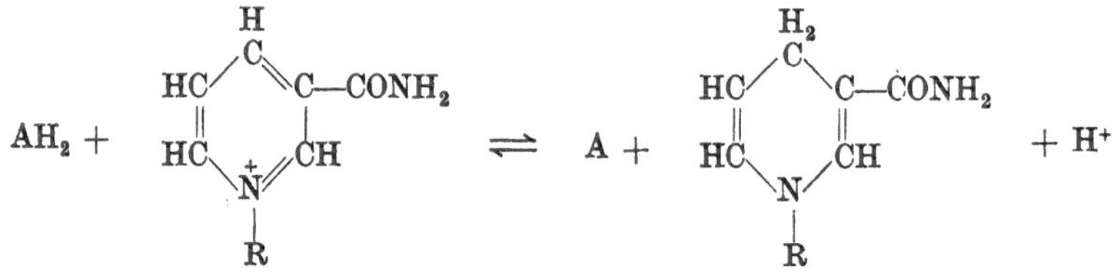

Unter der Katalyse der für den Wasserstoffdonator spezifischen Dehydro-genase wird das Coenzym durch den Metaboliten zu einem Dihydropyridinderivat im Sinne einer Gleichgewichtsreaktion reduziert.

Das Pyridinnucleotid läßt sich als Bestandteil der Dehydrogenase auffassen. Euler und Albers (1936) haben die Ansicht vertreten, daß das spezifische Pro-tein der Dehydrogenase den Träger des Pyridinnucleotids darstellt, so daß letzteres analog der Konstitution der Flavoproteine zur prosthetischen Gruppe der „Pyri-din-dehydrogenasen" wird. Der hochmolekulare Teil des dehydrierenden Fer-mentes wird „Apodehydrogenase" genannt, das Coferment „Codehydrogenase". Durch die Vereinigung beider Komponenten entsteht die wirksame Dehydro-genase oder die „Holodehydrogenase".

Hydrierung der Pyridinnucleotide und Reoxydation.

Die partielle, reversible Reduktion von DPN und TPN ist die biologisch be-deutsame Reaktion der beiden Nucleotide. Warburg, Christian und Griese machten 1935 die Beobachtung, daß dann, wenn Glucose-6-phosphat mit TPN in Gegenwart des substratspezifischen Proteins reagiert, im langwelligen Ultra-violett eine Absorptionsbande mit einem Maximum bei 340 mμ entsteht. Dieselbe charakteristische Absorption tritt auch auf, wenn andere Pyridiniumverbin-dungen chemisch z. B. mit Hyposulfit ($Na_2S_2O_4$) reduziert werden. Hieraus ergab sich die Wirkungsweise der Nucleotide als erste biologische Wasserstoff-acceptoren:

$$\text{Glucose-6-phosphat} + \text{TPN}^+ \rightleftarrows \text{6-Phospho-gluconsäurelacton} + \text{TPNH} + \text{H}^+.$$

Abb. 2 zeigt die Absorptionsspektren der oxydierten und der reduzierten Pyridinnucleotide. Die Spektren von DPN und TPN unterscheiden sich nicht, und auch die hydrierten Formen weisen praktisch die gleiche Absorption auf. Die reduzierten Coenzyme zeigen ein Absorptionsmaximum bei 340 mμ, in einem Bereich, in dem die oxydierten Verbindungen nicht absorbieren.

Die Verwendung der spektrophotometrischen Methode WARBURGS hat auf die Erforschung der Pyridin-dehydrogenasen einen entscheidenden Einfluß gehabt. In Kenntnis der molekularen Extinktion der reduzierten Form bietet sie die ideale Methode zur Bestimmung der Coenzyme (Zusammenfassung bei HASSE 1955). Aber auch der Ablauf enzymatischer Reaktionen, an denen Pyridinnucleotide beteiligt sind, läßt sich durch das Entstehen oder Verschwinden der Dihydropyridinbanden in einfacher Weise messen.

In dem enzymatischen Gleichgewicht hat das Pyridinnucleotid die Funktion einer Reaktionskomponente. Die Lage des Gleichgewichts ist abhängig von dem Redoxpotential des Systems. Stoffwechselprodukte der verschiedenen Körperklassen können DPN und TPN reduzieren. In der folgenden Tabelle sind eine Reihe von Substraten angeführt, die unter der Katalyse der spezifischen Apodehydrogenasen als Wasserstoffdonatoren fungieren können. Als dehydrierbare Angriffsstellen der Substrate fungieren insbesondere alkoholische Hydroxylgruppen und Aldehydgruppen. Aber auch Carbonsäuren lassen sich oxydativ decarboxylieren (Zusammenfassungen über Pyridindehydrogenasen bei SCHLENK 1951; SINGER und KEARNEY 1954, HASSE 1958).

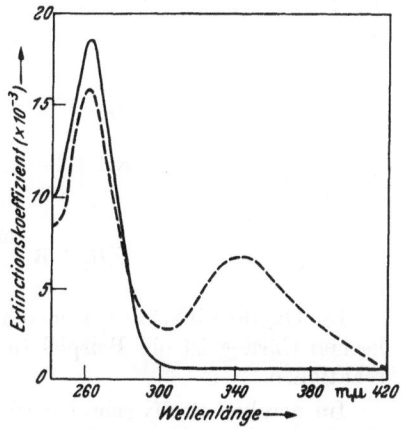

Abb. 2. Absorptionsspektren des oxydierten und reduzierten Diphospho-pyridinnucleotids.
—— Oxydierte Form; ---- reduzierte Form.

Auch in der Photosynthese der grünen Pflanze kommt den Pyridinnucleotiden eine wichtige Funktion zu. Als Folge der photolytischen Wasserspaltung werden in einem noch ungeklärten Chemismus die Pyridinnucleotide hydriert. Sie sind zur Zeit die ersten faßbaren Reduktionsprodukte dieses Vorgangs (VISHNIAC 1952a und b, TOLMACH 1951, ARNON 1951).

$$\text{PN}^+ + \text{H}_2\text{O} \xrightarrow{h\nu} 1/2\,\text{O}_2 + \text{PNH} + \text{H}^+.$$

Tabelle 1.

Äthylalkohol und Homologe	3-Phospho-glycerinaldehyd
Glycerin	Acetaldehyd und Homologe
Glykolsäure	L-Glutaminsäure
1-Phospho-glycerin	Dihydro-orotsäure
Milchsäure	Ameisensäure
Äpfelsäure	Brenztraubensäure
Isocitronensäure	α-Ketoglutarsäure
β-Hydroxyacyl-Co A	6-Phospho-gluconsäure
6-Phospho-glucose	Liponsäure

Das Gleichgewicht der Dehydrierungsreaktion läßt sich durch die Erhöhung der Konzentration an reduziertem Pyridinnucleotid im Sinne der Reduktion des Substrats verschieben, so daß das Enzym nunmehr als Reduktase wirkt. Eine

12*

Reihe von Dehydrogenasen sind nicht nur an der biologischen Oxydation der Nährstoffe beteiligt, sondern auch an der reduktiven Synthese dieser Verbindung. Je nach der Stoffwechsellage — ob viel oder wenig reduziertes Pyridinnucleotid zur Verfügung steht — verläuft die Reaktionsfolge im Sinne des Abbaus oder des Aufbaus. Beispiele hierfür sind die 3-Phospho-glycerinaldehyd-dehydrogenase im Kohlenhydratstoffwechsel, die β-Hydroxyacyl-CoA-dehydrogenase im Fettstoffwechsel und die L-Glutaminsäure-dehydrogenase auf dem Gebiete der Proteine.

Die Reoxydation des in einer enzymatischen Reaktion gebildeten hydrierten Pyridinnucleotids (Gl. 1) kann in einem zweiten PN-abhängigen Dehydrierungsvorgang (Gl. 2) mit geeigneter Gleichgewichtslage erfolgen. Durch die Kopplung der beiden Reaktionen wird die Funktion des Pyridinnucleotids zu dem eines Katalysators der Wasserstoffübertragung von einem Metaboliten auf einen anderen:

$$AH_2 + PN^+ \xrightleftharpoons{AH_2\text{-apodehydrogenase}} A + PNH + H^+. \tag{1}$$

$$B + PNH + H^+ \xrightleftharpoons{BH_2\text{-apodehydrogenase}} BH_2 + PN^+ \tag{2}$$

$$AH_2 + B \xrightleftharpoons[PN^+]{\substack{AH_2\text{-apodehydrogenase} \\ BH_2\text{-apodehydrogenase}}} A + BH_2$$

Die Oxydoreduktion zwischen Triosephosphat und Acetaldehyd in der alkoholischen Gärung ist ein Beispiel für das Zusammenwirken von 2 Dehydrierungsreaktionen.

Im aeroben Stoffwechsel wird die Reoxydation der hydrierten Formen der Pyridinnucleotide durch Flavoproteide bewirkt, die spezifisch auf DPNH und TPNH als Wasserstoffdonatoren eingestellt sind. Unter intermediärer Reduktion des Isoalloxazinsystems, der prosthetischen Gruppe dieser Fermente, wird der Wasserstoff (bzw. die Elektronen) auf spezifische Acceptoren übertragen.

Wasserstoffübertragung zwischen oxydierten und reduzierten Formen der Pyridinnucleotide.

In Bakterien und in tierischen Geweben finden sich „Pyridinnucleotid-transhydrogenasen", die den Wasserstofftransport von einem Pyridinnucleotid in der reduzierten Form auf ein solches in oxydiertem Zustand katalysieren (Colowick und Mitarbeiter 1952, Kaplan, Colowick und Neufeld 1952, 1953, Kaplan und Mitarbeiter 1953).

Für die reversible Reaktion

$$TPNH + DPN \rightleftharpoons TPN + DPNH$$

mit Enzymextrakten aus *Pseudomonas fluorescens* ist Adenosin-2'-phosphat ein Cofaktor. Mit solchen Enzymen ist auch eine Übertragung von Wasserstoff von DPNH und TPNH auf die Desamino-pyridinnucleotide und auf Nicotinsäureamid-mononucleotid möglich. Diese Transhydrogenasen spielen vielleicht eine wichtige Rolle in der Regulierung des Wasserstofftransportes in der pflanzlichen und tierischen Zelle.

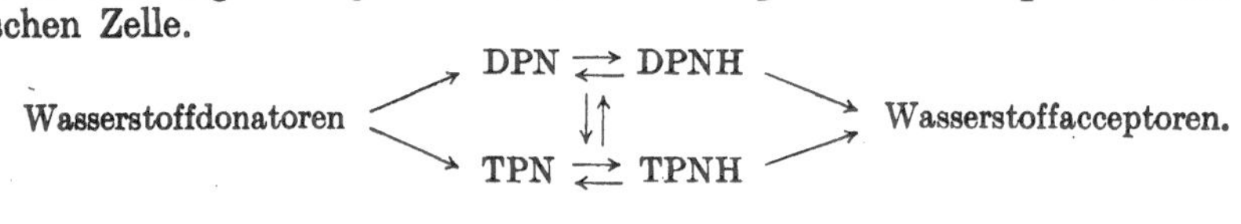

$$\text{Wasserstoffdonatoren} \diagup \begin{matrix} DPN \rightleftharpoons DPNH \\ \updownarrow \\ TPN \rightleftharpoons TPNH \end{matrix} \diagdown \text{Wasserstoffacceptoren}.$$

Biosynthese.

Nicotinsäure.

Während die höhere Pflanze zur Synthese der Nicotinsäure befähigt ist, hat diese Verbindung für viele Tiere Vitamincharakter. Auch für zahlreiche Mikroorganismen ist Nicotinsäure ein Wuchsstoff. *Haemophilus influenzae* und *H. para-influenzae* sind darüber hinaus nicht imstande, Nicotinsäureamid mit Ribose zu verbinden. Das Wachstum dieser Organismen ist abhängig von der Zufuhr eines Pyridinnucleotids oder von Nicotinsäureamid-ribosid (GINGRICH und SCHLENK 1944). Nicotinsäure ist ein Stoffwechselprodukt des Tryptophans. Diese Erkenntnis wurde am Tier und bei einigen Mikroorganismen gewonnen. ELVEHJEM und Mitarbeiter konnten 1938 zeigen, daß Nicotinsäuremangel die wesentliche Ursache der Pellagra ist. Dieser Arbeitskreis fand, daß neben Nicotinsäure auch Tryptophan das Auftreten der Mangelerscheinungen im Tierversuch verhindern kann (KREHL und Mitarbeiter 1945, KREHL, SARMA und ELVEHJEM 1946).

Die Umwandlung von Tryptophan in N^1-Methyl-nicotinsäure (HUFF und Mitarbeiter 1942) und später auch in Nicotinsäure (HEIDELBERGER und Mitarbeiter 1948) ließ sich mit markierten Verbindungen beweisen. Aus Fütterungsversuchen hatte sich schon früher ergeben, daß Kynurenin (MATSUOKA und YOSHIMATSU 1925, KOTAKE und IWAO 1931) und Anthranilsäure (KOTAKE und OTANI 1933) Abbauprodukte des Tryptophans darstellen.

Tabelle 2. *Abbauprodukte des Tryptophans.*

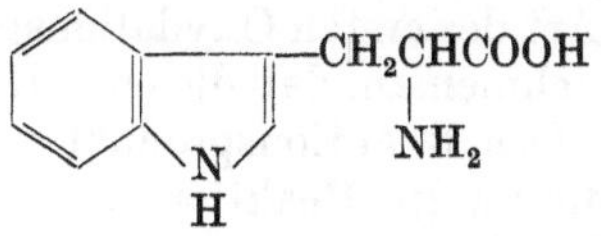

Tryptophan

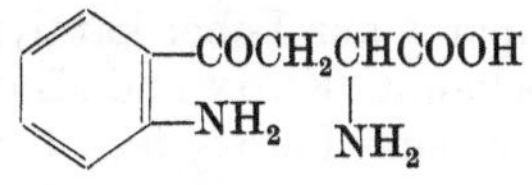

Kynurenin

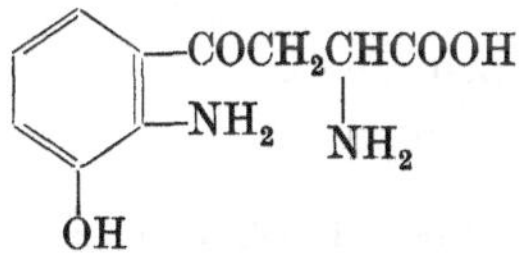

3-Hydroxy-Kynurenin

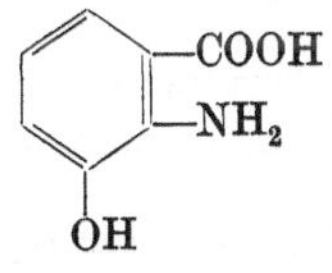

3-Hydroxy-Anthranilsäure

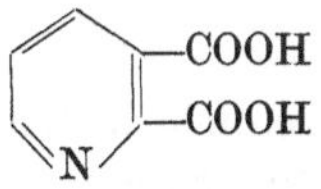

Chinolinsäure

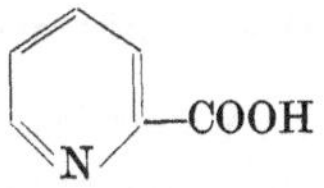

α-Picolinsäure

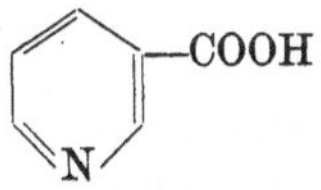

Nicotinsäure

Untersuchungen über den Wuchsstoffbedarf von *Neurospora*-Mutanten haben wesentlich zur Lösung des Chemismus der Nicotinsäuresynthese beigetragen. Durch Röntgen- und UV-Bestrahlung ließen sich Mutanten erzeugen, deren Nicotinsäurebedarf auch durch Tryptophan, Kynurenin (BEADLE, MITCHELL und NYC 1947) und durch 3-Hydroxy-anthranilsäure (MITCHELL und NYC 1948, BONNER und YANOFSKY 1949) gedeckt werden kann.

Phycomyces-Kulturen, die durch Zugabe von 2-Methyl-sulfochinon ihre Fähigkeit zur Nicotinsäurebildung verloren haben, finden diese wieder durch Zusatz von Tryptophan oder Kynurenin (SCHOPFER und BOSS 1949). Tryptophan und Kynurenin begünstigen auch die Synthese von Nicotinsäure in höheren Pflanzen.

NASON (1949) stellte fest, daß Maiskeimlinge nach 10tägiger Keimung in einer Nährlösung mit Tryptophan bis 60% mehr Nicotinsäure enthielten als die in tryptophanfreien Lösungen.

Auch Kohl- und Tomatenblätter, deren Stiele in eine Tryptophanlösung tauchten, enthielten mehr Nicotinsäure als die Kontrollen (GUSTAFSON 1949). In keimenden Getreidesamen beobachteten CHITRE, DESAI und RAUT (1955) eine Steigerung der Synthese durch Tryptophan und Kynurenin. 3-Hydroxy-anthranilsäure vermag ebenfalls Nicotinsäure in *Neurospora crassa* zu ersetzen (MITSCHELL und NYC 1948). YANOFSKY und BONNER (1951) zeigten in Isotopenversuchen, daß diese Verbindung die einzige Vorstufe für das Vitamin in *Neurospora*-Mutanten sein kann. Andererseits soll 3-Hydroxy-anthranilsäure die Synthese von Nicotinsäure in keimendem Samen nicht steigern (CHITRE, DESAI und RAUT 1955).

3-Hydroxy-anthranilsäure kann sowohl durch den tierischen Organismus als auch durch Mikroorganismen in Chinolinsäure übergeführt werden (HENDERSON und RAMASARMA 1949). Es ist noch ungeklärt, wie Nicotinsäure und Chinolinsäure im Stoffwechsel miteinander verknüpft sind. Mit Chinolinsäure anstelle der Nicotinsäure konnte in *Neurospora*-Mutanten keine Wachstumswirkung erzielt werden (BONNER und YANOFSKY 1949). Dagegen fanden HENDERSON und HIRSCH (1949) nach starker Überdosierung der Chinolinsäure bei dem gleichen Organismus eine sehr schwache Verwertung und bei Fütterung an Ratten nur eine geringe Erhöhung von Nicotinsäurederivaten im Harn. In keimenden Getreidesamen soll Chinolinsäure wirksam sein (CHITRE und Mitarbeiter 1955).

Die Bildung von Kynurenin aus Tryptophan läßt sich mit einem löslichen Enzympräparat aus Leber katalysieren. Die Art des ersten Oxydationsproduktes ist nicht bekannt. KNOX und MEHLER (1950) nehmen an, daß die erste Reaktionsphase durch eine Peroxydase bewirkt wird. Das Reaktionsprodukt wird dann durch eine Oxydase weiteroxydiert. Durch die zweite Reaktion wird das H_2O_2 gebildet, das die erste Reaktion benötigt. Die Enzymlösungen enthalten eine Formylase, die Formyl-Kynurenin zu Kynurenin und Ameisensäure hydrolytisch spaltet:

L-Tryptophan → „oxydiertes Tryptophan" →
Formyl-kynurenin → Kynurenin + Ameisensäure.

Kynurenin wird durch eine Kynureninase aus Leber (KOTAKE und NAKAYAMA 1941) und aus Bakterien (HAYAISHI 1952) zu Anthranilsäure und Alanin hydrolysiert.

Kynurenin + H_2O → Anthranilsäure + Alanin.

Anthranilsäure kommt jedoch als Vorstufe der Nicotinsäure nicht in Betracht. Diese Verbindung ist im Tierversuch und auch bei keimenden Samen (CHITRE und Mitarbeiter 1955) unwirksam. Die Kynureninase spaltet aber auch in analoger Weise 3-Hydroxy-kynurenin unter Bildung von 3-Hydroxy-anthranilsäure (WISS 1953).

3-Hydroxy-kynurenin + H_2O → 3-Hydroxy-anthranilsäure + Alanin.

Man muß deshalb annehmen, daß vor der Abspaltung des Alanins Kynurenin zu Hydroxy-kynurenin oxydiert wird. Aus Lebermitochondrien von Katze und Ratte läßt sich ein zellfreies Enzymsystem bereiten, das in Gegenwart von TPNH 3-Hydroxy-kynurenin aus L-Kynurenin bildet. DPNH ist weniger wirksam (DE CASTRO, PRICE und BROWN 1956).

MEHLER (1956) hat mit Leberextrakten 3-Hydroxy-anthranilsäure oxydativ in Chinolinsäure und α-Picolinsäure übergeführt. Von WISS (1956) wurde das ver-

mutlich erste Reaktionsprodukt durch das UV-Spektrum charakterisiert, und es
ließ sich eine Aldehydgruppe nachweisen. Das erste Oxydationsprodukt läßt sich
in nichtenzymatischer Reaktion durch Erwärmen quantitativ in Chinolinsäure
überführen. WISS (1956) gibt folgende Formulierung des Vorgangs:

Bisher ist es nicht gelungen, im Enzymversuch 3-Hydroxy-anthranilsäure in
Nicotinsäure umzuwandeln.

Über die Amidierung der Nicotinsäure in der Pflanze ist noch nichts bekannt.
QUAGLIARIELLO und PORCELLATI (1953) haben in Niere und Leber der Ratte
eine enzymatische Aktivität gegenüber Nicotinsäure gefunden. Diese Organe
bilden — besonders nach Zugabe von ATP — aus Ammoniumnicotat Nicotin-
amid.

Nicotinsäureamid-ribotid.

ROWEN und KORNBERG (1951) konnten aus Leber eine Phosphorylase gewinnen, unter deren Katalyse Nicotinsäureamid glykosidisch mit D-Ribose verbunden wird:

Nicotinsäureamid + Ribose-1-phosphat $\rightleftarrows$ Nicotinsäureamid-ribosid + H_3PO_4.

Ein Enzym mit ähnlichen Eigenschaften wurde auch aus Hefe in angereichertem
Zustand erhalten (HEPPEL und HILMOE 1952). Das Leberenzym reagiert nicht
mit DPN, TPN und Nicotinsäureamid-mononucleotid in Gegenwart von anorganischem Phosphat. Es bewirkt aber auch die Phosphorolyse von Inosin. Das
Enzym aus Hefe spaltet außer Nicotinsäureamid-ribosid auch Guanosin und
Inosin. Wahrscheinlich ist die „Nicotinsäureamid-ribosid-phosphorylase" identisch mit der Purinnucleosid-phosphorylase, die KALCKAR 1945 beschrieben hat.
Das Gleichgewicht der Reaktion liegt wie bei der Inosinsynthese sehr zugunsten
des Nucleosids.

Die Phosphorylierung des Nicotinsäureamid-ribosids zum Ribotid mit Adenosin-triphosphat wird durch Phosphokinasen katalysiert:

Nicotinsäureamid-ribosid + ATP $\rightarrow$ Nicotinsäureamid-ribotid + ADP.

ROWEN und KORNBERG (1951) haben die Phosphorylierung von Nicotinsäure-
ribosid dadurch indirekt nachgewiesen, daß sie mit dem partiell gereinigten
DPN-synthetisierenden Enzym aus Leber (s. S. 184) das aus NMN und ATP
sekundär entstehende DPN bestimmen konnten. Die Möglichkeit einer direkten
Synthese von NMN aus Nicotinsäureamid und Ribose-1,5-diphosphat oder
5-Phospho-ribosyl-pyrophosphat ist in Erwägung zu ziehen.

Ribose-5-phosphat wird mit ATP in Gegenwart von Enzymen aus Leber und wahrschein-
lich auch aus Bakterien in 5-Phospho-ribosyl-pyrophosphat verwandelt (LIEBERMAN und
KORNBERG 1955, KORNBERG, LIEBERMAN und SIMMS 1955). Dieses Triphosphat reagiert mit
Pyrimidin- und Purin-Derivaten unter der Katalyse von Pyrophosphorylasen zu Nucleotiden.

Diphospho-pyridinnucleotid.

Die Synthese von DPN durch eine Pyrophosphatase aus Nicotinsäureamid-ribotid und Adenosin-triphosphat hat Kornberg (1950a) erstmals beschrieben:

$$\text{Nicotinsäureamid-ribotid} + \text{ATP} \rightleftarrows \text{DPN} + \text{Pyrophosphat.}$$

DPN-Pyrophosphorylase findet sich in Hefe und in Leber. TPN und Flavin-adenin-dinucleotid können dem Enzym nicht als Substrat dienen. Doch entsteht auch DPNH in analoger Reaktion aus dem hydrierten Nicotinsäureamid-ribotid.

Triphospho-pyridinnucleotid.

Euler und Adler haben schon 1938 die Phosphorylierung von DPN mit Adenosintriphosphat in Hefeextrakten bewirkt. Da die Lokalisierung des 3. Phosphatrestes in der Struktur des TPN lange Zeit Schwierigkeiten bereitete, konnte die Reaktion erst später als die Katalyse einer Phosphokinase erkannt werden (Kornberg 1950b).

$$\text{DPN} + \text{ATP} \rightarrow \text{TPN} + \text{ADP.}$$

Die Kinase aus Hefe reagiert auch mit DPNH. Das Enzym aus Taubenleber ist inaktiv gegen DPNH und Desamino-DPN. Mit dieser Kinase lassen sich präparativ leicht größere Mengen des bisher schwer zugänglichen TPN aus DPN bereiten (Wang und Kaplan 1954, Wang, Kaplan und Stolzenbach 1954).

Enzymatischer Abbau.

DPN und TPN sind als Dinucleotide chemisch sehr labil. Sie werden sowohl durch Säuren als auch durch Alkalien hydrolytisch gespalten. Aber selbst das Erwärmen in einer neutralen Lösung führt zu partieller Zersetzung. Die Pyridinnucleotide sind jedoch auch in vielen biologischen Extrakten — gerade in physiologischem p_H-Bereich — unbeständig und werden durch eine Reihe von Enzymen abgebaut (Schlenk 1945).

Desaminierung. DPN, nicht jedoch TPN, wird durch die unspezifische Desaminase aus Takadiastase zu Desamino-DPN hydrolysiert (Kaplan, Colowick und Ciotti 1952):

$$\text{DPN} + \text{H}_2\text{O} \rightarrow \text{Desamino-DPN} + \text{NH}_3.$$

Dieses Enzym desaminiert auch die folgenden Spaltprodukte der Coenzyme: Adenosin-diphospho-ribose, 5'-AMP und Adenosin. Die spezifische 5'-Adenylsäure-desaminase aus Muskeln und die Adenosin-desaminase aus der Darmschleimhaut (Kalckar und Shafran 1947) greifen die Pyridinnucleotide nicht an.

Spaltung der Pyrophosphatbindung. In Pflanzen und auch in tierischen Geweben kommen weit verbreitet Pyrophosphatasen vor, die die Pyrophosphatbindung von DPN und TPN spalten. Aus DPN entstehen Nicotinsäureamid-ribotid und Adenosin-5'-phosphat, aus TPN neben NMN das Adenosin-2',5'-diphosphat:

$$\text{DPN} + \text{H}_2\text{O} \rightarrow \text{Nicotinsäureamid-ribotid} + \text{Adenosin-5'-phosphat.}$$

$$\text{TPN} + \text{H}_2\text{O} \rightarrow \text{Nicotinsäureamid-ribotid} + \text{Adenosin-2',5'-diphosphat.}$$

Die Pyrophosphatase aus Kartoffeln spaltet auch DPNH, Flavin-adenin-dinucleotid und ATP (Kornberg und Lindberg 1948, Kornberg und Pricer 1950a). Wahrscheinlich werden auch Uridin-diphospho-glucose und Coenzym A durch das gleiche Enzym hydrolysiert.

In Gegenwart von anorganischem Pyrophosphat kann die Pyrophosphat-bindung auch durch das DPN-synthetisierende Enzym pyrophosphorolytisch gespalten werden (KORNBERG 1950a) (s. S. 184).

Dephosphorylierung. Spezifische und unspezifische Phosphatasen verschie-dener Herkunft dephosphorylieren TPN und die Mononucleotide der Coenzyme. DPN selbst wird dagegen nicht angegriffen. Unspezifische Phosphatasen führen TPN in DPN über (EULER und ADLER 1938, KORNBERG 1950a, SANADI 1952).

$$\text{TPN} + \text{H}_2\text{O} \rightarrow \text{DPN} + \text{H}_3\text{PO}_4.$$

3'-Nucleotidase aus "rye grass" (SHUSTER und KAPLAN 1953) ist spezifisch für 3'-Nucleotide und greift dementsprechend TPN nicht, wohl dagegen Coenzym A an. Unspezifische Phosphatasen und 5'-Nucleotidasen hydrolysieren Nicotinsäure-amid-ribotid zu Nicotinsäureamid-ribosid und Adenosin-5'-phosphat zu Adenosin (HEPPEL und HILMOE 1951).

$$\text{Nicotinsäureamid-ribotid} + \text{H}_2\text{O} \rightarrow \text{Nicotinsäureamid-ribosid} + \text{H}_3\text{PO}_4.$$

$$\text{Adenosin-5'-phosphat} + \text{H}_2\text{O} \rightarrow \text{Adenosin} + \text{H}_3\text{PO}_4.$$

Adenosin-2',5'-diphosphat aus TPN wird durch 5'-Nucleotidasen (HEPPEL und HILMOE 1951) nicht angegriffen. Adenosin-5'-phosphatase (KORNBERG und PRICER 1951) aus Kartoffeln spaltet nur den Phosphatrest vom C_5 ab.

Spaltung der Nicotinsäureamid-Ribosebindung. Beide Pyridinnucleotide wer-den durch Nucleosidasen (,,DPN-ase") an der N-glykosidischen Bindung hydroly-siert zu Nicotinsäureamid und Adenosin-diphospho- bzw. -triphospho-ribose:

$$\text{N}^+\text{RPPRA} + \text{H}_2\text{O} \rightarrow \text{N} + \text{RPPRA} + \text{H}^+.$$
$$(= \text{DPN})$$

Derartige Enzyme sind in Mikroorganismen und verschiedenen tierischen Geweben nachgewiesen worden. Die DPN-ase aus *Neurospora crassa* (NASON, KAPLAN und COLOWICK 1951, KAPLAN, COLOWICK und NASON 1951) reagiert auch mit Desamino-DPN (KAPLAN, COLOWICK und CIOTTI 1952). Reduziertes DPN und TPN, NMN, Nicotinsäureamid-ribosid und das α-Isomere des DPN (KAPLAN und Mitarbeiter 1955) sind als Substrate unwirksam.

Im Gegensatz zu den *Neurospora*-Nucleosidasen haben die entsprechenden tierischen Enzyme auch die Eigenschaften von Transglykosidasen.

Durch die Verwendung von Nicotinsäureamid, das mit C^{14} markiert war, ließ sich eine Austauschreaktion zwischen freiem und an DPN gebundenem Nicotinsäureamid nachweisen (ZATMAN, KAPLAN und COLOWICK 1953, ZATMAN und Mitarbeiter 1955). Durch solche Trans-glykosidierungen mit Pyridinderivaten ließen sich verschiedene DPN-Derivate enzymatisch synthetisieren:

$$\text{X} + \text{N}^+\text{RPPRA} \rightarrow \text{X}^+\text{RPPRA} + \text{N}.$$
$$(= \text{DPN})$$

In Hefen findet sich eine Nucleosidase, die Nicotinsäureamid-ribosid hydrolytisch spaltet (HEPPEL und HILMOE 1952). Die Spaltung der Nicotinsäureamid-Ribosid-bindung kann auch phosphorolytisch in Umkehrung der von ROWEN und KORN-BERG (1951) gefundenen reversiblen Nicotinsäureamid-ribosid-Synthese aus Nicotinsäureamid und Ribose-1-phosphat erfolgen (s. S. 183).

Desamidierung von Nicotinsäureamid.

Verschiedene *Lactobacillus*-Arten enthalten ein Enzym, das die Umwandlung von Nicotinsäureamid in Nicotinsäure bewirkt (HUGHES und WILLIAMSON 1953):

$$\text{Nicotinsäureamid} + \text{H}_2\text{O} \rightarrow \text{Nicotinsäure} + \text{NH}_3.$$

Auch in *Staphylococcus albus* wurde dieses Enzym gefunden, jedoch nicht in zahlreichen anderen Bakterienarten. Die bakterielle „Nicotinsäureamid-des-amidase" besitzt ein p_H-Optimum von 5,8. Das Ferment greift keine anderen natürlichen Nicotinsäureamidderivate an.

Abbau der Nicotinsäure.

Über den Abbau der Nicotinsäure in der höheren Pflanze und in Mikroorganis-men ist nur wenig bekannt. Hingegen ist die Literatur über den Stoffwechsel dieser Verbindung im tierischen Organismus sehr umfangreich (Williams und Mitarbeiter 1950).

Verschiedene Stämme von *Pseudomonas fluorescens* können auf einem Medium wachsen, das nur anorganische Salze und Nicotinsäure als einzige Kohlenstoff- und Stickstoffquelle enthält (Allinson 1943). Gewaschene Zellsuspensionen

Tabelle 3. *Abbauprodukte des Nicotinsäureamids.*

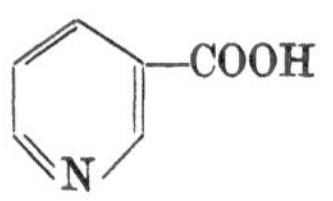

Nicotinsäure

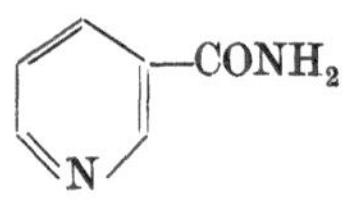

Nicotinsäureamid

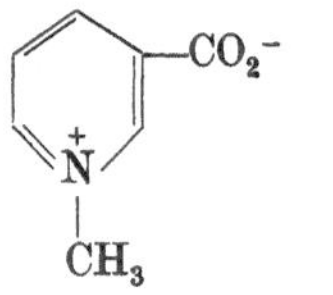

Trigonellin

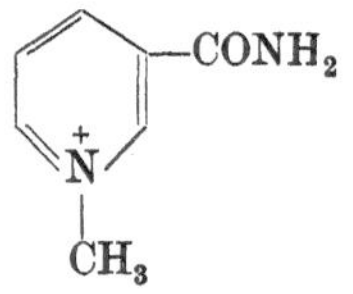

N^1-Methyl-nicotinsäureamid

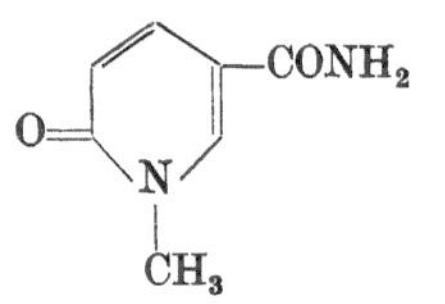

N^1-Methyl-6-pyridon-3-carbonsäuramid

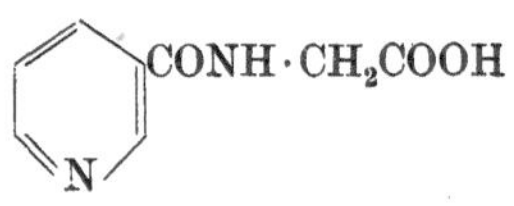

Nicotinursäure

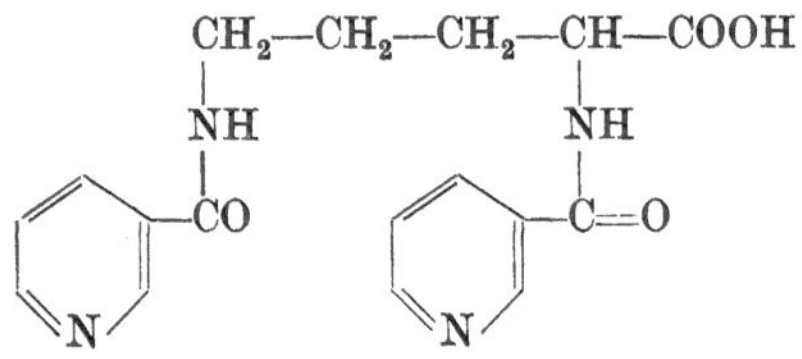

2,5-Dinicotinyl-ornithin

dieser Mikroorganismen oxydieren Nicotinsäure unter Bildung von Kohlendioxyd und Ammoniak. 6-Hydroxy-nicotinsäure läßt sich als partielles Oxydations-produkt isolieren (Hughes 1955). Es wird deshalb angenommen, daß der Pyridin-ring vor der Spaltung hydroxyliert wird, und daß 6-Hydroxy-nicotinsäure das primäre Oxydationsprodukt ist. *Pseudomonas fluorescens*, gewachsen auf Nicotin-säure, oxydiert 6-Hydroxy-nicotinsäure, jedoch nicht 2-Hydroxy-nicotinsäure (Hughes 1955).

Trigonellin, das Betain der Nicotinsäure, ist in den Samen von *Strophantus*, von Bohnen und Erbsen, in den Knollen von Kartoffeln und Dahlien u. a. aufgefunden worden (GUGGENHEIM 1951). Fäulnisbakterien greifen Trigonellin nur wenig an. Nach dem Fäulnisprozeß waren noch 80% unverändert (THIELMANN 1924). In Kulturen von *Proteus vulgaris* entsteht aus Nicotinsäure und Nicotinsäureamid N^1-Methyl-nicotinsäureamid (ELLINGER, FRAENKEL und ABDEL KADER (1947). Diese Pyridiniumverbindung wird durch Darmbakterien abgebaut (ELLINGER 1947). Für Ricinin (1-Methyl-3-nitrilo-4-methoxy-pyridon) und Ricinidin (1-Methyl-3-nitrilo-2-pyridon) wird angenommen, daß diese Verbindungen in den Pflanzen aus N^1-Methyl-pyridinium-Derivaten gebildet werden.

Im Organismus des Warmblütlers wird die mit der Nahrung zugeführte Nicotinsäure in das Amid und in N^1-Methyl-nicotinsäureamid übergeführt. N^1-Methyl-nicotinsäureamid ist bei den meisten Tieren das überwiegende Produkt des Nicotinsäure-Stoffwechsels. Nach Verabreichung großer Dosen von Nicotinsäure kann eine kleine Menge unverändert im Harn ausgeschieden werden. Ein kleiner Teil erscheint mit Aminosäuren (Glykokoll und Ornithin) durch Säureamidbindung verknüpft als Nicotinursäure und 2,5-Dinicotinyl-ornithin im Harn. Beim Menschen ist 1-Methyl-2-pyridon-5-carbonsäureamid das Hauptausscheidungsprodukt (KNOX und GROSSMAN 1946, HOLMAN und DE LANGE 1949, 1950a und b).

Durch Versuche mit markierter, injizierter Nicotinsäure wurde bewiesen, daß alle diese Pyridinverbindungen im Harn aus dem Vitamin hervorgegangen sind (LEIFER, ROTH und HOGNESS 1951). Ein Teil der injizierten Radioaktivität, die in dem Carboxyl der Nicotinsäure lokalisiert war, wurde als Kohlendioxyd ausgeatmet. Der Abbau der Nicotinsäure kann also noch weiter gehen.

Literatur.

ALLINSON, M. J. C.: Spezific enzymic method for the determination of nicotinic acid in blood. J. of Biol. Chem. **147**, 785—791 (1943). — ARNON, D. J.: Extracellular photosynthetic reaction. Nature (Lond.) **167**, 1008—1010 (1951).

BEADLE, G. W., H. K. MITCHELL and J. F. NYC: Kynurenine as an intermediate in the formation of nicotinic acid from tryptophan by *Neurospora*. Proc. Nat. Acad. Sci. U.S.A. **33**, 155—158 (1947). — BONNER, D. M., and C. YANOFSKY: Quinolinic acid accumulation in the conversion of 3-hydroxyanthranilic acid to niacin in *Neurospora*. Proc. Nat. Acad. Sci. U.S.A. **35**, 576—581 (1949).

CASTRO, F. T. DE, J. M. PRICE and R. R. BROWN: Reduced triphosphopyridinnucleotide requirement for the enzymatic formation of 3-hydroxykynurenin from l-kynurenin. J. Amer. Chem. Soc. 78, 2904—2905 (1956). — CHITRE, R. G., D. B. DESAI and V. S. RAUT: Die Biosynthese der Nicotinsäure in keimenden Getreidesorten und Hülsenfrüchten. Proc. Soc. Biol. Chem., India **13**, 17—18 (1955). Zit. nach Chem. Zbl. **1955**, 6429. — COLOWICK, S. P., N. O. KAPLAN, E. F. NEUFELD and M. CIOTTI: Pyridine nucleotide transhydrogenase. I. Indirect evidence for the reaction and purification of the enzyme. J. of Biol. Chem. **195**, 95—105 (1952).

ELLINGER, P.: Fate of nicotinamide methochloride and the effect of liver poisons on its elimination rate in the rat. Biochemic. J. **41**, 308—314 (1947). — ELLINGER, P., G. FRAENKEL and M. M. ABDEL KADER: Utilization of nicotinamide derivates and related compounds by mammals, insects and bacteria. Biochemic. J. **41**, 559—568 (1947). — ELVEHJEM, C. A., R. J. MADDEN, F. M. STRONG and D. W. WOOLLEY: The isolation and identification of the antiblacktongue factor. J. of Biol. Chem. **123**, 137—149 (1938). — EULER, H. v., u. E. ADLER: Über die gegenseitige enzymatische Umwandlung von Codehydrase I und Codehydrase II. Z. physiol. Chem. **252**, 41—48 (1938). — EULER, H. v., u. H. ALBERS: Über die Komponenten der Dehydrasesysteme. IX. Die Co-Dehydrasen: Co-Zymase und Co-Dehydrase II. Co-Zymase als Wasserstoffübertrger. Z. physiol. Chem. **238**, 233—260 (1936).

FOUTS, P. J., O. M. HELMER, S. LEPKOWSKI and T. H. JUKES: Die Behandlung menschlicher Pellagra mit Nicotinsäure. Proc. Soc. Exper. Biol. a. Med. **37**, 405—407 (1937). Zit. nach Chem. Zbl. **1938** I, 2576.

GINGRICH, W. D., and F. SCHLENK: Codehydrogenase I and other pyridinium compounds as V-factor for *Hemophilus influenzae* and *H. parainfluenzae*. J. Bacter. **47**, 535—550 (1944).— GUGGENHEIM, M.: Die biogenen Amine. Basel: S. Karger 1951. — GUSTAFSON, F. G.: Tryptophan as an intermediate in the synthesis of nicotinic acid by green plants. Science (Lancaster, Pa.) **110**, 279—280 (1949).

HARDEN, A., and W. J. YOUNG: Das alkoholische Ferment des Hefesaftes. Proc. Roy. Soc. Lond., Ser. B 77, 405—420 (1906). Zit. nach Chem. Zbl. **1906** I, 1625. — HASSE, K.: Codehydrasen I und II. In: Moderne Methoden der Pflanzenanalyse. Herausgeg. von K. PAECH u. M. V. TRACEY, Bd. 4, S. 320—344. Berlin-Göttingen-Heidelberg: Springer 1955. — Pyridinnucleotid-dehydrogenasen. In Handbuch der Pflanzenphysiologie. Herausgeg. von

W. Ruhland, Bd. 12. Berlin-Göttingen-Heidelberg: Springer 1958. — Hayaishi, O.: Kynureninase of *Pseudomonas fluorescens*. J. of Biol. Chem. **195**, 735—740 (1952). — Heidelberger, C., M. E. Gullberg, A. F. Morgan and S. Tepkowski: Concering the mechanism of the mammalian conversion of tryptophan into kynurenine, kynurenic acid and nicotinic acid. J. of Biol. Chem. **175**, 471—472 (1948). — Henderson, L. M., and H. M. Hirsch: Quinolinic acid metabolism. I. Urinary excretion by the rat following tryptophan and 3-hydroxyanthranilic acid administration. J. of Biol. Chem. **181**, 667—675 (1949). — Henderson, L. M., and G. B. Ramasarma: Quinolinic acid metabolism. III. Formation from 3-hydroxy-anthranilic acid by rat liver preparations. J. of Biol. Chem. **181**, 687—962 (1949). — Heppel, L. A., and R. J. Hilmoe: Purification and properties of 5-nucleotidase. J. of Biol. Chem. **188**, 665—676 (1951). — Phosphorolysis and hydrolysis of purine nucleosides by enzymes from yeast. J. of Biol. Chem. **198**, 683—694 (1952). — Holman, W. I. M., and D. J. de Lange: Determination of N-methyl-2-pyridone-5-carboxylamide and of N-methyl-2-pyridone-3-carboxylamide in human urine. Biochemic. J. **45**, 559—563 (1949). — Metabolism of nicotinic acid and related compounds by humans. Nature (Lond.) **165**, 604—605 (1950a). — The determination of N-methyl-2-pyridone-5-carboxylic acid in human urine. Biochemic. J. **46**, 47—49 (1950b). — Huff, J. W., W. A. Perlzweig, R. Forth and F. Spilman: Nicotinic acid metabolism. III. Metabolism and synthesis of nicotinic acid in the rat. J. of Biol. Chem. **142**, 401—416 (1942). — Hughes, D. E.: 6-Hydroxynicotinic acid as an intermediate in the oxidation of nicotinic acid by *Pseudomonas fluorescens*. Biochemic. J. **60**, 303—310 (1955). — Hughes, D. E., and W. H. Williamson: The deamidation of nicotinamide by bacteria. Biochemic. J. **55**, 851—856 (1953).

Kalckar, H. M.: Enzymic synthesis of a nucleoside. J. of Biol. Chem. **158**, 723—724 (1945). — Kalckar, H. M., and M. Shafran: The enzymic synthesis of purine ribosides. J. of Biol. Chem. **167**, 477—486 (1947). — Kaplan, N. O., M. M. Ciotti, F. E. Stolzenbach, and N. R. Bachner: Isolation of a DPN isomer containing nicotinamide riboside in the α-linkage. J. Amer. Chem. Soc. **77**, 815 (1955). — Kaplan, N. O., S. P. Colowick and M. M. Ciotti: Enzymatic desamination of adenosine derivates. J. of Biol. Chem. **194**, 579—591 (1952). — Kaplan, N. O., S. P. Colowick and A. Nason: *Neurospora* diphosphopyridine nucleotidase. J. of Biol. Chem. **191**, 473—483 (1951). — Kaplan, N. O., S. P. Colowick and E. F. Neufeld: Pyridine nucleotide transhydrogenase. II. Direct evidence for and mechanism of the transhydrogenase reaction. J. of Biol. Chem. **195**, 107—119 (1952). — Pyridine nucleotide transhydrogenase. III. Animal tissue transhydrogenases. J. of Biol. Chem. **205**, 1—15 (1953). — Kaplan, N. O., S. P. Colowick, E. F. Neufeld and M. M. Ciotti: Pyridine nucleotide transhydrogenase. IV. Effect of adenylic acid on the bacterial transhydrogenases. J. of Biol. Chem. **205**, 17—30 (1953). — Kaplan, N. O., S. P. Colowick, L. J. Zatman and M. M. Ciotti: Pyridine nucleotide transhydrogenase. V. Exchange reactions studied with C[14]. J. of Biol. Chem. **205**, 31—44 (1953). — Knox, W. E., and W. J. Grossman: A new metabolite of nicotinamide. J. of Biol. Chem. **166**, 391—392 (1946). — Knox, W. E., and A. H. Mehler: The conversion of tryptophan to kynurenine in liver. I. The coupled tryptophan peroxidase-oxidase system forming formylkynurenine. J. of Biol. Chem. **187**, 419—430 (1950). — Kornberg, A.: Reversible enzymatic synthesis of diphosphopyridine nucleotide and inorganic pyrophosphate. J. of Biol. Chem. **182**, 779—793 (1950a). — Enzymatic synthesis of triphosphopyridine nucleotide. J. of Biol. Chem. **182**, 805—813 (1950b). — Kornberg, A., J. Lieberman and E. S. Simms: Enzymatic synthesis of purine nucleotides. J. of Biol. Chem. **215**, 417—427 (1955). — Kornberg, A., and O. Lindberg: Diphosphopyridine nucleotide pyrophosphatase. J. of Biol. Chem. **176**, 665—677 (1948). — Kornberg, A., and W. E. Pricer jr.: Nucleotide pyrophosphatase. J. of Biol. Chem. **182**, 763—778 (1950a). — The structure of triphosphopyridine nucleotide. J. of Biol. Chem. **186**, 557—567 (1950b). — Kotake, Y., u. J. Iwao: Studien über den intermediären Stoffwechsel des Tryptophans. I. Mitt. Über das Kynurenin, ein intermediäres Stoffwechselprodukt des Tryptophans. Z. physiol. Chem. **195**, 139—147 (1931). — Kotake, Y., u. Y. Nakayama: Studien über den intermediären Stoffwechsel des Tryptophans. XXXIV. Mitt. Über die Anthranilsäurebildung aus Kynurenin durch Organsaft. Z. physiol. Chem. **270**, 76—83 (1941). — Kotake, Y., u. S. Otani: Studien über den intermediären Stoffwechsel des Tryptophans. XII. Mitt. Über das Kynurenin, ein intermediäres Stoffwechselprodukt des Tryptophans. Z. physiol. Chem. **214**, 1—5 (1933). — Krehl, W. A., P. S. Sarma and C. A. Elvehjem: The effect of protein on the nicotinic acid and tryptophan requirement of the growing rat. J. of Biol. Chem. **162**, 403—411 (1946). — Krehl, W. A., L. J. Tepley, P. S. Sarma and C. A. Elvehjem: Growth-retarding effect of corn in nicotinic acid-low rations and its counteraction by tryptophan. Science (Lancaster, Pa.), N. S. **101**, 489—490 (1945).

Leifer, E., L. J. Roth and D. S. Hogness: The metabolism of radioactive nicotinic acid and nicotinamide. J. of Biol. Chem. **190**, 595—602 (1951). — Lieberman, J., A. Kornberg and E. S. Simms: Enzymatic synthesis of pyrimidine nucleotides. Orotidine-5'-phosphate and uridine-5'-phosphate. J. of Biol. Chem. **215**, 403—415 (1955).

MATSUOKA, Z. und N. YOSHIMATSU: Über eine neue Substanz, die aus Tryptophan im Tierkörper gebildet wird. Z. physiol. Chem. 143, 206—217 (1925). — MEHLER, A. H.: Formation of picolinic and quinolinic acids following enzymatic oxidation of 3-hydroxy-anthranilic acid. J. of Biol. Chem. 218, 241—254 (1956). — MITCHELL, H. K., and J. F. NYC: Hydroxyanthranilic acid as a precursor of nicotinic acid in Neurospora. Proc. Nat. Acad. Sci. U.S.A. 34, 1—5 (1948). — MYRBÄCK, K.: Co-Zymase (zusammenfassende Darstellung der bisherigen Ergebnisse über die Cozymase). In Ergebnisse der Enzymforschung. Herausgeg. von F. F. NORD u. R. WEIDENHAGEN, Bd. 2, S. 139—168. Leipzig 1933.

NASON, A.: Existence of a tryptophan-niacin relationship in corn. Science (Lancaster, Pa.) 109, 170—171 (1949). — NASON, A., N. O. KAPLAN and S. P. COLOWICK: Changes in enzymatic constitution in zink-deficient Neurospora. J. of Biol. Chem. 188, 397—406 (1951).

QUAGLIARIELLO, G., e G. PORCELLATI: Über die enzymatische Amidierung der Nicotinsäure. Boll. Soc. ital. Biol. sper. 29, 273—275 (1953). Zit. nach Chem. Zbl. 1955, 5097.

ROWEN, J. W., and A. KORNBERG: The phosphorolysis of nicotinamide riboside. J. of Biol. Chem. 193, 497—507 (1951).

SANADI, D. R.: Enzymic dephosphorylation of triphosphopyridine nucleotide (TPN). Arch. of Biochem. a. Biophysics 35, 268—277 (1952). — SCHLENK, F.: Enzymatic reactions involving nicotinamide and its related compounds. In: Advances in enzymology. Herausgeg. von F. F. NORD u. C. H. WERKMAN, Bd. 5, S. 207—236. New York: Interscience Publ. 1945. — Co-dehydrogenase I and II and apoenzymes. In: The enzymes. Chemistry and mechanism of action. Herausgeg. von J. B. SUMNER u. K. MYRBÄCK, Bd. II/1, S. 250—315. New York: Academic Press 1951. — SCHLENK, F., u. H. v. EULER: Cozymase. Naturwiss. 24, 794—795 (1936). — SCHOPFER, W. H., and M. L. BOSS: Tryptophan-nicotinic acid relationship. Production of nicotinic acid avitaminosis in an organism by vitamin K_3 (2-methyl-1,4-naphthoquinone). Helvet. physiol. Acta 7, C 20—22 (1949). — SHUSTER, L., and N. O. KAPLAN: A specific b nucleotidase. J. of Biol. Chem. 201, 535—546 (1953). — SINGER, T. P., and E. B. KEARNEY: Chemistry, metabolism and scope of action of the pyridine nucleotide coenzymes. In: Advances in enzymology and related subjects. Herausgeg. von F. F. NORD, Bd. 15, S. 79—139. New York: Interscience Publ. 1954.

THIELMANN, F.: Ein Fäulnisversuch mit Trigonellin. Z. Biol. 81, 208—210 (1924). — TOLMACH, L. J.: The influence of triphosphopyridine nucleotide (TPN) and other physiological substances upon oxygen evolution from illuminated chloroplasts. Arch. of Biochem. a. Biophysics 33, 120—142 (1951).

VISHNIAC, W., and S. OCHOA: Fixation of carbon dioxide coupled to photochemical reduction of pyridine nucleotides by chloroplast preparations. J. of Biol. Chem. 195, 75—93 (1952a). — Phosphorylation coupled to photochemical reduction of pyridine nucleotides by chloroplast preparations. J. of Biol. Chem. 198, 501—506 (1952b).

WANG, T. P., and N. O. KAPLAN: Kinases for the synthesis of coenzyme a and triphosphopyridine nucleotide. J. of Biol. Chem. 206, 311—325 (1954). — WANG, T. P., N. O. KAPLAN and F. E. STOLZENBACH: Enzymatic preparation of triphosphopyridine nucleotide from diphosphopyridine nucleotide. J. of Biol. Chem. 211, 465—472 (1954). — WARBURG, O.: Chemische Konstitution von Fermenten. In: Ergebnisse der Enzymforschung. Herausgeg. von F. F. NORD u. R. WEIDENHAGEN, Bd. 7, S. 210—245. Leipzig: Akademische Verlagsgesellschaft 1938. — WARBURG, O., u. W. CHRISTIAN: Über Aktivierung der ROBISONschen Hexose-Mono-Phosphorsäure in roten Blutzellen und die Gewinnung aktivierender Fermentlösungen. Biochem. Z. 242, 206—227 (1931). — Pyridin, der wasserstoffübertragende Bestandteil von Gärungsfermenten (Pyridinnucleotide). Biochem. Z. 287, 291—333 (1936). — WARBURG, O., W. CHRISTIAN u. A. GRIESE: Wasserstoffübertragendes Co-Ferment, seine Zusammensetzung und Wirkungsweise. Biochem. Z. 282, 157—205 (1935). — WARBURG, O., W. CHRISTIAN u. W. SCHÖLLER: Co-Fermentproblem. Biochem. Z. 275, 464 (1935). — WILLIAMS, R. J., R. E. EAKIM, E. BEERSTECKER jr. and W. SHIVE: The biochemistry of B vitamin. New York: Reinbold 1950. — WISS, O.: Der enzymatische Abbau des Kynurenins und 3-Oxy-kynurenins im tierischen Organismus. Z. physiol. Chem. 293, 106—121 (1953). — Über die Umwandlung der 3-Hydroxy-anthranilsäure, Chinolinsäure und Nicotinsäure im tierischen Organismus. Z. physiol. Chem. 304, 221—231 (1956).

YANOFSKY, C., and D. M. BONNER: Studies on the conversion of 3-hydroxyanthranilic acid in Neurospora. J. of Biol. Chem. 190, 211—218 (1951).

ZATMAN, L. J., N. O. KAPLAN and S. P. COLOWICK: Inhibition of spleen diphosphopyridine nucleotidase by nicotinamide, an exchange reaction. J. of Biol. Chem. 200, 197—212 (1953). — ZATMAN, L. J., N. O. KAPLAN, S. P. COLOWICK and M. M. CIOTTI: The isolation and properties of the isonicotinic acid hydrazide analogue of diphosphopyridine nucleotide. J. of Biol. Chem. 209, 467—484 (1955).

Riboflavin nucleotides*.

By

B. J. D. Meeuse.

With 5 figures.

Introduction.

The recognition that certain flavin derivatives can act as co-enzymes is based on the pioneer work done by WARBURG and CHRISTIAN on the so-called "old yellow enzyme" (OYE), a hemin-free respiratory agent which they isolated from bottom yeast in 1932 (WARBURG and CHRISTIAN 1932 [1], [2], [3]). In 1933, they were able to show that OYE consists of a highmolecular carrier protein to which is attached, as the "active group", a low-molecular yellow pigment with a powerful green fluorescence (WARBURG and CHRISTIAN 1933 [4], [5], [6]). Warming the enzyme with methanol/water led to a splitting-off of the active group; a photo-derivative of the latter, lumiflavin, could be obtained in crystalline form from chloroform, and the empirical formula could be determined. The structure of the original (non-irradiated) pigment was, thereafter, established unequivocally by KUHN, REINEMUND et al. (1935) as well as by KARRER et al. (1935); total synthesis showed it to be a nucleoside, 6,7-dimethyl-9-(D-1-ribityl)-isoalloxazine (see formula).

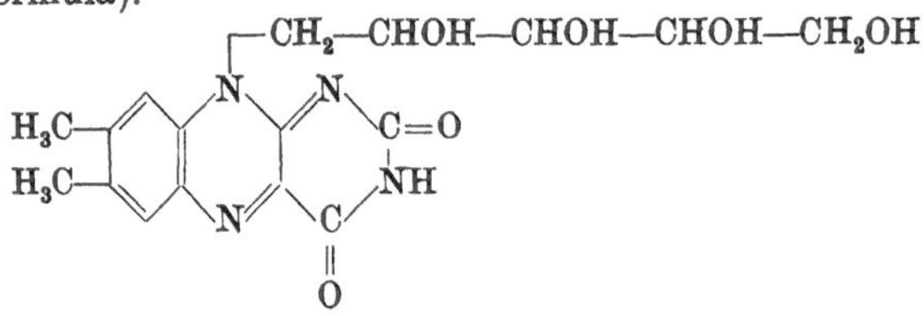

Riboflavin (lactoflavin)

It was soon recognized that there is no real difference between this compound and the various "flavins" found by KUHN, GYÖRGYI and WAGNER-JAUREGG (1933 [1], [2], [3], 1934) in biological materials of widely different origin: lacto-flavin from milk, ovoflavin from eggs, hepatoflavin from liver, etc. Neither is there a difference with vitamin B_2, BLEYER and KALLMANN's "lactochrom" [from milk (1925)], SZENT-GYÖRGYI's "cytoflav" [from heart muscle (1932)], or ELLINGER and KOSCHARA's "lyochromes" [from various sources (1933)]. The accepted name nowadays is riboflavin. According to KUHN et al. (1934), the photoderivative lumiflavin, mentioned above, is 6,7,9-trimethyl-iso-alloxazine. Lumichrome, another photodegradation product, is 6,7-dimethyl-iso-alloxazine (see formulae). However, from THEORELL's beautiful work on OYE in 1937

---

* The following abbreviations will be used: Rbf = riboflavin; FMN = riboflavin-5′-phosphate or flavin mononucleotide; FAD = flavin-adenine dinucleotide; OYE = old yellow enzyme; AMP, ADP and ATP = adenosine mono-, di- and triphosphate; PP = pyrophosphate; DPN = cozymase or co-enzyme I or diphosphopyridine nucleotide; TPN = WARBURG's co-enzyme or co-enzyme II or triphosphopyridine nucleotide.

No serious attempt will be made to respect the artificial boundary between the biochemistry of higher plants and that of other forms such as yeasts, bacteria and animals.

it became evident that the nucleoside, riboflavin, does not itself represent the original active group of the flavoprotein; it must be considered a breakdown

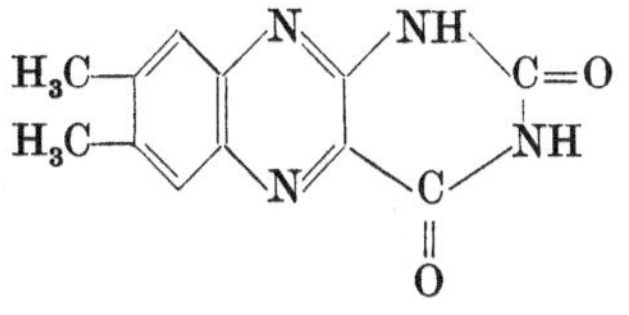

$$H_3C- \quad N \quad NH \quad C=O \quad NH \quad C \quad \| \quad O$$

lumichrome

$$CH_3 \quad H_3C- \quad N \quad N \quad C=O \quad NH \quad C \quad \| \quad O$$

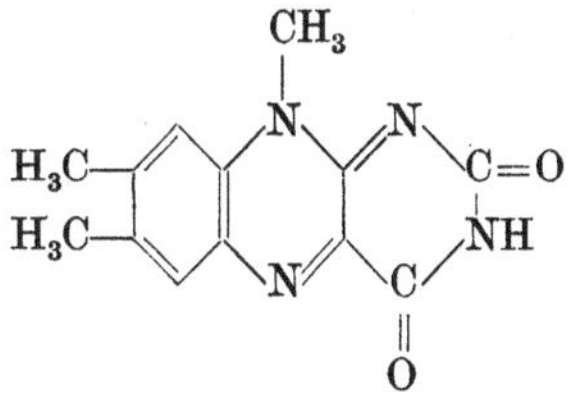

lumiflavin

product of a naturally occurring flavin nucleotide. Riboflavin-5'-phosphate (FMN, for flavin mononucleotide; see formula) was indicated by THEORELL as the original active group. In 1938, WARBURG and CHRISTIAN (1938 [7]—[10])

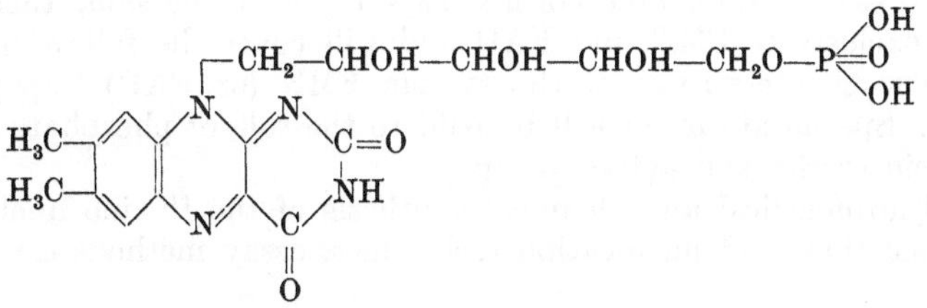

$$-CH_2-CHOH-CHOH-CHOH-CH_2O-P \underset{OH}{\overset{OH}{=}} O$$

$$H_3C- \quad N \quad N \quad C=O \quad NH \quad C \quad \| \quad O$$

Riboflavin-5'-phosphoric acid

could bring excellent evidence for the idea that the co-enzyme of another flavo-protein, D-amino acid oxidase, is flavin-adenine-dinucleotide or FAD (see formula). This was supported by ABRAHAM (1939), who identified both FMN and adenosine-5'-phosphate as hydrolysis products of FAD. The discovery of the coenzyme-role of FAD has induced WARBURG *and coworkers* to engage in some speculation concerning the possibility that FMN is itself an artifact, produced through a breakdown of FAD. This aspect will be discussed later, in connection with recent work done on OYE in THEORELL's laboratory.

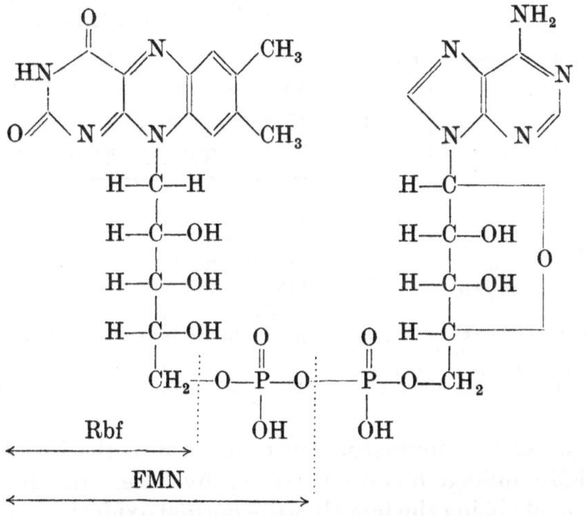

Flavine adenine dinucleotide (FAD).
(After CHRISTIE, KENNER and TODD 1952.)

Substances of at least potential importance in a study of flavoprotein co-enzymes are the so-called pseudo-flavins, studied by HUENNEKENS *et al.* (1954). These compounds, possibly related to toxoflavin, isolated by VAN VEEN and MERTENS (1933) from cultures of *Pseudomonas cocovenenans (Bacterium bongkrek)*, can function as co-enzyme in processes usually catalyzed by flavoproteins. Thus, a pseudoflavin is active in the oxidation of reduced DPN (cozymase) by oxygen and a pig heart DPNH-oxidase (HUENNEKENS and BASFORD 1954); the oxidation of reduced co-enzyme II (TPNH) by oxygen (or methemoglobin) and erythrocyte methemoglobin reductase is also pseudoflavin-catalyzed (HUENNEKENS and GABRIO 1954). However, since the chemical structure of toxoflavin and the pseudo-flavins still awaits elucidation, and since little can be said about a possible role of phosphate, the pseudoflavins will not be discussed here. Likewise, no special mention will be made of the role of metals in combination with flavin enzymes. (For an excellent discussion of the so-called metalloflavoproteins, see MAHLER and GREEN 1954.) The discussion, then, will concern almost exclusively FMN and FAD and will cover the following points:

1) The characteristics of the system FMN (or FAD) + apoprotein ⇌ flavoprotein. Special attention will be paid to the role of phosphate in the coupling of protein carrier and active group.

2) The biological and chemical synthesis of the flavine nucleotides.

3) Isolation- and purification techniques, assay methods *etc.*

I. The system FMN + apoprotein ⇌ flavoprotein.

(Studied in the case of OYE.)

In 1937, THEORELL (1937 [1], [2]) succeeded in obtaining a highly purified preparation of OYE, which he dialyzed against a large volume of dilute HCl at low temperature and in the presence of some ammonium sulfate; this resulted in the splitting-off of FMN without concomitant denaturation of the protein carrier. Active holo-enzyme could therefore be reconstituted from protein and FMN, which combined in a 1:1 ratio. If carrier protein was present in excess, the activity turned out to be directly proportional to the quantity of FMN added; by continued addition of the nucleotide, the quantity of protein component could be made the limiting factor (see Fig. 1).

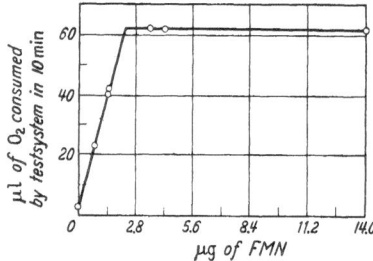

Fig. 1. Resynthesis of the "old yellow enzyme" from carrier protein and active group. The amount of carrier protein in this experiment was kept constant. (According to THEORELL 1937 [2].)

Table 1. E_0' at p_H 7 and 30° C.

O₂	0.81 V
cytochrome c	0.262
cytochrome b	0
OYE	−0.123 } 0.063 V
FMN	−0.186
alcoholdehydrogenase-DPN⁺	−0.240 } 0.078 V
DPN⁺	−0.318
H₂	−0.420

The importance of the iminogroup in position 3 of the iso-alloxazine nucleus for the protein-FMN linkage has been stressed by KUHN and BOULANGER (1936), one of their arguments being the fact that the normal oxidation-reduction potential (at p_H 7) of riboflavin and FMN is much lower than that of OYE. They believed the E_0' (FMN, Rbf) to be between −0.185 and −0.191 Volt, and the E_0' (OYE) to be between −0.059 and −0.066 Volt. These figures have recently been challenged by VESTLING (1955). The Table 1 gives his tentative redox potential values for FMN and OYE, within the proper framework of other biologically active compounds.

It is clear that the difference between the redox potentials of FMN and OYE is not as spectacular as believed before; the effect of binding FMN to its apo-enzyme is very similar to that of binding DPN⁺ to yeast alcohol dehydrogenase, as calculated by THEORELL and BONNICHSEN (1951).

The importance of the phosphate group of FMN (and presumably of FAD) for the coupling with the specific protein has been highlighted in a beautiful series of investigations carried out recently in THEORELL's laboratory. Working with crystalline OYE of almost 100% purity, and basing their approach on the fact that the fluorescence of FMN disappears when it combines with the specific protein to form OYE, THEORELL and NYGAARD (1954 [1]) have con-structed an extremely sensitive fluorescence recorder enabling the investigators to carry out, for the first time in history, a detailed study of the kinetics of the reversible reaction between a co-enzyme and its apo-enzyme. With a view to the fundamental nature of this study, and the far-reaching consequences it will undoubtedly have, it will here be discussed in some detail.

In a short advance note (1954 [2]) THEORELL and NYGAARD come to the conclusion that FMN and protein are attached through electrostatic forces, the FMN furnishing the negative and the protein the positive charge. The old idea that there are 2 linkages between the FMN and the protein, one through the phosphate group of the FMN and the other through the NH-group in position 3 of the iso-alloxazine ring, is confirmed. From the dissociating effects which various anions have on the FMN-protein system, is it concluded that the latter can be considered as an anionic exchanger of the weakly basic type; the order of effectiveness in causing dissociation can be represented, namely, as: $Br^- > NO_3^- > Cl^- > SO_4^= > PO_4^=$ > acetate, and it can be seen that there is a definite correlation with the acidic strengths of the acids from which the salts are derived. In general, k_1 (the velocity constant for the association-reaction) is decreased, k_2 (the constant for the dissociation) is increased, but chloride (e.g.) affects k_2 more than 1000 times as strongly as acetate does.

In water, there is a tremendous affinity between FMN and apo-enzyme; the dissociation constant K for the system is too low to be accurately measured and is assumed to be in the vicinity of 10^{-12} M. In comparison, the affinity between Rbf and apo-enzyme (which can also be determined with the fluorescence recorder because the fluorescence of Rbf, like that of FMN, is quenched by the coupling) is very weak, $K = 5 \times 10^{-6}$ M.

Experiments on the p_H-influence (THEORELL and NYGAARD 1954 [1, 3]; Figs. 2, 3 and 4) provide more detailed information on the mechanism of coupling through the phosphate group; with increasing acidity, the association velocity constant k_1 decreases along a curve almost identical with the second dissociation step of the FMN phosphate group ($pK' \sim 6.0$). For the rapid combination of FMN and protein to occur, then, it is necessary that both the acidic groups of the phosphate be dissociated. Between p_H 8.5 and 9, k_1 increases, most probably because the iminogroup in position 3 dissociates with a pK of 10.2; it can be calculated that, at p_H 8.5, 2% of the FMN molecules are present in trivalent form and at p_H 9.2, 9%. If it is assumed that the association velocity is many times greater for the trivalent FMN anions than it is for the others, the rise in k_1 can thus be accounted for. It is true that above a p_H of about 9 a sudden drop in k_1 can be noticed, but this can be explained satisfactorily by assuming that there occurs a loss of positive charges on the protein at this p_H. Primary amino groups are assumed to serve as binding sites for FMN; they probably act in such a fashion that one FMN phosphate combines with 2 different -NH₂ groups in close proximity to each other, *e.g.* on the same side of a "helix" in two consecutive turns. Such a situation could account, at least in part, for the high affinity between FMN and apoprotein, and also for the specific binding of only one FMN to the large protein molecule.

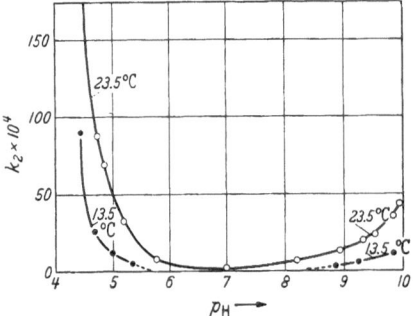

Experiments with Rbf have implicated tyrosine hydroxyl groups of the apo-enzyme as agents in the coupling, confirming an earlier suggestion by Weber (1950). Between p_H 4.3 and 9, hydrogen bonds might also be involved; as suggested by Geissman (1949), the presence of a hydrogen bond from the protein to nitrogen atom 10 in the leucoform of isoalloxazine might account for the low dissociation of the reduced holo-enzyme as compared with that of the oxidized complex (see below).

Fig. 2. k_2 for OYE as a function of p_H in 0.4 M NaCl. (After Theorell and Nygaard 1954.)

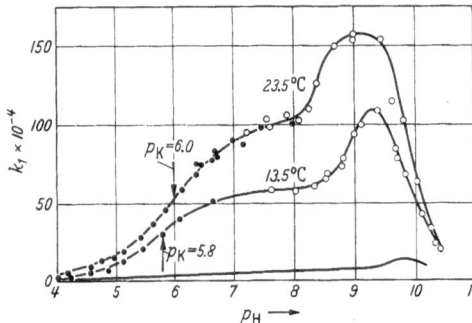

Fig. 3. k_1 for FMN; + protein as a function of p_H; ● in 0.1 M acetate buffer; o in 0.1 M glycine buffer. The lower curve, experimental points omitted, is k_1 for riboflavin + protein in 0.1 M acetate at 13.5°. (After Theorell and Nygaard 1954.)

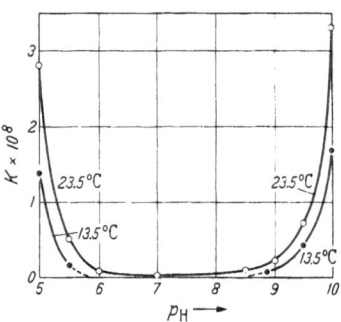

Fig. 4. Dissociation constant, K or OYE as a function of p_H in 0.4 M NaCl. (After Theorell and Nygaard 1954.)

On the basis of their findings, Nygaard and Theorell give the following diagram for the FMN-protein complex:

In later work (1955) the complete reliability of the new fluorometric method was demonstrated by a comparison with the spectrophotometric method. Combining sites of the protein could be identified by studying the effects on k_1, k_2 and K of specific inhibitions (modifications) of the functional groups thought to be active in the coupling. The primary aminogroups involved in the latter process must be more reactive than other aminogroups; for it turns out that the specific acetylation of only a small proportion of the 60 primary aminogroups present per apoprotein molecule causes a strong rise in k_2 and a severe drop in k_1 (see Fig. 5). Strongly acetylated apoenzyme (40/60 acetylation), in the presence of an excess of FMN, was still able to bind one mole of FMN per mole of protein. Likewise: the specific substitution with iodine of some of the 24 tyrosyl groups gave a drop in k_1 and a rise in k_2, again without abolishing the ability of the apoprotein to combine with FMN in the presence of an excess of that compound. Atabrine, which has been used as a structural analogue in the study of flavins, and an incompletely defined riboflavin diphosphate were found to combine loosely with the apoprotein. This was also the case with FAD, although WARBURG and CHRISTIAN have previously reported (1938) that the apo-enzyme of OYE is nearly as active, enzymatically, with FAD as it is with FMN.

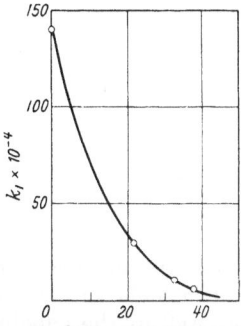

Fig. 5. The effect of acetylation of apoenzyme on k_1, as measured in 0.1 M glycine p_H 9.0. 1 μl portions of acetic anhydride were added to 3 ml of a 27 μM solution of the apoenzyme in 2 M NaAc, p_H 6.5, 0° C. (After NYGAARD and THEORELL 1955.)

The latter authors probably used a large excess of FAD; NYGAARD and THEORELL showed that at neutral p_H values, the K was nearly 200,000 times as large for the FAD apoprotein complex as it was for OYE. The following table allows a comparison between FMN, FAD and Rbf.

These data certainly militate against the idea that FMN is an "artifact" produced from the "original" active group, FAD.

VESTLING (l.c.), on the basis of his E_0' values for OYE and FMN (see before), and accepting a K value of 10^{-8} M for the FMN-OYE system, has arrived at a value of 6×10^{-11} M for K_R, the dissociation constant for

Table 2. *Reaction of the apoenzyme of OYE with FMN, FAD or riboflavin, in 0.1 M glycine, p_H 9.0, at 23° C.*
(From NYGAARD and THEORELL 1955.)

	K	k_1	k_2
FMN	$<10^{-12}$ M	1.4×10^6	$<10^{-6}$
FAD	2×10^{-7} M	0.2×10^6	0.04
Rbf	2×10^{-6} M	0.1×10^6	0.2

the reduced OYE complex. This means that there is an extremely tight binding of reduced FMN to its apo-enzyme. A similar situation may obtain for complexes between reduced FAD and apo-enzymes.

II. The system FAD + apoprotein ⇌ flavoprotein.
(Studied in the case of D-amino acid oxidase.)

BURTON (1951), working with the apo-enzyme of sheep kidney D-amino acid oxidase, found that it was to a certain extent protected from thermal inactivation by FAD, substrates, or inhibitors such as L-leucine and sodium benzoate. Several substances related to D-amino acids or FAD, or resembling these compounds in structure (such as quinine, mepacrine, adenosine, adenine and caffeine) gave no protection. On the other hand, AMP and ADP competed with FAD for the apoenzyme; the affinity between apo-enzyme and these compounds was about twenty times as great as that shown by ATP, inosine-5'-phosphate, adenosine-3'phosphate, guanosine-3'-phosphate or adenosine.

More information on the coupling between FAD and its apo-protein(s) has been produced by Walaas and Walaas (1954). These authors, working with D-amino acid oxidase from hog kidney and FAD purified by ionophoresis on paper, studied the inhibitory effects on the activity of the holo-enzyme by salts as well as by structural analogues of FAD. Salts inhibited in the following order: $I^->Br^->NO_3^->Cl^->PO_4^\equiv>SO_4^=>$acetate$\geq F^-$, indicating, like in the case of OYE and FMN, that the anions of strong acids are most effective. In the presence of high FAD concentrations, the inhibition was reversible, suggesting that there is a competition between inhibitor and prosthetic group. AMP, ATP and FMN (structural analogues of FAD) gave 50% inhibition in concentrations of about 5×10^{-4} M; adenosine, adenine and hypoxanthine did the same at a concentration of 6×10^{-3} M. DPN had a slightly weaker inhibitory activity than the members of the first group. Cytosin and uracil showed little effect, and riboflavin none at all. The conclusion is, that the phosphoric ester groups as well as the adenine part of the FAD molecule are essential in the coupling with the apo-enzyme. It is worthy of note that the coupling of FAD and apo-enzyme causes no quenching of the fluorescence; in this respect, the enzyme is similar to Straub's diaphorase.

III. Some data on the biological and chemical synthesis of Rbf, FMN and FAD.

Data on the biological synthesis are scarce. Nevertheless, 2 different approaches have at least begun to yield promising results. They are, in the first place, feeding experiments with normal organisms, and secondly, the use of so-called biochemical mutants.

In 1952 McLaren, working with the mold *Eremothecium Ashbyii*, which is an important commercial producer of riboflavin, discovered that certain purine derivatives such as xanthine and adenine promote Rbf production in a manner independent of the growth of the mold; uracil inhibits markedly, probably because it acts as a competitive inhibitor. On the basis of these facts, McLaren has postulated his "purine pathway" of Rbf synthesis, which can be pictured as follows:

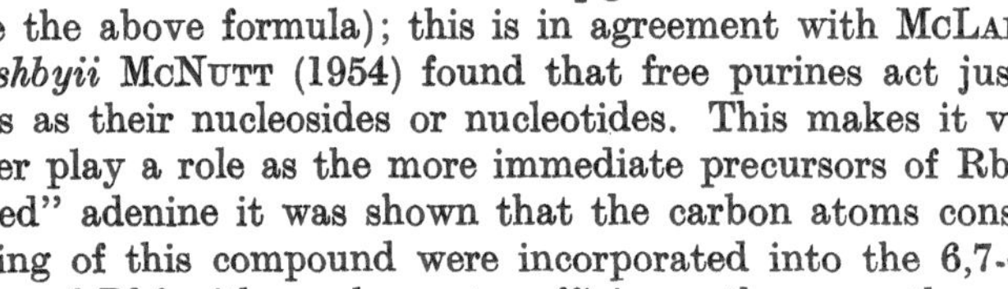

Plaut, in 1953, then reported that in the biogenesis of Rbf by *Ashbya gossypii* formate gives rise to carbon atom 2 and CO_2 gives rise to carbon atom 4 of the Rbf ring (see the above formula); this is in agreement with McLaren's views.

For *E. Ashbyii* McNutt (1954) found that free purines act just as well in Rbf synthesis as their nucleosides or nucleotides. This makes it very unlikely that the latter play a role as the more immediate precursors of Rbf. With the aid of "labeled" adenine it was shown that the carbon atoms constituting the pyrimidine ring of this compound were incorporated into the 6,7-dimethyliso-alloxazine ring of Rbf with much greater efficiency than was the ureido-carbon atom of the imidazole ring. From this it was concluded that the immediate riboflavin precursors formed from adenine probably lack this carbon atom, number 8. In 1956, McNutt was able to confirm this conclusion; adenine thus serves as a precursor in the biogenesis of Rbf through the contribution of an intact pyrimidine ring.

In 1954, Goodwin and Pendlington were able to collect evidence showing that in *E. Ashbyii* xanthine and adenine act as "receptors" for certain amino acids on the way to Rbf. Threonine and serine provide the same, or at least similar, building units. The amino acids form the "aromatic" part of the Rbf molecule, as illustrated here for threonine:

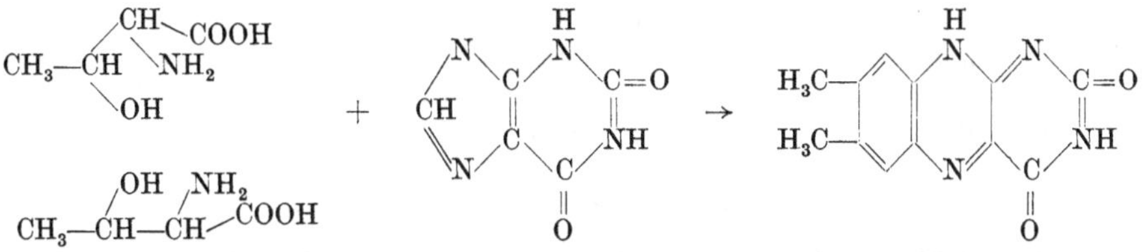

As to the use of Rbf-requiring mutants: a limited number of these has been obtained in *Neurospora*; *e.g.*, Mitchell and Houlahan (1946) have described such a "Rbf-less" mutant which was temperature-sensitive. Their article is valuable in that it also gives data on the natural occurrence of Rbf-derivatives such as lumichrome and lumiflavin. Unfortunately, none of the mutants obtained thusfar has permitted the isolation or identification of intermediates on the pathway to Rbf; neither have cross-feeding experiments been successful (McNutt, private communication).

Chemical synthesis of FMN was achieved by Kuhn *et al.* (1936), by following the reaction scheme: Rbf→5'Trityl Rbf→2',3',4'-acetyl-5' Trityl-Rbf→2',3',4'-Acetyl Rbf→2',3',4'-Acetyl-phospho Rbf→5'-phospho Rbf. A simpler method, namely the direct phosphorylation of Rbf with $POCl_3$ in dry pyridine plus a trace of water, has been worked out by Forrest and Todd (1950). Viscontini *et al.* (1952) have applied metaphosphoric acid, and Flexser and Farkas (1951) have used substituted phosphorylchlorides for the phosphorylation of Rbf.

The process of FMN biosynthesis has been elucidated by Kearney and Englard (1951 [1], [2]) who were able to isolate, from brewers' yeast, an enzyme flavokinase, catalyzing the following reaction:

$$\text{Rbf} + \text{ATP} \rightleftharpoons \text{FMN} + \text{ADP}.$$

The enzyme seems to possess a certain specificity (although riboflavin analogues such as arabityl flavin and 6,7-dichloro-flavin are also converted to their mononucleotide form; see Kearney 1952). It has a temperature optimum of 38° C and a p_H optimum of 7.8—8.5, and requires certain metallic ions for full activity. Mg^{++} is more active than Co^{++} or Mn^{++}, Zn^{++} is active, Ca^{++} inhibitory; Fe^{++}, Ni^{++} and Al^{+++} are inactive. ADP is about half as active as ATP as a phosphate donor. AMP is a competitive inhibitor, inosine triphosphate is inactive.

The dissociation constants for the complex of enzyme and Rbf, ATP, ADP, or AMP are low, and the FMN-forming reaction goes virtually to completion. Flavokinase is thus ideally suited for the manufacture of the trace-substance, FMN.

Enzymatic breakdown of FMN to Rbf, at a very slow rate, was found to occur in bull seminal plasma, by Heppel and Hilmoe (1951). There is no convincing proof that the agent responsible is identical with the seminal 5-nucleotidase which splits off orthophosphate from adenosine-5'-phosphate, inosine-5'-phosphate, uridine-5'-phosphate, cytidine-5'-phosphate, nicotinamide-ribose-5'-phosphate and ribose-5'-phosphate; after 50 fold purification of the 5-nucleotidase, the FMN-splitting activity was very far from being boosted to the same extent.

The complete chemical synthesis of FAD (*i.e.*, the first total synthesis of any dinucleotide!) has been achieved by Christie *et al.* (1952) according to the following scheme: monosilver salt of riboflavin-5'-phosphate + 2',3'-isopropylidine adenosine-5'-benzyl chlorophosphonate → mono benzylester of 2',3' isopropylidine FAD → FAD.

The process of FAD biosynthesis, recognized by Klein and Kohn (1940), was elucidated by Schrecker and Kornberg (1950) who demonstrated that it is very similar to the enzymatic synthesis of DPN (Kornberg 1950). From brewers' yeast, an FAD-synthesizing enzyme, FAD-synthetase, could be extracted and partially purified. In the presence of Mg^{++}-ions, it catalyzes the following reaction:

$$FMN + ATP \rightleftharpoons FAD + PP \text{ (pyrophosphate)}.$$

In this, ATP could not be replaced by ADP or adenosine-5'-phosphate; neither could the PP be replaced by ortho- or metaphosphate. A small amount of FAD was formed from ATP and Rbf, but this is probably due to the initial formation of FMN by another enzyme. The corresponding reaction for the synthesis of DPN (catalyzed by an enzyme found in yeast and liver) is:

$$NMN \text{ (nicotinamide mononucleotide)} + ATP \rightleftharpoons DPN + PP.$$

The FAD-synthesizing activity in yeast autolyzates is only about 1% of that reported for the synthesis of DPN. However, the dissociation constants of the complexes formed by the synthetase with FMN and ATP are relatively low in comparison with those formed by NMN and ATP with the DPN-synthesizing enzyme: 1.4×10^{-6} and 1.2×10^{-5} M for the FAD synthetase, versus 1.5×10^{-4} and 4.6×10^{-4} M for the DPN synthetase. Also, the concentration of FAD in fresh yeast is very much lower than that of DPN (Warburg and Christian 1938 [11]; Schlenk 1942), so that the reported amount of FAD-synthetase must be considered adequate.

It has later been found (Novelli *et al.* 1954) that the biosynthesis of co-enzyme A follows a very similar pattern. Todd (1951) has called attention to the probable similarities between the pathways of biological and chemical FAD synthesis.

The enzymatic breakdown of FAD to FMN + AMP has been studied by Kornberg and Pricer (1950) with the aid of nucleotide pyrophosphatase. This enzyme could be purified 750-fold from potatoes; the authors present some evidence showing that it is also found in animal tissues and in yeast. It splits the PP linkages of DPN, $DPNH_2$, TPN, FAD, ATP, ADP, and thiamin pyrophosphate. The dissociation constants of the enzyme-substrate complexes and some kinetics of inhibition have been studied. The importance of the enzyme for pyrophosphate metabolism in animal tissues and fungi is indicated by the following diagram (from Kornberg 1950):

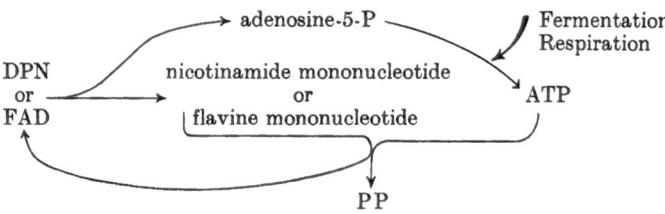

KEARNEY and ENGLARD (1951), on the basis of their findings as well as on that of the results just reported, arrive at the following diagram for flavine metabolism:

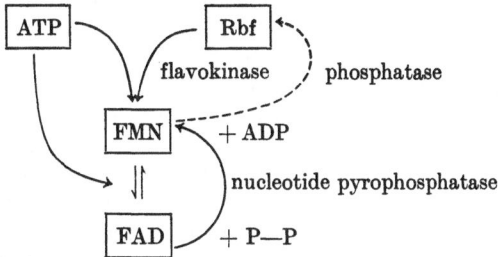

The two key enzymes in this scheme (flavokinase and synthetase) had, at the time their article was written, been found in yeast only, the others also in animals. However, KEARNEY and ENGLARD quote literature showing that FAD synthesis from Rbf also occurs in animals (KLEIN and KOHN 1940, TRU-FANOV 1941, 1942). Although the mechanism has not been elucidated, there is a good chance that the synthesis occurs through the same cycle. The flavo-kinase activity of yeast autolysate is of the same low order of magnitude as its content of FAD-synthetase, but (just as in the case of the latter enzyme) it must be considered quite adequate.

IV. Methods for the isolation and assay of flavins and flavin nucleotides.

Though it is not the present writer's intention to give elaborate descriptions of isolation and assay procedures, he feels that at least the governing principles should be indicated. The difficulty is that there is such a great choice, for among the various characteristics displayed by flavin derivatives there are at least 4 that can be used as "guides":

1) The very reliable and specific SNELL and STRONG method (1939) for the microbiological assay of riboflavin and derivatives with the aid of *Lactobacillus casei* is based on the characteristic that certain flavins can act as vitamins.

2) The typical flavin absorption spectrum can also be used as a basis, especially since WHITBY has produced precise values for the molecular extinction coefficients of Rbf, FMN and FAD at the absorption peak wavelenghts (275, 375 and 450 mμ).

3) As we mentioned before, the fact that FMN and FAD have such a strong affinity for their carrier proteins can be used with advantage. Even in quantities lower than 1 γ, FAD can be assayed with the apo-enzyme of D-amino acid oxidase (see there) and FMN with that of TPNH-cytochrome c-reductase.

4) Assay methods based on the strong yellow-green fluorescence have been used in higher plants (pea seedlings) by VAN HERK, as early as 1935. More recently (1949) BESSEY et al. have worked out good quantitative fluorescence methods for Rbf, FMN and FAD; WEBER (1950) has thoroughly studied the quenching effect on isoalloxazine fluorescence of a variety of substances, including the effect which the adenine moiety of FAD molecule exerts on the other moiety (FMN).

The fluorescence of flavins is also an excellent way to find the position of these compounds on paper chromatograms, and it is therefore not amazing that chromatographic methods play such a predominant part in recent flavin studies (CRAMMER 1948, HUMMEL and LINDBERG 1949, FORREST and TODD

1950, Whitby 1950, 1952, Dimant *et al.* 1952, Forter and Karrer 1955, Huenne-kens *et al.* 1953). For the large scale separation of flavins, partition columns consisting of ion exchange resins or celite have been used successfully by many workers; Huennekens *and coworkers, e.g.*, have been able to achieve a separation between FAD and FAD-X, a cyclic form of FAD and possibly an artifact, on celite columns with phenol-butanol-water as the developing solvent. For a survey of these methods, we refer to Huennekens' excellent short review (1956). An interesting variation is to be found in a recent paper by Siliprandi and Bianchi (1955), who applied chromatography on a carboxylic resin followed by electrophoresis on a column packed with cellulose powder to prepare pure FAD. The claim is made that FAD of 98% purity was obtained. Even though this figure is based on the assumption that the commercial sample used for comparison was 60% pure, this may well be the purest FAD ever prepared.

Table 3. *List of various FMN- and FAD-flavoproteins found in plants.*

Enzyme	Source	Active group[1]	Authority
TPNH-cytochrome c reductase	Yeast	FMN	Haas, Horecker and Hogness (1940)
DPNH-cytochrome c reductase	Bacteria	FAD	Brodie and Gots (1952)
Cytochrome c reductase (DPN and TPN ?) . . .	Soybean leaf, leaves of 15 other plant species	uncertain	Evans (1955)
Glycolic acid oxidase . . .	Spinach	FMN	Zelitch and Ochoa (1953)
Oxalic acid oxidase. . . .	Moss	FMN	Datta and Meeuse 1955)
Glucose oxidase (notatin) .	*Penicillium*	FAD	Keilin and Hartree (1946, 1948); Coulthard *et al.* (1942, 1945)
D-amino acid oxidase . . .	*Penicillium, Aspergillus*	FAD	Warburg and Christian (for horse kidney!) (1938 [9], [11])
Nitrate reductase.	Soybean leaf	FAD	Evans and Nason (1953)
TPNH-nitrate reductase. .	*Neurospora crassa*	FAD	Nicholas *et al.* (1953); Nason and Evans (1953)
Nitrite reductase	*Neurospora*	FAD	Zucker and Nason (1955)
Fumaric hydrogenase . . .	Yeast	FAD	Fischer *et al.* (1942)
Old yellow enzyme	Yeast	FMN	Warburg and Christian (1933 [7])
New yellow enzyme . . .	Yeast	FAD	Haas (1938)
Lactic oxidase	Bakers' yeast	FMN	Appleby and Morton (1954)
L-lactic oxidase	*Mycobacterium phlei*	FAD	Edson (1947)
L-Amino acid oxidase. . .	*Neurospora crassa*	FAD	Burton (1951)
DPNH peroxidase	*Streptococcus faecalis*	FAD	Dolin (1953)
TPNH-hydroxyl amine reductase		FAD	Zucker and Nason (1954)
DPNH-menadione-reductase	*Streptococcus faecalis*	FMN	Dolin (1954)

[1] With omission of the non-flavine components sometimes present.

Literature.

Abraham, E. P.: Experiments relating to the constitution of alloxazine-adenine dinucleotide. Biochemic. J. **33**, 543–548 (1939). — Appleby, C. A., and R. K. Morton: Crystalline cytochrome b₂ and lactic dehydrogenase of yeast. Nature (Lond.) **173**, 749–752 (1954).

Bessey, O., O. H. Lowry and R. H. Love: The fluorometric measurement of the nucleotides of riboflavin and their concentration in tissue. J. of Biol. Chem. **180**, 755–769 (1949). —

BLEYER, B., u. O. KALLMANN: Beiträge zur Kenntnis einiger Inhaltsstoffe der Milch. II. Biochem. Z. 155, 54—79 (1925). — BRODIE, A. F.: A bacterial diphosphopyridine nucleotide-linked cytochrome c reductase. J. of Biol. Chem. 199, 835–844 (1952). — BRODIE, A. F., and J. F. GOTS: Bacterial DPN-cytochrome c reductase. Federat. Proc. 11, 191 (1952). — BURTON, K.: [1] Stabilization of D-amino acid oxidase by flavin-adenine dinucleotide, substrates, and competitive inhibitors. Biochemic. J. 48, 458–467 (1951). — [2] The L-amino-acid oxidase of Neurospora. Biochemic. J. 50, 258–268 (1951).

CHRISTIE, S. M. H., G. W. KENNER and A. R. TODD: Total synthesis of flavine-adenine-dinucleotide. Nature (Lond.) 170, 924 (1952). — COULTHARD, C. E., R. MICHAELIS, W. F. SHORT, G. SYKES, G. E. H. SKRIMSHIRE, A. F. B. STANDFAST, J. H. BIRKINSHAW and H. RAISTRICK: [1] Notatin: an antibacterial glucose-aerodehydrogenase from Penicillium notatum, WESTLING. Nature (Lond.) 150, 634–635 (1942). — [2] Notatin: an antibacterial glucose-aerodehydrogenase from Penicillium notatum, WESTLING and Penicillium resticulosum sp. nov. Biochemic. J. 39, 24–36 (1945). — CRAMMER, J. L.: Paper chromatography of flavine nucleotides. Nature (Lond.) 161, 349–350 (1948).

DATTA, P. K., and B. J. D. MEEUSE: Moss oxalic acid oxidase, a flavoprotein. Biochim. et Biophysica Acta 17, 602–603 (1955). — DIMANT, E., D. R. SANADI and F. M. HUENNE-KENS: The isolation of flavin-nucleotides. J. Amer. Chem. Soc. 74, 5440–5444 (1952). — DOLIN, M. I.: [1] The oxidation and peroxidation of DPNH$_2$ in extracts of Streptococcus faecalis, 10 C 1. Arch. of Biochem. a. Biophysics 46, 483–485 (1953). — [2] The flavin requirement for DPNH-menadione reductase in Streptococcus faecalis. Biochim. et Biophysica Acta 15, 153–154 (1954).

EDSON, N. L.: The oxidation of lactic acid by Mycobacterium phlei. Biochemic. J. 41, 145–151 (1947). — ELLINGER, PH., u. W. KOSCHARA: Über eine neue Gruppe tierischer Farbstoffe (Lyochrome). 1. Vorl. Mitt. Ber. dtsch. chem. Ges. 66, 315–317 (1933). — EVANS, H. J.: Studies on cytochrome reductase in higher plants. Plant Physiol. 30, 437–444 (1955). — EVANS, H. J., and A. NASON: Pyridine nucleotide-nitrate reductase from extracts of higher plants. Plant Physiol. 28, 233–254 (1953).

FISCHER, F. G., A. ROEDIG u. K. RAUCH: Über Fumarat-hydrase als gelbes Ferment. Biochemische Hydrierungen. IX. Liebigs Ann. 552, 203–242 (1942). — FLEXSER, L. A., and W. G. FARKAS: Abstracts. 12. Internat. Congr. of Pure and Appl. Chemistry, New York, p. 71, 1951. — FORREST, H. S., and A. R. TODD: Nucleotides. Part V. Riboflavin-5'-phosphate. J. Chem. Soc. (Lond.) 1950, 3295—3299. — FORTER, W., u. P. KARRER: Das Verhalten einiger Flavine im Papierchromatogramm. Helvet. chim. Acta 36, 1530–1531 (1953).

GEISSMAN, T. A.: A theory of the mechanism of enzyme action. Quart. Rev. Biol. 24, 309–327 (1949). — GOODWIN, T. W., and S. PENDLINGTON: Nitrogen metabolism and flavino-genesis in Eremothecium Ashbyi. Biochemic. J. 57, 631–641 (1954).

HAAS, E.: Isolierung eines neuen gelben Ferments. Biochem. Z. 298, 378–390 (1938). — HAAS, E., B. L. HORECKER and T. R. HOGNESS: The enzymatic reduction of cytochrome c. Cytochrome c reductase. J. of Biol. Chem. 136, 747–774 (1940). — HEPPEL, L. A., and R. J. HILMOE: Purification and properties of 5-nucleotidase. J. of Biol. Chem. 188, 665–676 (1951). — HERK, A. W. H. VAN: Die Konzentrationsänderungen der Co-zymase, des Z-faktors und des Flavins während der Keimung der Erbsen. Ark. Kem., Mineral. u. Geol., Ser. A 11, Nr 22 (1935). — HUENNEKENS, F. M.: Flavin nucleotides and flavoproteins. Experientia (Basel) 12, 1–6 (1956). — HUENNEKENS, F. M., and R. E. BASFORD: Oxidation of DPNH by a-ketoglutaric dehydrogenase. Federat. Proc. 13, 232 (1954). — HUENNEKENS, F. M., and B. W. GABRIO: Methemoglobin reductase. Federat. Proc. 13, 232 (1954). — HUENNE-KENS, F. M., D. R. SANADI, E. DIMANT and A. I. SCHEPARTZ: The occurrence of a new flavin dinucleotide (FAD-X). J. Amer. Chem. Soc. 75, 3611–3612 (1953). — HUMMEL, J. P., and LINDBERG O.: Studies on the mechanism of aerobic phosphorylation. J. of Biol. Chem. 180, 1–11 (1949).

KARRER, P., K. SCHÖPP u. F. BENZ: Synthesen von Flavinen. IV. Helvet. chim. Acta 18, 426–429 (1935). — KARRER, P., K. SCHÖPP, F. BENZ u. K. PFAEHLER: Synthesen von Flavinen. III. Helvet. chim. Acta 18, 69–79 (1935). — KEARNEY, E. B.: The interaction of yeast flavokinase with riboflavin analogues. J. of Biol. Chem. 194, 747–754 (1952). — KEARNEY, E. B., and S. ENGLARD: [1] Enzymatic synthesis of flavin mononucleotide. Arch. of Biochem. 32, 222–223 (1951). — [2] The enzymatic phosphorylation of riboflavin. J. of Biol. Chem. 193, 821–834 (1951). — KEILIN, D., and E. F. HARTREE: Prosthetic group of glucose oxidase. Nature (Lond.) 157, 801 (1946). — KLEIN, J. R., and H. I. KOHN: The synthesis of flavin-adenine dinucleotide from riboflavin by human blood cells in vitro and in vivo. J. of Biol. Chem. 136, 177–189 (1940). — KORNBERG, A.: Reversible enzymatic synthesis of diphosphopyridine nucleotide and inorganic pyrophosphate. J. of Biol. Chem. 182, 779–793 (1950). — KORNBERG, A., and W. E. PRICER: Nucleotide pyrophosphatase. J. of Biol. Chem. 182, 763–778 (1950). — KUHN, R., u. P. BOULANGER: Beziehungen zwischen Reduktions-Oxydations-Potential und chemischer Konstitution der Flavine. Ber. dtsch.

chem. Ges. B **69**, 1557–1556 (1936). — Kuhn, R., P. Györgyi u. T. Wagner-Jauregg: [1] Über eine neue Klasse von Naturfarbstoffen. Vorl. Mitt. Ber. dtsch. chem. Ges. **66**, 317–320 (1933). — [2] Über Ovoflavin, den Farbstoff des Eiklars. Ber. dtsch. chem. Ges. **66**, 576–580 (1933). — [3] Über Lactoflavin, den Farbstoff der Molke. Ber. dtsch. chem. Ges. **66**, 1034–1038 (1933). — Kuhn, R., K. Reinemund, H. Kaltschmitt, R. Ströbele u. H. Frischmann: Synthetisches 6,7-Dimethyl-9-d-riboflavin. Naturwiss. **23**, 260 (1935). — Kuhn, R., K. Reinemund u. F. Weygand: Synthese des Lumi-lactoflavins. Ber. dtsch. chem. Ges. **67**, 1460–1462 (1934). — Kuhn, R., K. Reinemund, F. Weygand u. R. Strö-bele: Über die Synthese des Laktoflavins. Vitamin B_2. Ber. dtsch. chem. Ges. **68**, 1765–1774 (1935). — Kuhn, R., u. H. Rudy: Katalytische Wirkung der Lactoflavin-5′-phosphorsäure; Synthese des gelben Ferments. Ber. dtsch. chem. Ges. **69**, 1974–1977 (1936). — Kuhn, R., H. Rudy u. F. Weygand: Synthese der Lactoflavin-5′-phosphorsäure. Ber. dtsch. chem. Ges. **69**, 1543–1547 (1936). — Kuhn, R., u. T. Wagner-Jauregg: Lactoflavin (Vitamin B_2) aus Leber. Ber. dtsch. chem. Ges. **67**, 1770–1773 (1934).

Mahler, H. R., and D. E. Green: Metallo-flavoproteins and electron transport. Science (Lancaster, Pa.) **120**, 7–12 (1954). — McLaren, J. A.: The effects of certain purines and pyrimidines upon the production of riboflavin by *Eremothecium Ashbyii*. J. Bacter. **63**, 233–241 (1952). — McNutt jr., W. S.: [1] The incorporation of the pyrimidine ring of adenine into the alloxazine ring of riboflavin. J. of Biol. Chem. **219**, 365–373 (1956). — [2] The direct contribution of adenine to the biogenesis of riboflavin by *Eremothecium Ashbyii*. J. of Biol. Chem. **210**, 511–519 (1954). — Mitchell, H. K., and M. B. Houlahan: *Neurospora*. IV. A temperature-sensitive riboflavinless mutant. Amer. J. Bot. **33**, 31–35 (1946).

Nason, A., and H. J. Evans: Triphosphopyridine nucleotide-nitrate reductase in *Neurospora*. J. of Biol. Chem. **202**, 655–673 (1953). — Nicholas, D. J. D., A. Nason and W. D. McElroy: Effect of molybdenum deficiency on nitrate reductase in cell-free extracts of *Neurospora* and *Aspergillus*. Nature (Lond.) **172**, 34 (1953). — Novelli, G. D., F. J. Schmetz and N. O. Kaplan: Enzymatic degradation and resynthesis of coenzyme A. J. of Biol. Chem. **206**, 533–545 (1954). — Nygaard, A. P., and H. Theorell: Kinetics and equilibria in flavoprotein systems. III. The effects of chemical modifications of the apoprotein on the dissociation and reassociation of the old yellow enzyme. Acta chem. scand. (Copenh.) **9**, 1587–1599 (1955).

Plaut, G. W. E.: Mechanism of riboflavin biosynthesis. Federat. Proc. **12**, 254–255 (1953).

Schlenk, F.: In: A symposium on respiratory enzymes. Madison: Wisconsin 1942. — Schrecker, A., and A. Kornberg: Reversible enzymatic synthesis of flavin adenine dinucleotide. J. of Biol. Chem. **182**, 795–803 (1950). — Siliprandi, N., and P. Bianchi: A new method for preparing flavin-adenine-dinucleotide. Biochim. et Biophysica Acta **16**, 424–428 (1955). — Snell, E. E., and F. M. Strong: A microbiological assay for riboflavin. Ind. Engng. Chem., Analyt. Edit. **11**, 346–350 (1939). — Szent-Györgyi, A. v., u. I. Banga: Über Co-Fermente, Wasserstoffdonatoren und Arsenvergiftung der Zellatmung. Biochem. Z. **246**, 203–214 (1932).

Theorell, H.: [1] Das gelbe Ferment, seine Chemie und Wirkungen. Erg. Enzymforsch. **6**, 111–138 (1937). — [2] Die freie Eiweißkomponente des gelben Ferments und ihre Kupplung mit Lactoflavinphosphorsäure. Biochem. Z. **290**, 293–303 (1937). — Theorell, H., u. R. Bonnichsen: Liver alcohol dehydrogenase. I. Equilibria and initial reaction velocities. Acta chem. scand. (Copenh.) **5**, 1105–1126 (1951). — Theorell, H., and A. P. Nygaard: [1] Kinetics and equilibria in flavoprotein systems. I. A fluorescence recorder and its application to a study of the dissociation of the old yellow enzyme and its resynthesis from riboflavin phosphate and protein. Acta chem. scand. (Copenh.) **8**, 877–888 (1954). — [2] The combination of flavin mononucleotide and riboflavin with the protein of the old yellow enzyme. Acta chem. scand. (Copenh.) **8**, 1104–1105 (1954). — [3] Kinetics and equilibria in flavoprotein systems. II. The effect of p_H, anions and temperature on the dissociation and reassociation of the old yellow enzyme. Acta chem. scand. (Copenh.) **8**, 1649–1658 (1954). — Todd, A. R.: The nucleotides: some recent chemical research and its biological implications. Harvey Lect. **1951**. — Trufanov, A. V.: [1] Synthesis of flavine-adenine-dinucleotide. Biochimija **6**, 301–311 (1941). Quoted from Chem. Abstr. **35**, 7499³ (1941). — [2] Enzymic synthesis of flavine-adenine-nucleotide. Biochimija **7**, 188–200 (1942). Quoted from Chem. Abstr. **38**, 131⁵ (1944).

Veen, A. G. van, u. W. K. Mertens: On the isolation of a toxic bacterial pigment. Proc. Kon. Akad. Wetensch. Amsterdam **36**, 666–670 (1933). — Vestling, C. S.: Kinetics and equilibria in flavoprotein systems. IV. The standard potential of the old yellow enzyme of yeast. Acta chem. scand. (Copenh.) **9**, 1600–1609 (1955). — Viscontini, M., C. Ebnöther u. P. Karrer: Einfache Synthese kristallisierter Lactoflavin-5′-phosphorsäure (Coferment des Flavinenzyms). Helvet. chim. Acta **35**, 457–459 (1952).

WALAAS, O., u. E. WALAAS: Studies on the inhibition of D-amino acid oxidase. Acta chem. scand. (Copenh.) 8, 1104 (1954). — WARBURG, O., u. W. CHRISTIAN: [1] Über ein neues Oxydationsferment und sein Absorptionsspektrum. Biochem. Z. 254, 438–458 (1932). — [2] Ein zweites sauerstoffübertragendes Ferment und sein Absorptionsspektrum. Naturwiss. 20, 688 (1932). — [3] Über das neue Oxydationsferment. Naturwiss. 20, 980 (1932). — [4] Über das gelbe Oxydationsferment. Biochem. Z. 258, 496–498 (1933). — [5] Über das gelbe Oxydationsferment. Biochem. Z. 263, 228–229 (1933). — [6] Über das gelbe Ferment und seine Wirkungen. Biochem. Z. 266, 377–411 (1933). — [7] Co-Ferment der d-Aminosäure-Deaminase. Biochem. Z. 295, 261 (1938). — [8] Coferment der d-Alanin-Oxydase. Biochem. Z. 296, 294 (1938). — [9] Coferment der d-Alanin-Oxydase. Biochem. Z. 298, 150 (1938). — [10] Isolierung der prosthetischen Gruppe der d-Aminosäure-Oxydase. Biochem. Z. 298, 158–168 (1938). — [11] Bemerkung über gelbe Fermente. Biochem. Z. 298, 368–377 (1938). — WEBER, G.: Fluorescence of riboflavin and flavine-adenine dinucleotide. Biochemic. J. 47, 114–121 (1950). — WHITBY, L. G.: [1] Enzymic formation of a new riboflavin derivative. Nature (Lond.) 166, 479–480 (1950). — [2] Riboflavinyl glucoside: a new derivative of riboflavin. Biochemic. J. 50, 433–438 (1952). — [3] A new method for preparing flavin-adenine dinucleotide. Biochemic. J. 54, 437–442 (1953). — [4] Preparation of flavin-adenine dinucleotide (FAD). Biochim. et Biophysica Acta 15, 148–149 (1954).

ZELITCH, I., and S. OCHOA: Oxidation and reduction of glycolic and glyoxylic acids in plants. I. Glycolic acid oxidase. J. of Biol. Chem. 201, 707–718 (1953). — ZUCKER, M., and A. NASON: Enzymatic reduction of hydroxylamine to ammonia by reduced pyridine nucleotides. Federat. Proc. 13, 328 (1954). — A pyridine nucleotide-hydroxylamine reductase from Neurospora. J. of Biol. Chem. 213, 463–478 (1955).

The formation and degradation of thiaminpyrophosphate.

By

J. R. P. O'Brien.

Thiamine is essential to the life of most plants, animals and micro-organisms. Its catalytic effect upon the oxygen consumption of tissues and on the oxidation of pyruvic acid in particular was first established by PETERS and his colleagues (PETERS 1936). It was early suspected that it was not thiamine itself that was metabolically active but a compound derived from it (PETERS, RYDIN and THOMPSON 1935; WESTENBRINK and POLAK 1937). The nature of this compound emerged from studies upon yeast carboxylase, an enzyme promoting the decarboxylation of pyruvic acid to acetaldehyde. The enzyme lost its activity if washed with phosphate buffer, and recovered it if complemented with a thermolabile substance in yeast extract called co-carboxylase (AUHAGEN 1931, 1933). The isolation from yeast and identification of this compound was achieved by LOHMANN and SCHUSTER (1937). It proved to be thiamine pyrophosphate:

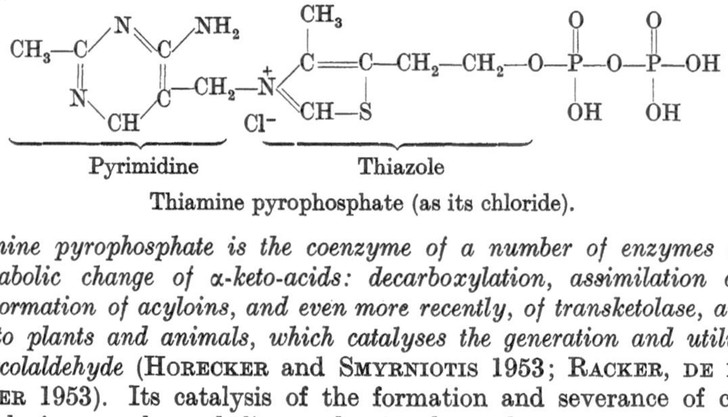

Thiamine pyrophosphate (as its chloride).

Thiamine pyrophosphate is the coenzyme of a number of enzymes concerned with metabolic change of α-keto-acids: decarboxylation, assimilation of carbon dioxide, formation of acyloins, and even more recently, of transketolase, an enzyme common to plants and animals, which catalyses the generation and utilisation of active glycolaldehyde (HORECKER and SMYRNIOTIS 1953; RACKER, DE LA HABA and LEDER 1953). Its catalysis of the formation and severance of carbon to carbon links in several metabolites makes it a key substance in animal and plant economy. Its significance is not better illustrated than by its early established indispensability in the decarboxylation of pyruvic acid, the end product of glycolysis. Deprived of thiamine, man succumbs to the grievious ailment of beri-beri and the experimental animal to crippling disorders of the central and peripheral nervous system; both lose their power to oxidise pyruvic acid and only when fed thiamine regain it and their health (PETERS 1936). Its role in yeast fermentation has already been mentioned. The mechanism of its action is still obscure and none of the speculative schemes for its action have gained general approval. This may in part be due to indecision about the form co-carboxylase takes in reactions, *i.e.* whether it retains its thiazolium ring intact or assumes a thiol form by opening of the thiazolium ring or whether it forms conjugates with lipoic acid (see WILLIAMS, EAKIN, BEERSTECHER and SHIVE 1950; REED 1953; JOHNSON 1955).

A discussion of the biological synthesis and degradation of thiamine pyro-phosphate is equivalent to one of these processes in regard to thiamine. Yet at the moment knowledge of these matters is thin and perhaps with reason. Animals are unable to synthesise thiamine though they can phosphorylate it; hence they need a continuous provision of it to balance losses by its degradation and its excretion. In contrast numerous species of plants and micro-organisms are blessed with the capacity in varying degrees, and should consequently be the ideal material for synthetic studies. Most of the facts about thiamine and hence carboxylase synthesis have however come from investigations upon the growth requirements of micro-organisms and molds and only few from the higher plants.

The very nature of plants and their extraordinary synthetic powers exclude an easy study of the way in which they form the vitamin. They cannot be depleted of their vitamin by a simple procedure and in themselves are self-sustaining as regards their thiamine needs. But their parts are not equally effective in synthesising the vitamin. From the more efficient parts, thiamine is translocated to those which need it. Thiamine is to be found in most organs of a plant, the leaves, the stems, fruit, flowers, roots and seeds. It is, however, the mature leaf which seems to be an active site of thiamine formation and a source of supply to roots and apical growing points. This has been especially well demonstrated in the tomato by several methods including the technique of girdling (see BONNER and BONNER 1948). Beyond this little is known, scarcely anything about metabolites the leaf may use to construct the pyrimidine and thiazole parts of the vitamin molecule. The experimental difficulties in the way of resolving this problem however may be ultimately met by combining radio-isotopic and chromatographic techniques with that of plant culture, a promising but not fully explored approach to the problems of biosynthesis. The latter has been especially well developed for the study of the growth re-quirements of isolated roots and has yielded some interesting information upon the thiamine needs and thiamine synthesis by pea and tomato roots which is mentioned later (BONNER 1937; ROBBINS and BARTLEY 1937).

Most of the little known about *thiamine synthesis* has come from the abundant data of work upon the growth needs of micro-organisms. Many yeasts and many bacteria are certainly autotrophic for thiamine (WILLIAMS *et al.* 1950; SCHÖPFER 1943). On the contrary many molds, bacteria, yeasts and algae are heterotrophic in respect to the vitamin or to one or both its components. Thus *Phycomyces Blakesleeanus*, some strains of *Saccharomyces cerevisiae* and *Staphylo-coccus aureus* grow well given thiamine or both the thiazole and pyrimidine parts of its molecule (SCHÖPFER 1943; SCHULTZ, ATKIN and FREY 1937a, b, 1938; KNIGHT 1937). This behaviour suggests that in living organisms the components of thiamine are first formed and subsequently united. This certainly seems to be the case for *Phycomyces* and *Rhodotorula rubra* (SCHÖPFER 1943). A similar ability is possessed by excised pea and tomato roots, the growth of which is stimulated by thiamine, but the synthetic power depends to some extent upon the strain of the plant. Some tomato and pea roots grow well given the thiazole and the pyrimidine components; some tomato strains and flax roots manage well with thiazole alone and can presumably synthesise the pyrimidine part (BONNER and BONNER 1948). In the case of the pea root, it has been proved by microbiological tests with *Phycomyces* and *Phytophthora* that thiamine is formed from its pyrimidine and thiazole components (BONNER and BUCHMAN 1938). There are some comments to be made on these findings. From a con-siderable amount of work upon the growth effects of analogues of the pyrimidine and thiazole parts of thiamine it has become clear that micro-organisms and pea roots are sensitive to small structural changes in the pyrimidine component and to a lesser extent to any in the thiazole part (Bonner and Buchman 1938; Schöpfer 1943; Bonner and Bonner 1948). This response suggests that the synthesising enzyme system has a certain degree of specificity. There is still a long way to go before the conditions which influence the weak synthetic capacity of these partially heterotrophic organisms are understood, and among these are medium composition and temperature. One micro-organism may lose irretrievably its power to synthesise thiamine from its components and revert to absolute heterotrophism; another may under "training" strengthen its weak synthetic powers (Schöpfer 1943). The results so far mentioned give some idea of what may be the last stage in thiamine synthesis but no clues to the nature of the precursors of the thiazole and pyrimidine components. Nevertheless although little has been said about the origin of the pyrimidine, a suggestion has been made that the thiazole is formed by the fermentative degradation of the corresponding α-amino-propionic acid, α-amino-β-(4-methylthiazole-5-) propionic acid, which might be formed by a reaction of methionine, acetaldehyde and ammonia (Harington and Moggridge 1939, 1940; Buchman and Richardson 1939). The amino acid is broken down to the thiazole of thiamine by yeast and is transformed into thiamine by pea roots supplied with the vitamin pyrimidine, but not by *Staphylococcus aureus* or *Phycomyces* similarly nourished. The meaning of these results is not at once obvious; it may be that yeast and pea roots can metabolise the α-amino acid more to their advantage than can bacteria and molds. Pea roots are certainly versatile; they can form the vitamin thiazole from simple compounds although unable to form the vitamin pyrimidine. Supplied with a mixture of vitamin pyrimidine, thioformamide, and either chloroacetopropyl alcohol or acetopropyl alcohol, the roots grow as well and contain as much thiamine, measured by *Phycomyces* assay, as those supplied with the vitamin (Bonner and Buchman 1938). At present, the evidence is far too slight to consider that these reactions are common to many organisms and plants; indeed *Phycomyces* cannot use these thiazole precursors in place of vitamin thiazole.

Another form of microbiological analysis, the use of mutants, indicates that the biosynthesis of thiamine is not so straightforward as nutritional studies upon micro-organisms lead one to suppose. Wild *Neurospora* synthesise thiamine but some mutant strains cannot form the thiazole part or combine the thiazole and pyrimidine parts of the vitamin and hence do not grow. From the products which accumulate during the growth of different mutants of *Neurospora* it has been concluded that thiamine synthesis by this mold takes the following line [the numbers indicate mutant strains and ↕ metabolic blocks (Tatum and Bell 1946)].

The accumulations of pyrimidine by mutant 18558 and of thiazole and pyrimidine by 9185 were proved by bioassay; the first, therefore, cannot synthesise thiazole, and the second is unable to form thiamine from its pyrimidine and thiazole parts. Furthermore the mutants 17084 and 1090 produce a substance as active as thiamine towards mutants 9185 and 18558. These findings fit substantially with the view, already described, namely that in thiamine synthesis, the thiazole and pyrimidine parts are first formed and subsequently condensed to the vitamin. But there is the fact that although they accumulate a thiamine-like substance, mutants 17084 and 1090 do not use it and require thiamine for growth. This is explained by a theory of HARRIS (1953, 1955) who, from a study of the utilisation and production of the pyrimidine and the thiazole components of thiamine and of thiamine itself postulates alternative pathways for thiamine synthesis:

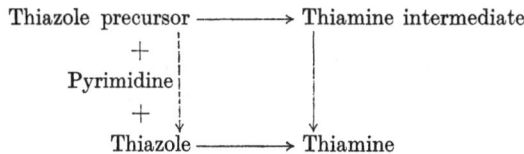

A thiazole precursor reacts with pyrimidine to form a thiamine intermediate which is transformed into thiamine. Mutants 17084 and 1090 presumably form this vitamin intermediate but cannot effect its transformation. Alternatively they can circumvent the blockage of thiamine synthesis from its precursor by condensing the pyrimidine, they can synthesise, with exogenous thiazole which acts competitively with the endogenous thiazole precursor for any available pyrimidine. The more abundant the supplies of exogenous pyrimidine, the less restrictive to growth is this competition.

Some idea of the nature of the thiazole precursor has come from studies of growth inhibition of a mutant strain of *Neurospora crassa U.T. 77A.* by threonine (DOUDNEY and WAGNER 1953). This inhibition is reversed competitively by homocysteine and non-competitively by methionine and choline and by a mixture of choline with thiazole or thiamine. It obviously involves a failure of synthesis of methionine and of the thiazole part of thiamine from homocysteine. In fact threonine and choline alone or together suppress thiazole formation during the growth phase of the mold, an effect which is checked by homocysteine. It is considered that an intermediary in the conversion of homocysteine to methionine is utilized in the synthesis of the thiazole part:

$$\text{Homocysteine} \xrightarrow{\underset{\substack{\text{Threonine} \\ \text{inhibition}}}{\big|}} \quad \underset{\substack{\downarrow \\ \text{Thiazole}}}{\text{X}} \quad \rightleftharpoons \quad \text{methionine}$$

In what manner homocysteine is concerned in this synthesis is undecided, but there are at least two possible ways. It might help in the incorporation of a formyl group into the thiazole, or it might be completely utilized in building up this molecule. The latter view is, of course, similar to that suggested by HARINGTON and MOGGRIDGE (1939, 1940) and BUCHMAN and RICHARDSON (1939).

In animal tissue but not to the same extent in plants, thiamine is present largely as co-carboxylase. Since LOHMANN and SCHUSTER (1937) first demonstrated that brain and intestine could synthesise co-carboxylase particularly if supplied with adenosine triphosphate (A.T.P.) and thiamine, the number of tissues with the same power has increased. Yeast, several bacteria, red cells, liver, brain, kidney and muscle can phosphorylate thiamine (see OCHOA 1942). The synthesis of co-carboxylase is most effectively demonstrated in the liver of the thiamine deficient pigeon. The amount of coenzyme synthesised rarely exceeds that found in the tissues of normal animals, and consequently it is not easy to observe synthesis in tissues of normal animals. There have been attempts to isolate the enzyme which phosphorylates thiamine (WEIL-MALHERBE 1939;

VAN THOAI and CHEVILLARD 1949), but only recently has it been obtained in a relatively pure condition from rat liver (LEUTHARDT and NIELSEN 1952). This enzyme has an optimum p_H of 6.8–6.9 and is activated by magnesium ions.

The synthesis of thiamine pyrophosphate by tissues and micro-organisms depends upon aerobic respiration or glycolysis whence supplies of A.T.P. are forthcoming. The phosphorylation of thiamine by A.T.P. in presence of its phosphorylase is still poorly understood. Apparently the reaction consists in the transfer of a pyrophosphoryl group from A.T.P. to thiamine (LEUTHARDT and NIELSEN 1952). This is supported by the fact that in the presence of respiring kidney particles, P^{32} labelled phosphate is not readily incorporated into co-carboxylase nor is co-carboxylase formed by orthophosphate phosphorylation of thiamine monophosphate or thiamine (BARTLEY 1954). Injections of thiamine and A.T.P. labelled with P^{32} into rats lead to the presence in the liver of labelled thiamine di- and tri-phosphates, but of little radioactive monophosphate. This suggests that P^{32} is localised mainly in the terminal phosphate of the esters of thiamine. Information upon the phosphorylation of thiamine still remains scanty and vague. Possibly a new approach to the problem may yield further details such as a use of those thiamine analogues which are known to inhibit phosphorylation of thiamine (WOOLLEY 1951; EICH and CERECEDO 1954).

The nature of the *degradation* of thiamine and its active metabolic form, co-carboxylase, is more obscure in the plant than it is in the animal and so are the conditions determining it. Most animal tissues but especially the liver and the kidney contain *phosphatases* which hydrolyse thiamine pyrophosphate. The free thiamine circulates in the plasma, whence it may be removed by tissues needing it or be excreted in the urine. In experiments in which rats have been injected with labelled thiamine, about two-thirds of the administered material appears in the urine. Following the intraperitoneal injection of 1 mg. thiamine labelled with radio C^{14} in position 2 in the thiazole ring eight substances have been identified chromatographically and microbiologically; thiamine, co-carboxylase, three, apparently, conjugates of *lipoic acid* and three derivatives of thiazole (IACONO, WOLF and JOHNSON 1953). In another experiment a reasonable dose of 50 μg/rat of thiamine labelled with S^{35} was given intramuscularly to normal rats. About 2% of the radio-sulfur appeared in the urine as sulfate indicating that part of the thiazole had undergone oxidation, and 63% as neutral sulfur. Of this neutral sulfur roughly 60% was probably thiamine and about 40% was break-down products of the labelled thiamine (McCARTHY, CERECEDO and BROWN 1954). These investigations with labelled thiamine provide interesting useful information but their differences in techniques and in procedure forbid any general conclusions.

There are two other reactions of thiamine which deserve brief mention since they may have a physiological significance not yet apparent (FUJITA 1954). An enzyme, *thiaminase*, is present in shell fish, crustacea and fresh water fish, and some plants, chiefly the fern. It promotes the decomposition of thiamine into thiazole and pyrimidine parts. Nitrogenous compounds such as aniline and pyridine catalyse the reaction by combining with the pyrimidine part of the vitamin. What the natural counterparts to these bases are in this base exchange reaction is a problem yet to be resolved. The second reaction is that of thiamine with thiol compounds. Thus thiamine reacts with an extract of garlic to yield *alli-thiamine*, a disulfide of allyl mercaptan and the thiol form of thiamine. These two reactions may account for some of the unidentified degradation products of thiamine.

Literature.

AUHAGEN, E.: Cocarboxylase, ein neues Co-enzym der alkoholischen Gärung. Hoppe-Seylers Z. **204**, 149–167 (1931). — Über Co-carboxylase. Biochem. Z. **258**, 330–339 (1933).

BARTLEY, W.: Metabolism of thiamine phosphates in washed suspensions of kidney particles. Biochemic. J. **56**, 379–387 (1954). — BONNER, J.: Vitamin B_1 a growth factor for higher plants. Science (Lancaster, Pa.) **85**, 183–184 (1937). — BONNER, J., and H. BONNER: The B_1 vitamins as plant hormones. Vitamins a. Hormones **6**, 225–275 (1948). — BONNER, J., and E. R. BUCHMAN: Syntheses carried out in vivo by isolated pea roots. I. Proc. Nat. Acad. Sci. U.S.A. **24**, 431–438 (1938). — BUCHMAN, E. R., and E. M. RICHARDSON: Thiamine analogues. I. B_1 (4-methyl thiazolyl-5)-alanine. J. Amer. Chem. Soc. **61**, 891–893 (1939).

DOUDNEY, C. O., and R. P. WAGNER: A relationship of homocysteine metabolism to thiamine serine, and adenine biosynthesis in a mutant strain of *Neurospora*. Proc. Nat. Acad. Sci. U.S.A. **39**, 1043–1052 (1953).

EICH, S., and C. R. CERECEDO: Studies on thiamine analogues. III. Effects on enzyme systems. J. of Biol. Chem. **207**, 295–303 (1954).

FUJITA, A.: Thiaminase. Adv. Enzymol. **15**, 389–421 (1954).

HARINGTON, C. R., and R. C. G. MOGGRIDGE: α-Amino-β-(4-methylthiazole-5-)-propionic acid, a possible precursor of aneurin. J. Chem. Soc. Lond. **1939**, 443–446. — Experiments on the biogenesis of vitamin B_1. Biochemic. J. **34**, 685–689 (1940). — HARRIS, D. L.: Biosynthesis of thiamine in *Neurospora*. Federat. Proc. **12**, 214–215 (1953). — Alternative pathways of thiamine biosynthesis in *Neurospora*. Arch. of Biochem. a. Biophysics **57**, 240–251 (1955). — HORECKER, B. L., and P. Z. SMYRNIOTIS: The coenzyme function of thiamine pyrophosphate in pentose phosphate metabolism. J. Amer. Chem. Soc. **75**, 1009 (1953).

IACONO, J. M., G. WOLF and B. C. JOHNSON: Metabolism of radioactive thiamine in the rat. Federat. Proc. **12**, 223 (1953).

JOHNSON, B. C.: Water-soluble vitamins. Part III. Annual Rev. Biochem. **24**, 419–455 (1955).

KNIGHT, B. C. J. G.: The nutrition of *Staphylococcus aureus*. The activities of nicotinamide, aneurin (vitamin B_1) and related compounds. Biochemic. J. **31**, 967–973 (1937).

LEUTHARDT, F., u. H. NIELSEN: Phosphorylation biologique de la thiamine. Helvet. chim. Acta **35**, 1196–1209 (1952). — LOHMANN, K., u. P. SCHUSTER: Untersuchungen über die Co-carboxylase. Biochem. Z. **294**, 188–214 (1937).

McCARTHY, P. T., L. R. CERECEDO and E. V. BROWN: Fate of thiamine-S^{35} in the rat. J. of Biol. Chem. **209**, 611–618 (1954).

OCHOA, O.: The Biological Action of the Vitamins. Edit. by E. A. EVANS, p. 17. Chicago: The University Chicago Press 1942.

PETERS, R. A.: The biochemical lesion in vitamin B_1 deficiency. Lancet **1936** I, 1161 to 1164. — PETERS, R. A., H. RYDIN and R. H. S. THOMPSON: Brain respiration, a chain of reactions, as revealed by experiments upon catatorulin effect. Biochemic. J. **29**, 53–62 (1935).

RACKER, E., G. DE LA HABA and I. G. LEDER: Thiamine pyrophosphate, a coenzyme of transketolase. J. Amer. Chem. Soc. **75**, 1010 (1953). — REED, L. J.: Metabolic functions of thiamine and lipoic acid. Physiologic. Rev. **33**, 544–559 (1953). — ROBBINS, W. J., and M. A. BARTLEY: Vitamin B_1 and the growth of excised tomato roots. Science (Lancaster, Pa.) **85**, 246–247 (1937). — ROSSI-FANELLI, A., N. SILLIPRANDI, P. FASELLA, D. SILLIPRANDI u. F. L. SALVETTI: On phosphorylation of thiamine in the living animal. Experientia (Basel) **10**, 73–74 (1954).

SCHÖPFER, W. H.: Plants and Vitamins. Waltham, Mass., U.S.A.: Chronica Botanica Co. 1943. — SCHULTZ, A. S., L. ATKIN and C. N. FREY: A fermentation test for vitamin B_1. J. Amer. Chem. Soc. **59**, 948–949 (1937a). — A fermentation test for vitamin B_1. J. Amer. Chem. Soc. **59**, 2457–2460 (1937b). — Thiamine, pyrimidine and thiazole as bios factors. J. Amer. Chem. Soc. **60**, 490 (1938).

TATUM, E. L., and T. T. BELL: *Neurospora*. III. Biosynthesis of thiamine. Amer. J. Bot. **33**, 15–20 (1946). — THOAI, N. VAN, et L. CHEVILLARD: Sur la synthèse enzymatique de la cocarboxylase. Purification du proteine phosphorylant la thiamine. Bull. Soc. Chim. biol. Paris **31**, 204–212 (1949).

WEIL-MALHERBE, H.: The enzymic phosphorylation of vitamin B_1. Biochemic. J. **33**, 1997–2007 (1939). — WESTENBRINK, H. G. K., u. J. J. POLAK: Some experiments on the catatorulin effect. Rec. Trav. chim. Pays-Bas (Amsterd.) **56**, 315–329 (1937). — WILLIAMS, R. J., R. E. EAKIN, E. BEERSTECHER jr. and W. SHIVE: The Biochemistry of B_1 Vitamins. New York (U.S.A.): Reinhold Publishing Co. 1950. — WOOLLEY, D. W.: An enzymatic study of the mode of action of pyrithiamine (neopyrithiamine). J. of Biol. Chem. **191**, 43–54 (1951).

Coenzyme A.

By

Harry G. Albaum.

In 1945 FRITZ LIPMANN described a cell-free system from pigeon liver which had the ability to acetylate sulfanilamide and several other substances including p-aminobenzoic acid, sulfathiazole, and sulfadiazine. This study was undertaken because it appeared to offer some insight into the mechanism of acetate "activation". The system had the ability to respire aerobically during which time acetylation occurred. The reaction was able to take place anaerobically if adenosine triphosphate, (ATP) was added. The function of the aerobic portion of the system appeared, therefore, to be the generation of high energy phosphate required for the acetylation reaction. While acetate was the most effective substrate for this system, acetoacetic acid and pyruvic acid were also active presumably because they were able to yield acetate. Active enzyme preparations responsible for the above reaction were freely soluble and could be separated from the particulate matter of the liver by centrifugation. Active solutions could also be obtained by extraction from acetone powders. The latter preparations were, however, readily inactivated by autolysis. This inactivation could be partly or wholly reversed through the addition of boiled liver juice. Boiled liver juice therefore appeared to contain a thermostable component necessary for the acetylation reaction. Before its structure became known, this co-factor was named co-enzyme A.

In the interval since the original discovery of coenzyme A and the present time a large number of investigations have been carried out on its structure, its biosynthesis, and its role in the organism in acetate activation and acyl group transfer. Very little has been done on this unique compound in plant tissues. Most of the investigations have centered about animal tissues and microorganisms. The discussion which follows is based mainly, therefore, on the latter studies with applications to the tissues of higher plants given at the end.

Structure.

Coenzyme A was isolated and purified from *Streptomyces fradiae* through a complex series of steps involving adsorption and elution from charcoal, reduction with zinc and hydrochloric acid and chromatography through a Duolite CS-100 resin column (DE VRIES et al.). This procedure resulted in a coenzyme A preparation with a purity of well over 90% (GREGORY, NOVELLI and LIPMAN). Coenzyme A was found to contain panthothenic acid, adenine, ribose, phosphorus, and sulfur, the latter present in β-mercaptoethylamine. The molar ratio of pantothenic acid to adenine to total phosphorus to sulfur was approximately 1 to 1 to 3 to 1. On the basis of degradation studies, the following structural formula was suggested:

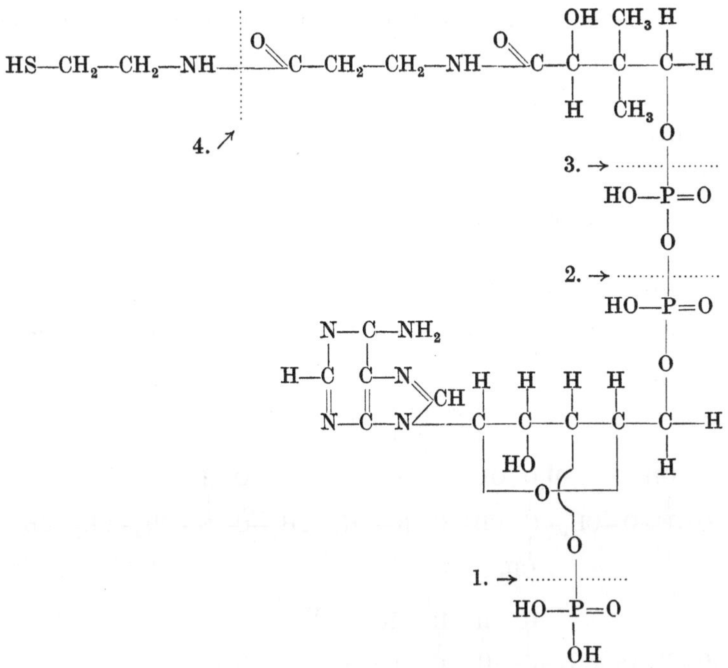

$$HS-CH_2-CH_2-NH-\overset{O}{\underset{\vdots}{C}}-CH_2-CH_2-NH-\overset{O}{C}-\overset{OH}{\underset{H}{C}}-\overset{CH_3}{\underset{CH_3}{C}}-\overset{H}{C}-H$$

4. ↗

3. → $\overset{O}{\underset{O}{HO-P=O}}$

2. → $\overset{O}{HO-P=O}$

$$\begin{matrix} N-C-NH_2 \\ H-C\ \ C-N \\ N-C-N\ \ CH \end{matrix}\ \overset{H\ \ H\ \ H\ \ H}{\underset{HO}{C-C-C-C-C-H}}$$

1. → $HO-P=O$ $\overset{}{\underset{OH}{}}$

Structure of coenzyme A.

The phosphate in position number 1 could be removed by a phosphomonoesterase prepared from prostate. The resulting fragment called dephospho-CoA was inactive in the acetylation reaction. The presence of a pyrophosphate linkage at position number 2 was verified by the use of a nucleotide pyrophosphatase from potato or snake venom. Splitting at this point gave rise to diphosphoadenosine and phosphopantetheine; both fragments were enzymatically inactive. These products were isolated by paper chromatography and shown to be monophospho-esters. When coenzyme A was treated with both snake venom and prostate phosphatase, the pyrophosphate bridge at position number 2 was broken and the resulting monoesters were dephosphorylated giving rise to three moles of orthophosphate and one mole each of adenosine and pantetheine. β-mercaptoethylamine linked through a peptide bond was readily split from either CoA, dephospho-CoA, phosphopantetheine or pantetheine by an enzyme from the liver of the pigeon, hog, or kidney. These degradation studies appeared to establish the structure of coenzyme A as shown above.

Biosynthesis.

Starting with synthetic pantetheine it has been possible to bring about the biosynthesis of coenzyme A (NOVELLI and HOAGLAND). This involves the participation of 3 enzymes which have been partially purified from hog liver and other sources. The methods used in purifying the enzymes from hog liver supernatant solutions are summarized by NOVELLI. The first enzyme, pantetheine kinase, converts pantetheine into phosphopantetheine using ATP as an energy source. The second enzyme, called "condensing enzyme", in the presence of a second mole of ATP yields dephospho CoA and inorganic pyrophosphate. For this reaction to occur the substrate must be in the reduced form. Magnesium ions are also required. The third enzyme, dephospho CoA kinase in the presence of a third mole of ATP yields coenzyme A and ADP. The reactions are summarized below.

$$
\begin{array}{c}
\text{H}_3\text{C}\quad\text{OH}\quad\text{O}\quad\text{H}\qquad\qquad\qquad\text{O}\quad\text{H}\\
\text{HO—CH}_2\text{—C—CH—C—N—CH}_2\text{—CH}_2\text{—C—N—CH}_2\text{—CH}_2\text{—SH}\\
\text{CH}_3
\end{array}
$$

Panthetheine

Panthetheine kinase | + ATP ↓

$$
\begin{array}{c}
\text{OH}\qquad\quad\text{H}_3\text{C}\quad\text{OH}\quad\text{O}\quad\text{H}\qquad\qquad\text{O}\quad\text{H}\\
\text{O=P—O—CH}_2\text{—C—CH—C—N—CH}_2\text{—CH}_2\text{—C—N—CH}_2\text{—CH}_2\text{—SH}\\
\text{OH}\qquad\quad\text{CH}_3\qquad\qquad\qquad\qquad\qquad + \text{ADP}
\end{array}
$$

Phosphopanthetheine   + ATP ‖ Condensing enzyme ↓

$$
\begin{array}{c}
\text{OH}\qquad\quad\text{H}_3\text{C}\quad\text{OH}\quad\text{O}\quad\text{H}\qquad\qquad\text{O}\quad\text{H}\\
\text{O=P—O—CH}_2\text{—C—CH—C—N—CH}_2\text{—CH}_2\text{—C—N—CH}_2\text{—CH}_2\text{—SH}\\
\text{O}\qquad\qquad\text{CH}_3\qquad\qquad\qquad\qquad\qquad + \text{PP}
\end{array}
$$

$$
\begin{array}{c}
\text{H}\quad\text{H}\quad\text{H}\quad\text{H}\qquad\text{H}\\
\qquad\qquad\qquad\qquad\quad\text{C}\\
\text{O=P—O—CH}_2\text{—C—C—C—C—N}^{/}\ ^{\searrow}\text{N NH}_2\\
\text{OH}\qquad\quad\text{OH OH}\qquad\text{C=C—C}\\
\text{O}\qquad\qquad\qquad\qquad\text{N═══C—N}\\
\qquad\qquad\qquad\qquad\qquad\qquad\text{H}
\end{array}
$$

Dephospho CoA

Dephospho CoA kinase | + ATP ↓

$$
\begin{array}{c}
\text{OH}\qquad\quad\text{H}_3\text{C}\quad\text{OH}\quad\text{O}\quad\text{H}\qquad\qquad\text{O}\quad\text{H}\\
\text{O=P—O—CH}_2\text{—C—CH—C—N—CH}_2\text{—CH}_2\text{—C—N—CH}_2\text{—CH}_2\text{—SH}\\
\text{O}\qquad\qquad\text{CH}_3\qquad\qquad\qquad\qquad\qquad + \text{ADP}
\end{array}
$$

$$
\begin{array}{c}
\text{H}\quad\text{H}\quad\text{H}\quad\text{H}\qquad\text{H}\\
\qquad\qquad\qquad\qquad\quad\text{C}\\
\text{O=P—O—CH}_2\text{—C—C—C—C—N}^{/}\ ^{\searrow}\text{N NH}_2\\
\text{OH}\qquad\qquad\text{OH}\qquad\text{C=C—C}\\
\text{O}\qquad\qquad\qquad\qquad\text{N═══C—N}\\
\qquad\qquad\qquad\qquad\qquad\qquad\text{H}\\
\text{O}\\
\text{HO—P—OH}\\
\text{O}
\end{array}
$$

Coenzyme A

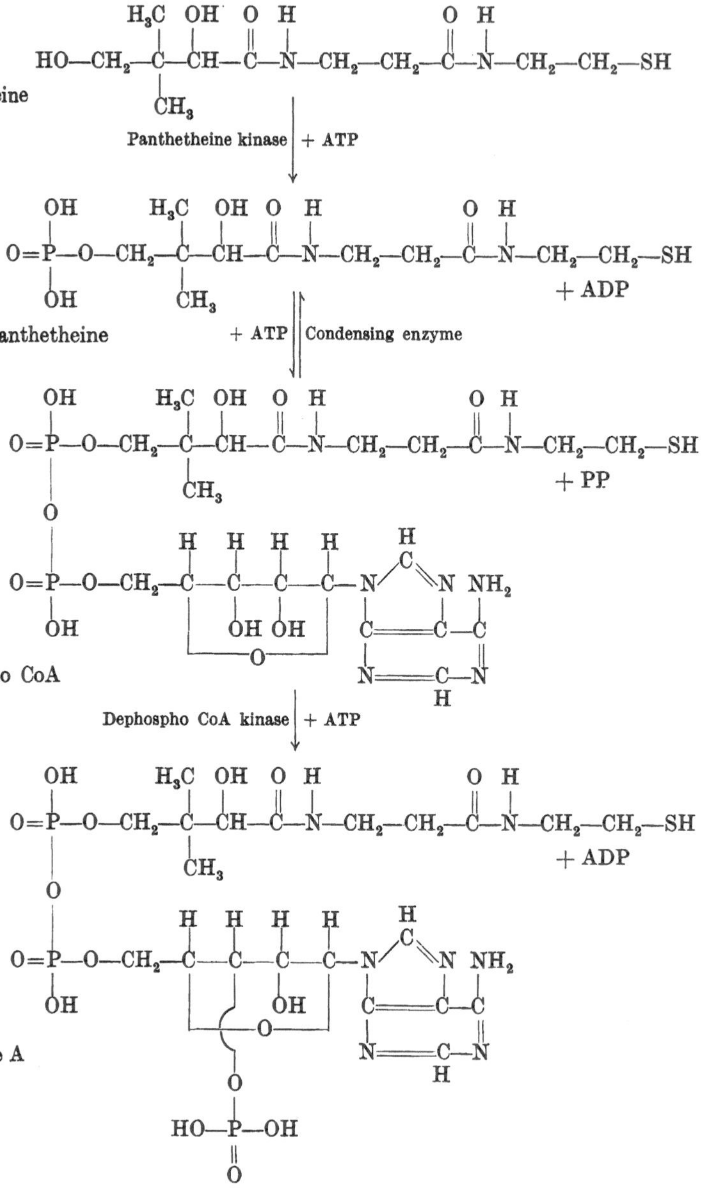

"Active acetate".

That the degradation of acetic by yeast occurs by way of the citric cycle and that the acetate is introduced into the cycle by conversion into "active acetate" was first suggested by LYNEN in 1941 (1941, 1942). "Active acetate" was identified in 1951 by LYNEN *and his coworkers* (1951). Active acetate was shown to be acetyl CoA whose structure is shown below.

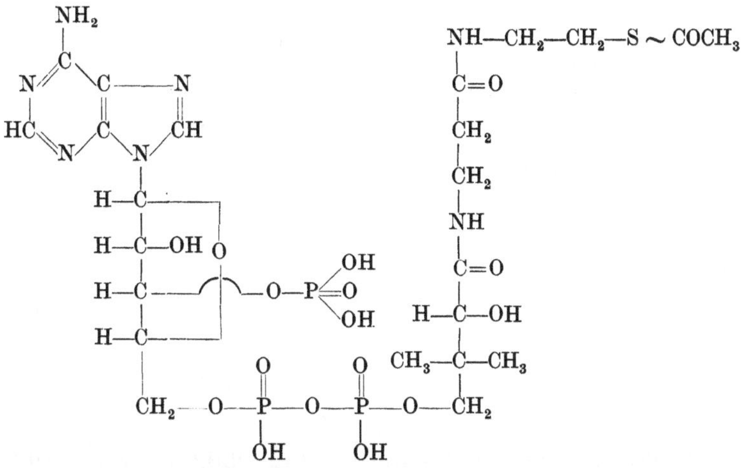

"Active acetate" (acetyl CoA).

It is now generally agreed that acetyl CoA is an intermediate not only in the utilization of acetate by yeast and in the reactions described by LIPMANN, but also in the oxidation of fatty acids in general in animal tissues and many micro-organisms. The acetyl CoA linkage involving the sulfhydryl group is especially interesting. The bond, it is generally believed, is of the high energy type and equivalent to the high energy bond which occurs in ATP. This was emphasized by the discovery that acetyl CoA in the presence of an appropriate enzyme system reversibly yields acetate, CoA and ATP. This reaction was carried out with an enzyme preparation from yeast (JONES).

$$\text{ATP} + \text{CoA} + \text{acetate} \rightleftharpoons \text{acetyl CoA} + \text{AMP} + \text{pyrophosphate.}$$

Recent tests from LIPMANN's laboratory using radioactive acetate and pyrophosphate make the following mechanism possible for this reaction (JONES, LIPMANN, HILZ and LYNEN).

1. $\text{E} + \text{Ad—P} \sim \text{PP} \rightleftharpoons \text{E} \sim \text{P—Ad} + \text{PP.}$
2. $\text{E} \sim \text{P—Ad} + \text{HSCoA} \rightleftharpoons \text{E} \sim \text{SCoA} + \text{Ad—P.}$
3. $\text{E} \sim \text{SCoA} + \text{CH}_3\text{COOH} \rightleftharpoons \text{E} + \text{CH}_3\text{CO} \sim \text{SCoA.}$

$\text{E} = \text{enzyme, Ad—P} \sim \text{PP} = \text{ATP, PP} = \text{inorganic pyrophosphate, Ad—P} = \text{AMP.}$

The first reaction shown above is between ATP and the enzyme to yield an AMP enzyme complex, in which the energy of the second pyrophosphate bond of ATP is retained in the complex, and inorganic pyrophosphate. The AMP enzyme complex then reacts with coenzyme A in a second reaction forming an enzyme CoA complex and AMP. In a third reaction the coenzyme A enzyme complex reacts with acetate giving rise to acetyl CoA and enzyme which is now available to participate once more in reaction number 1. As LYNEN (1953) has suggested, these results would indicate that the enzyme itself has an active SH group. According to this notion the pyrophosphate bond of ATP would be replaced by the equivalent sulfur linkage which now binds AMP to the enzyme. Removal of the AMP and its substitution by CoA on the enzyme surface would now lead to the activation of the CoA which could, in the third step, take on acetate, forming acetyl CoA.

Fatty acid activation.

The activation of acetate through the mechanism described above is not unique. Other fatty acids may be activated by a similar mechanism forming, for example, in the case of butyric acid, butyryl CoA or the CoA derivative of either higher or lower fatty acids. Butyryl CoA, as STADTMAN has pointed out using enzyme preparations derived from *Clostridium kluyveri*, may be formed by another mechanism. Starting with acetyl CoA and butyrate an exchange reaction may take place giving rise to butyryl CoA and acetate. Once the butyryl CoA complex is formed it may be degraded through the following system (MAHLER):

$$CH_3CH_2CH_2CO \sim SCoA + FAD \rightleftharpoons CH_3CH = CHCO \sim SCoA + FAD \cdot 2H$$
Acyl CoA dehydrogenase

$$CH_3CH = CHCO \sim SCoA + H_2O \rightleftharpoons CH_3CHOHCH_2CO \sim SCoA$$
Crotonase

$$CH_3CHOHCH_2CO \sim SCoA + DPN^+ \rightleftharpoons CH_3COCH_2CO \sim SCoA + DPNH$$
β-Keto dehydrogenase

$$CH_3COCH_2CO \sim SCoA + HSCoA \rightleftharpoons 2 CH_3CO \sim SCoA$$

The first step involves an enzyme whose prosthetic group is flavin adenine dinucleotide (FAD). The enzyme, acyl CoA dehydrogenase, would produce crotonyl CoA and reduced FAD. The next step catalyzed by the enzyme crotonase would yield β-hydroxy butyryl CoA which in the presence of a third enzyme which is DPN mediated, β-keto dehydrogenase, would form acetoacetyl CoA and reduced DPN. Acetacetyl CoA reacting with a second molecule of CoA would then yield two molecules of acetyl CoA. For higher fatty acids, for example hexanyl CoA or octanoyl CoA, similar oxidations at the β-carbon would take place yielding in the case of the former, first butyryl CoA and hexanyl CoA in the latter, and ultimately resulting in the overall conversion of the initial fatty acids into acetyl CoA. In animal tissues and in microorganisms the acetyl CoA would then react with oxaloacetate to form citrate and regenerate CoA. The citric acid would then be metabolized via the citric acid cycle.

The relationship of CoA to lipoic acid.

As LIPMANN had pointed out earlier pyruvic acid may be used as a substrate in the enzyme system from pigeon liver as an acetate donor. The participation of pyruvate in this reaction required coenzyme A, DPN, thiamine pyrophosphate, and an additional factor which has recently been identified as lipoic acid. The structure of the latter is shown below.

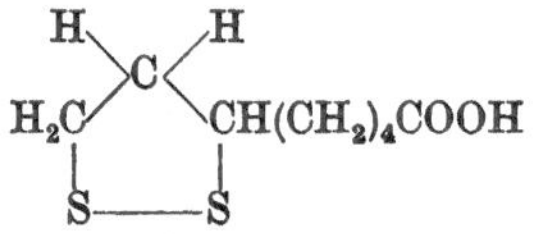

DL-Thioctic acid (Lipoic acid).

In the initial step the pyruvate is split yielding carbon dioxide and an acetyl lipoic acid complex. This complex then reacts with CoA to produce acetyl CoA and the reduced form of lipoic acid. The latter in the presence of an appropriate enzyme system reduces DPN, thus regenerating oxidized lipoic acid (GUNSALUS).

The reaction between pyruvate CoA and lipoic acid may be summarized as follows:

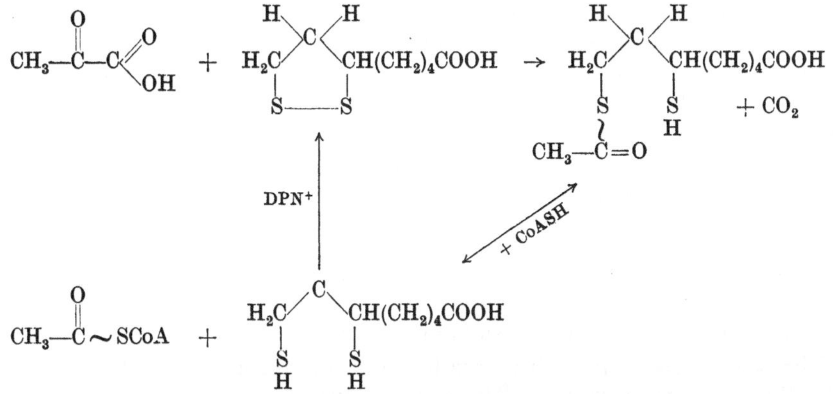

A similar mechanism operates at one other point in the citric acid cycle (KAUF-MAN), the conversion of α-keto-glutaric acid into succinate. Here instead of forming acetyl CoA as an intermediate, the product is succinyl CoA. The energy in the succinyl CoA bond may then be transferred to ADP leading to a regeneration of ATP.

Coenzyme A in the tissues of higher plants.

The occurrence of coenzyme A in the tissues of higher plants has been studied by SEIFTER. The work includes a comprehensive survey of various families of flowering plants and more detailed analyses on spinach, mung bean, pea, wheat germ, and corn. The material was prepared for assay in several ways. In the case of seeds, dried plant parts or materials high in coenzyme A, weighed samples were homogenized with equal weights of water at 80^0 C in a Waring Blendor for 2 minutes. After this treatment, the material was boiled for 5 minutes, filtered and clarified by centrifugation at $4600 \times$ g. The sediment was re-extracted once and the first and second supernatants combined. These could be directly assayed for coenzyme A.

For the assay procedures employed extracts must contain a minimum of 5 units of coenzyme A per ml. In cases where this quantity of material was not present, the extracts prepared above were concentrated at reduced pressure, below zero degrees, to one-fourth of their original volume. In some experiments, plant tissues were first dried and defatted with cold acetone. After the acetone treatment, the tissue residues were dried in vacuo over P_2O_5 and these powders were extracted and assayed by the direct method described above.

Two assay procedures were employed. The first of these employed the method of KAPLAN and LIPMANN. It is based on the observation that when extracts of acetone powders of pigeon livers are aged, the CoA is destroyed and ability to acetylate sulfanilamide is lost. It is regained, however, upon the addition of increasing quantities of CoA. Within limits the amount of sulfanilamide acetylated is proportional to the amount of CoA added. This reaction in the current investigations was carried out at p_H 7.1 at 37^0 C for 125 minutes.

Acetylated sulfanilamide fails to react in the BRATTON and MARSHALL method for the determination of free sulfanilamide. The quantity of CoA can therefore be determined by the disappearance of the free sulfanilamide. The latter is measured by reacting sulfanilamide with nitrous acid and l-naphthyl ethylenediamine. The colored complex formed is read in a photocell colorimeter or spectrophotometer at wave length 5300 to 5400 Å. For all assay experiments a standard sulfanilamide reaction mixture with the following composition was prepared:

10.0 ml 0.004 M sulfanilamide, 2.5 ml M magnesium acetate, 8.0 ml 0.05 M potassium adenosine, triphosphate (ATP), 10.0 ml 0.2 M potassium citrate.

The mixture was kept in 5 ml vials in a frozen state. It was not kept for more than one week before use. For the assay of CoA the following were used:

0.3 ml standard sulfanilamide reaction mixture, 0.1—0.3 ml solution being analyzed, 0.1 ml 0.15 M neutralized cysteine solution, 0.25 ml pigeon liver enzyme preparation water to make volume 1 ml.

After incubation the reaction was stopped by the addition of 4 ml of 5% trichloroacetic acid. The reaction mixture was then centrifuged and 1 ml aliquots were taken for analysis of free sulfanilamide.

The second assay procedure was based on the reaction described by STADT-MAN, NOVELLI, and LIPMANN. This reaction utilizes the enzyme transacetylase which catalyses the reversible reaction shown below.

1. Acetyl phosphate + CoA $\rightleftharpoons$ acetyl CoA + phosphate.
2. Acetyl CoA + arsenate $\rightarrow$ acetate + CoA + arsenate.

The enzyme is also capable of catalyzing the irreversible reaction shown. In a system supplied with acetyl phosphate, CoA, and arsenate, the transacetylase quantitatively converts acetyl phosphate into acetate, arsenate and phosphate. STADTMAN has shown that this arsenolysis within very wide ranges is proportional to the coenzyme A when the system is supplied with ample enzyme and arsenate. Since the disappearance of acetyl phosphate is a function of CoA concentration, the reaction may be followed quantitatively by the method of LIPMANN and TUTTLE for the determination of acyl phosphates. This method reacts an acid anhydride of an organic acid in an acidic solution with hydroxylamine. The reaction product is the corresponding hydroxamic acid. This compound combines with ferric chloride to form a complex possessing a characteristic color, which is then read in the photoelectric colorimeter. This method measures the disappearance of acetyl phosphate. The enzyme transacetylase used in the above essays is a commercial preparation made from *Streptococcus hemolyticus*. The acetylation

Table 1[1]. *Comparison of two methods of assay for coenzyme A.*

Source	Preparation	Units per gm dry wt	
		Acetylation	Arsenolysis
Wheat germ.	Direct[2]	41	40
Corn germ	Direct	38	39
Corn meal	Direct	22	21
Pea meal	Direct	44	43
Tomato fruit	Conc.[3]	11.2	12

[1] Taken from SEIFTER.   [2] Direct = hot water extract.   [3] Conc. = hot water extract, concentrated.

and arsenolysis procedures give comparable results when tested on the same material. For plant materials such a comparison is shown in Table 1. Table 2, taken from SEIFTER's paper, shows the distribution and concentration of CoA in some representative plants and plant tissues. The distribution of coenzyme A in the different tissues of 5 day old mung bean seedlings is presented in Table 3.

Table 2[1]. *Coenzyme a content of samples of representative plants ants and plant tissues.*

Source	Tissue	Preparation[2]	Units per gm dry wt	Test used[3]
Apple	Fruit	VC	3	Ac
Brussels sprouts	Bud	D	20.6	Ac
Bryophyllum calycinum	Leaf	VC	6.2	Ac
B. crenatum	Leaf	VC	5.1	Ac
Buckwheat	Seed	D	30	Ac
Cabbage	Leaf	VC	10.4	Ac
Cabbage	Bud	VC	10.4	Ac
Canada balsam	Pollen	D	38	Ac
Carrot	Xylem	VC	16.7	Ar
Carrot	Phloem	VC	21.6	Ar
Cattail	Pollen	D	27.3	Ac
Corn	Pollen	D	42.2	Ac
Cryptostegia grandiflora	Seed	D	32.7	Ac
C. magagascariensis	Seed	D	30	Ac
Cryptostegia hybrid (largely sterile) . .	Seed	D	13	Ac
Hevea	Seed	D	41	Ac
Kok Saghyz	Seed	D	28.5	Ac
Kok Saghyz	Root	D	19	Ac
Kok Saghyz	Leaves	D	9.3	Ac
Lettuce	Leaf	VC	14.2	Ac
Mustard	Seed	D	39.7	Ar
Onion	Bulb	AP	18.1	Ac
Orange	Seed	D	14.2	Ac
Orange	Rind	VC	2.1	Ar
Orange	Pulp	VC	36	Ar
Peanut	Seed	D and AP	47	Ac and Ar
Pine	Seed	D	34	Ar
Polypodium sp.	Leaf	VC	4.1	Ar
Polypodium sp.	Spores	D	23.5	Ar
Potato	Buds	D	17.6	Ar
Potato	Tuber	AP	5.3	Ar
Spinach	Leaf	VC	8.2	Ac
Valeriana sp.	Root	D	24	Ac
Valeriana sp.	Leaves	VC	13.2	Ac
Yam	Root	AP	13	Av

[1] Taken from SEIFTER. [2] AP = Acetone powder, D = Direct, VC = Vacuum concentrated. [3] Ac = Acetylation, Ar = Arsenolysis.

It is clear from the results presented that coenzyme A, as one might expect, is widely distributed among different plants and in the tissues of these plants. In

Table 3[1]. *Coenzyme a content of tissues of mung bean (Phaseolus aureus) seedlings[2].*

Tissue	Units per gm dry wt	Tissue	Units per gm dry wt
Plumule and first leaves .	22.8	Radicle	24
Cotyledons	15.1	Seed coat	2.4
Hypocotyl	14.7	Whole seed[3]	42

[1] Taken from SEIFTER. [2] Five day old seedlings; acetone powders were prepared and twice extracted with water at 85° C. [3] Ungerminated.

general the coenzyme appears to be present in highest concentrations in the seed and rather low concentrations in leaves. The lowest value detected was 2.1 units per gm dry weight in orange rind and the highest was found in the peanut seed—47 units per gm dry weight.

In the material presented above for animal tissues and bacteria, it was shown that the oxidation of fatty acids proceeds via the tricarboxylic acid cycle and that CoA is required in "activating" the fatty acid. In a study carried out on the ability of microsomal particles in germinating peanut cotyledons to oxidize fatty acids, data are presented to show that for palmitic acid, which is oxidized to carbon dioxide, the citric acid cycle may not be involved (HUMPHREYS, NEWCOMB, BOK-MAN and STUMPF). This conclusion is strengthened by the observation that palmityl CoA was found to be inactive as a substrate for this enzyme system. Whether this conclusion has relevance for the oxidation of fatty acids by plant tissues in general is not known at present.

Literature.

BRATTON, C., and E. K. MARSHALL: A new coupling component for sulfanilamide determination. J. of Biol. Chem. 128, 537–550 (1939).

DeVRIES, W. H., W. M. GOVIER, J. S. EVANS, J. D. GREGORY, G. D. NOVELLI, M. SOODAK and F. LIPMANN: Purification of coenzyme A from permentation sources and its further partial identification. J. Amer. Chem. Soc. 72, 4838 (1950).

GREGORY, J. D., G. D. NOVELLI and F. LIPMANN: The composition of coenzyme A. J. Amer. Chem. Soc. 74, 854 (1952). — GUNSALUS, I. C.: Oxidative and transfer reactions of lipoic acid. Federat. Proc. 13, 715–722 (1954).

HUMPHREYS, T. E., E. H. NEWCOMB, A. H. BOKMAN and P. K. STUMPF: Fat metabolism in higher plants. II. Oxidation of palmitate by a planut particulate system. J. of Biol. Chem. 210, 941–948 (1954).

JONES, M. E.: In symposium on chemistry and functions of coenzyme A. Federat. Proc. 12, 708–710 (1953). — JONES, M. E., F. LIPMANN, H. HILZ and F. LYNEN: On the enzymatic mechanism of coenzyme A acetylation with adenosine triphosphate and acetate. J. Amer. Chem. Soc. 75, 3285–3286 (1953).

KAPLAN, N. O., and F. LIPMANN: The assay and distribution of coenzyme A. J. of Biol. Chem. 174, 37–44 (1948). — KAUFMAN, SEYMOUR: Succinyl coenzyme A and its role in phosphorylation. Federat. Proc. 12, 704–708 (1953).

LIPMANN, F.: Acetylation of sulfanilamide by liver homogenates and extracts. J. of Biol. Chem. 160, 173–190 (1945). — LIPMANN, F., and L. C. TUTTLE: A specific micromethod for the determination of acyl phosphates. J. of Biol. Chem. 159, 21–28 (1945). — LYNEN, F.: Aerobic phosphate requirements of yeast-pasteur reaction. Liebigs Ann. 546, 120–141 (1941). — Biological degradation of AcOH. I. Induction period with impoverished yeast. II. Action of malonic acid on the degradation of AcOH by yeast. Liebigs Ann. 552, 270–306 (1942). — Functional group of coenzyme a and its metabolic relations in the fatty acid cycle. Federat. Proc. 12, 683–691 (1953). — LYNEN, F., u. E. REICHERT: Chemical structure of activated acetic acid. Z. angew. Chem. 63, 47, 490 (1951). — LYNEN, F., E. REICHERT u. L. RUEFF: Biological degradation of AcOH. VI. Isolation and chemical nature of activated AcOH. Liebigs Ann. 574 1–32 (1951).

MAHLER, H. R.: Role of coenzyme A in fatty acidmetabolism. Federat. Proc. 12, 694–702 (1953).

NOVELLI, G. D.: Enzymatic synthesis and structure of CoA. Federat. Proc. 12, 675–681 (1953). — NOVELLI, G. D., and M. B. HOAGLAND: The enzymatic degradation and resynthesis of coenzyme A. 123rd Meeting, Amer. Chem. Soc., Los Angeles, March 1953, Abstr. 26C.

SEIFTER, ELI: The occurrence of coenzyme a in plants. Plant Physiol. 29, 403–406 (1954). — STADTMAN, E. R.: Coenzyme A-dependent transacetylation and transphorylation. Federat. Proc. 9, 233 (1950). — In symposium on chemistry and functions of coenzyme A. Federat. Proc. 12, 692–693 (1953). — STADTMAN, E. R., G. D. NOVELLI and F. LIPMANN: Coenzyme A-function in and acetyl transfer by the phosphotransacetylase system. J. of Biol. Chem. 191, 365–376 (1951).

Pyridoxal phosphate.

By

W. W. Umbreit.

Vitamin B_6, like other vitamins, functions in the form of a coenzyme. This coenzyme, which has been variously called codecarboxylase, cotransaminase, *etc.*, is the 5-phosphate of pyridoxal:

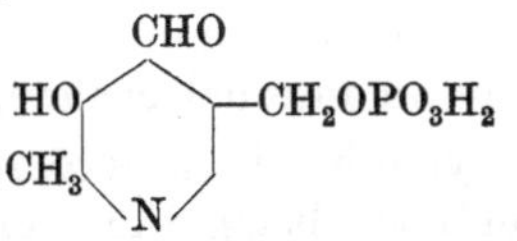

$$\text{CHO}$$

$$\text{HO} \quad \text{—CH}_2\text{OPO}_3\text{H}_2$$
$$\text{CH}_3 \quad \text{N}$$

Various members of the vitamin B_6 group (pyridoxine, pyridoxal, pyridoxamine and their respective phosphates) owe their vitamin activity to the ability of most organisms to convert them into the enzymatically active form, pyridoxal phosphate (BELLAMY, UMBREIT and GUNSALUS 1945). There are enzymes capable of phosphorylating pyridoxal with ATP (HURWITZ 1952) which reaction in fact led to the discovery and original synthesis of the coenzyme (GUNSALUS, BELLAMY and UMBREIT 1944). This enzyme system is inhibited by 2-ethyl-3-amino-4-ethoxymethyl-5-amino methyl pyridine which competes with the pyridoxal for the system, but the enzyme is capable of phosphorylating a variety of analogues of pyridoxal all containing a free 5-hydroxymethyl group (HURWITZ 1952). This reaction, forming desoxypyridoxine phosphate from desoxypyridoxine is the basis of the action of desoxypyridoxine as a vitamin B_6 antagonist (UMBREIT and WADDELL 1949). This same enzyme system, in a partially purified state, also phosphorylates pyridoxine, and the pyridoxine phosphate competes with (and inhibits) the action of pyridoxal phosphate in isolated enzyme systems. However, in the whole cell pyridoxine (and pyridoxamine) are converted to pyridoxal phosphate but it is not known which of the two available paths (pyridoxine $+ \text{ATP} \rightarrow$ pyridoxine phosphate $\rightarrow$ pyridoxal phosphate or pyridoxine $\rightarrow$ pyridoxal $[+ \text{ATP}] \rightarrow$ pyridoxal phosphate) are actually employed. Pyridoxamine and pyridoxamine phosphate both serve as substrates for a transaminase which converts them to the corresponding pyridoxal derivatives (UMBREIT, O'KANE, GUNSALUS 1948, MEISTER and TICE 1951, GUNSALUS and TONZETICH 1952). Thus there are also two routes for the formation of pyridoxal phosphate from pyridoxamine (pyridoxamine $\rightarrow$ pyridoxal $\xrightarrow{[+\text{ATP}]}$ pyridoxal phosphate or pyridoxamine $\xrightarrow{+\text{ATP}}$ pyridoxamine phosphate $\rightarrow$ pyridoxal phosphate).

In all of the enzymatic systems in which pyridoxal phosphate is the coenzyme, none of the other members of the vitamin B_6 group are active (unless they are converted to pyridoxal phosphate) with one exception. Pyridoxamine phosphate, under certain specified conditions, can replace pyridoxal phosphate in the transaminase system (MEISTER, SOBER and PETERSON 1952, 1954).

When the coenzyme was first synthesized in 1944 (GUNSALUS, BELLAMY and UMBREIT 1944), the impure synthetic material was compared with the impure naturally occurring substance, and with the demonstration that the two were interchangeable in several biological systems they were thus presumed to be identical (UMBREIT, BELLAMY and GUNSALUS 1945). Upon purification of the synthetic substance, the material obtained had 1 phosphorus per mole of pyridoxal and differed from pyridoxal in lacking an absorption maxima at 300 mμ under alkaline conditions (GUNSALUS, UMBREIT, BELLAMY and FOUST 1945). This would indicate that the phosphate was not at position 3, which conclusion was also supported by other types of evidence. While a few investigators were misled as to the nature of the activity and supposed for sometime that the active material was the 3-phosphate (KARRER, VISCONTINI and FORSTER 1948), chemical proof of structure (HEYL, LUZ, HARRIS and FOLKERS 1951, HEYL and HARRIS 1951) completely established the 5-phosphate as the coenzyme form. A 4-phosphate has been postulated as a possible tautomer (BADDILEY, THAIN and RODWELL 1954) with the 5-phosphate as the predominant form. However, other tautomeric arrangements, particularly an internal chelate, seem to be more likely.

Synthesis of the coenzyme has been accomplished on a reasonable scale, the most commonly used methods being those of WILSON and HARRIS (1951) or PETERSON and SOBER (1954).

The coenzyme is involved in a variety of enzymatic reactions including transamination, amino acid racemization, amino acid decarboxylation, β-hydroxy acid dehydration, cysteine desulfhydration, tryptophanase reaction, homocysteine desulfhydration, and the reversible cleavage of β-hydroxy amino acids. Many of these reactions may be conducted in chemical systems where they are catalyzed by pyridoxal and metal salts. On this basis a chemical hypothesis for mode of action, based largely upon chelation involving the α-amino group of the amino acid and the 4-aldehyde group of the pyridoxal phosphate, has been proposed (METZLER, IKAWA and SNELL 1954) which hypothesis is able to account for most of the known facts of the action of pyridoxal phosphate. The breakdown of pyridoxal phosphate has not been adequately studied. It is attacked by both acid and alkaline phosphatases and the 4-aldehyde group may be oxidized to yield what is apparently pyridoxic acid phosphate. The 4-aldehyde group may also be animated (to form pyridoxamine phosphate) and there is some indication that it may be reduced to pyridoxine phosphate. Much of this information is only incidential to other studies and in many cases the reaction conditions, the biological systems involved, or even an adequate characterization of the products has not been properly studied.

Literature.

BADDILEY, J., E. M. THAIN and A. W. RODWELL: A possible structure for codecarboxylase. Nature (Lond.) **167**, 556–557 (1951). — BELLAMY, W. D., W. W. UMBREIT and I. C. GUNSALUS: The function of pyridoxine; conversion of members of the vitamin B$_6$ group into codecarboxylase. J. of Biol. Chem. **160**, 461–472 (1945).

GUNSALUS, C. F., and J. TONZETICH: Transaminase for pyridoxamine and purines. Nature (Lond.) **170**, 162 (1952). — GUNSALUS, I. C., W. D. BELLAMY and W. W. UMBREIT: A phosphorylated derivative of pyroxidal as the coenzyme of tyrosine decarboxylase. J. of Biol. Chem. **155**, 685–686 (1944). — GUNSALUS, I. C., W. W. UMBREIT, W. D. BELLAMY and C. E. FOUST: Some properties of synthetic codecarboxylase. J. of Biol. Chem. **161**, 743–744 (1945).

HEYL, D., and S. A. HARRIS: Phosphates of the vitamin B$_6$ group. II. 3-Pyridoxal phosphoric acid. J. Amer. Chem. Soc. **73**, 3434–3436 (1951). — HEYL, D., E. LUZ, S. A. HARRIS and K. FOLKERS: Phosphates of the vitamin B$_6$ group. I. The structure of codecarboxylase. III. Pyridoxamine phosphate. J. Amer. Chem. Soc. **73**, 3430–3433, 3436–3437 (1951). — HURWITZ, J.: The enzymic phosphorylation of vitamin B$_6$ derivatives and their effects on tyrosine decarboxylase. Biochem. et Biophysica Acta **9**, 496–498 (1952).

KARRER, P., M. VISCONTINI and D. FORSTER: Pyridoxal-3-phosphate as coenzyme of L-amino decarboxylase. Helvet. chim. Acta **31**, 1004–1016 (1948).

MEISTER, A., H. A. SOBER and E. A. PETERSON: Activation of purified glutamic-aspartic apotransaminase by crystalline pyridoxamine phosphate. J. Amer. Chem. Soc. **74**, 2385–2386 (1952). — Studies of the coenzyme activation of glutamic-aspartic apotransaminase. J. of Biol. Chem. **206**, 89–100 (1954). — MEISTER, A., and S. V. TICE: Transaminat on from glutamine to α-keto acids. J. of Biol. Chem. **187**, 173–187 (1951). — METZLER, D. E., M. IKAWA and E. E. SNELL: A general mechanism for vitamin B -catalyzed reactions. J. Amer. Chem. Soc. **76**, 648–652 (1954).

PETERSON, E. A., and H. A. SOBER: Preparation of crystalline phosphorylated derivatives of vitamin B_6. J. Amer. Chem. Soc. **76**, 169–175 (1954).

UMBREIT, W. W., W. D. BELLAMY and I. C. GUNSALUS: The function of pyridoxine derivatives: A comparison of natural and synthetic codecarboxylase. Arch. of Biochem. **7**, 185–199 (1945). — UMBREIT, W. W., D. J. O'KANE and I. C. GUNSALUS: Function of the vitamin B_6 group: Mechanisms of transamination. J. of Biol. Chem. **176**, 629–637 (1948). — UMBREIT, W. W., and J. G. WADDELL: Mode of action of desoxypyridoxine. Proc. Soc. Exper. Biol. a. Med. **70**, 293–299 (1949).

WILSON, A. N., and S. A. HARRIS: Phosphates of the vitamin B_6 group. V. A synthesis of codecarboxylase. J. Amer. Chem. Soc. **73**, 4693–4694 (1951).

Transphosphorylation
(phosphotransferases, mutases and kinases).

By

Bernard Axelrod.

A. Introduction.

"Transphosphorylation" seems at first glance a somewhat arbitrary basis for categorizing apparently diverse groups of enzymes. However, a consideration of some of the reactions catalyzed by these enzymes may reveal some common features which are more than formalities. If some liberties are taken in occasionally discussing enzymes not of plant origin it is because many of the reactions in the plant kingdom are not unique with plants, and in many cases the more thorough-going information derived from a study of non-plant sources is of great value in interpreting the behavior of plant enzymes. The term "phosphotransferase" has been used to describe the activity of phosphatases upon phosphate esters wherein phosphorus is transferred to organic acceptors (AXELROD 1948a). It has also been employed to designate enzymes which can only cleave a phosphate ester when there is a specific acceptor present, as, for example, by STUMPF (1950) to refer to pyruvate kinase. The closely related term "transphosphatase" has been incorporated into the names of many kinases by HOFFMANN-OSTENHOF (1953) in his rational system of nomenclature. In this section the term "phosphotransferase" is considered in the broad sense to imply the transfer of phosphate (or more properly, the phosphoryl group) from an ester to an acceptor compound, be it water or an organic compound; or to the original ester, as with mutases. These three categories do not necessarily imply three classes of enzymes. For instance, it is possible that certain phosphatases can catalyze all of these types of reactions.

B. Enzymes transferring P from ATP[1].

The following group of "kinases" has been shown to utilize ATP as a source of P. However, in view of the ubiquity of the nucleotide diphosphate transferring enzymes, and the increasing instances when nucleotides derived from purines other than adenine can function in this way, one must recognize that ATP may not necessarily be the only donor, and not even the primary donor. It must be admitted that the plant preparations which have been studied, in general, have been relatively impure.

a) Pyruvate kinase.

This enzyme catalyzes the following reaction:

$$CH_2{=}C{-}COOH + ADP \rightarrow CH_3COCOOH + ATP.$$
$$\underset{OPO_3^=}{|}$$

---

[1] The following abbreviations are used in this section. ADP, adenosine diphosphate; AMP, adenosine monophosphate; ATP, adenosine triphosphate; P_i, inorganic phosphate; PEPA, phosphorylenolpyruvic acid; PGA, phosphoglyceric acid.

The presence of this enzyme in plants has been inferred from the cumulative observations of a number of workers who either isolated phosphorylated intermediates known to arise from the EMBDEN-MEYERHOF-PARNAS glycolytic pathway or actually isolated some of the enzymes which are involved in the scheme (see Volume 12 II, D, Volume 6, IV and Volume 5, III, E, d). The actual isolation of this enzyme from higher plants has not been achieved. The first reasonably direct indication of its presence was made by STUMPF (1950) who noted that 3-PGA treated with a crude extract of pea seeds (*Pisum sativum*, dwarf telephone) formed pyruvic acid if 5'-AMP were present. Acid-labile phosphate was also formed. When ATP was omitted, PEPA accumulated. It was further claimed that AMP was converted to ADP in the presence of PEPA. (Presumably the product may have been in part ATP, since the analytical method served merely to determine acid labile PO_4.) The fact that AMP could be used successfully implied the presence of adenyl kinase and a trace of ADP. Pyruvate kinase crystallized from mammalian tissue by NEGELEIN (cited in KUBOWITZ and OTT 1944) was shown by LARDY and ZIEGLER (1945) to require both K^+ and Mg^{++} for the reverse reaction. KACHMAR and BOYER (1953) established that K^+, NH_4^+ or Rb^+ were essential also for the forward reaction. It is not known whether the plant enzyme possesses these requirements. A clarification of this reaction in plants awaits the purification of this enzyme. With impure extracts the problem is complicated by the possible presence of at least two enzymes which can act on PEPA. BANDURSKI and GREINER (1953) and BANDURSKI (1955) have found a very active PEPA-carboxylase in plant extracts which can decompose PEPA as follows, if CO_2 is not excluded:

$$PEPA + CO_2 \rightarrow COOHCOCH_2COOH + P_i.$$

The oxalacetate, in turn, can, both spontaneously and under the influence of oxalacetic decarboxylase, form pyruvic acid. These workers have also found that leaves of higher plants contain a phosphatase which is very effective in hydrolyzing PEPA. (Unpublished results cited in AXELROD, BANDURSKI, GREINER and JANG 1953.)

b) Phosphoglyceryl kinase.

This enzyme has been shown to occur in a number of higher plants (AXELROD and BANDURSKI 1953). It can be readily determined by a sensitive method directly on crude plant extracts. Its occurence has been demonstrated in the following plants, *inter alia*: Tomato (stem, leaf, fruit), *Bryophyllum* (aerial portion), *Silene* (aerial portion), *Xanthium* (aerial portion), pea seedling (leaves), potato (tuber), sweet potato, mung bean (seed), soybean (seed), maize (seed), wheat germ. It was partially purified from pea seed meal, and unlike the crude extract was completely inactive unless certain cations were present. Mn^{++}, Mg^{++} and Co^{++} were equally effective. Ca^{++} and Fe^{+++} were also effective, but to a considerably lesser degree. The inhibition of glycolysis by F^- is normally laid to its marked effect on enolase. It is noteworthy, therefore, that F^- also inhibited plant phosphoglyceryl kinase, although higher concentrations were required. The assay procedure used, based upon the results of LIPMANN and TUTTLE (1945) depends on the presence of a high concentration of hydroxylamine (to which, fortunately, the enzyme is insensitive), to remove 1,3-diphosphoglycerate as it is formed and thus drive the reaction to completion.

$$
\begin{array}{llll}
\text{COOH} & & \text{COOPO}_3^= & \\
| & & | & \\
\text{H--C--OH} + \text{ATP} & \rightleftarrows & \text{HC--OH} + \text{ADP} \\
| & & | & \\
\text{H}_2\text{COPO}_3^= & & \text{H}_2\text{COPO}_3^=
\end{array}
$$

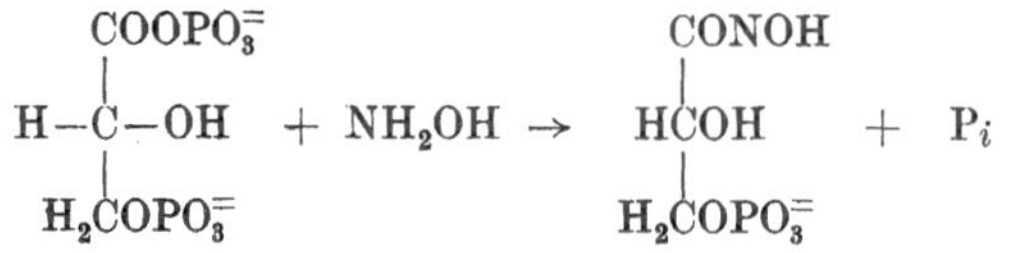

$$\begin{array}{c} \text{COOPO}_3^= \\ | \\ \text{H}-\text{C}-\text{OH} \\ | \\ \text{H}_2\text{COPO}_3^= \end{array} + \text{NH}_2\text{OH} \rightarrow \begin{array}{c} \text{CONOH} \\ | \\ \text{HCOH} \\ | \\ \text{H}_2\text{COPO}_3^= \end{array} + \text{P}_i$$

The hydroxamic acid of 3-PGA is readily measured by the color which it forms with Fe^{+++}. Unlike the glutamo-transferase enzyme, this enzyme was not able to use NH_3 in place of NH_2OH (AXELROD, unpublished results).

c) Phosphofructokinase.

The presence of this enzyme in plants was foreshadowed by TANKO'S (1936) discovery that pea seed brei esterified inorganic phophate and formed considerable quantities of fructose-1,6-diphosphate. AXELROD, SALTMAN, BANDURSKI and BAKER (1952) effected a modest purification of this enzyme from peas and used it to study the isolated reaction:

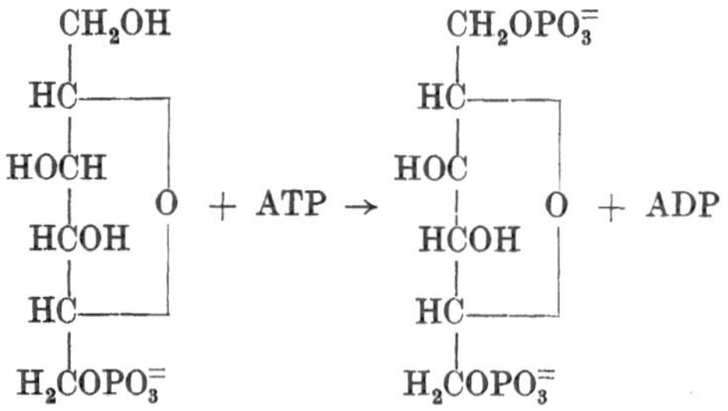

$$\begin{array}{c} \text{CH}_2\text{OH} \\ | \\ \text{HC}\!-\!\!-\!\!- \\ | \\ \text{HOCH} \quad | \\ | \quad\quad \text{O} \\ \text{HCOH} \quad | \\ | \\ \text{HC}\!-\!\!-\!\!- \\ | \\ \text{H}_2\text{COPO}_3^= \end{array} + \text{ATP} \rightarrow \begin{array}{c} \text{CH}_2\text{OPO}_3^= \\ | \\ \text{HC}\!-\!\!-\!\!- \\ | \\ \text{HOC} \quad | \\ | \quad\quad \text{O} \\ \text{HCOH} \quad | \\ | \\ \text{HC}\!-\!\!-\!\!- \\ | \\ \text{H}_2\text{COPO}_3^= \end{array} + \text{ADP}$$

The enzyme, like its animal counterpart, required Mg^{++}. It was sensitive to anti-sulfhydryl reagents. It was more readily denatured by heat than the animal enzyme. The presence of this enzyme has been established in the following higher plants: oat seeds, sunflower fruit, tomato seed, pollen of *Pinus ponderosa*, pea leaf, pea seed, radish seed, soy bean seed, broad bean seed and maize seed.

d) Adenyl kinase.

The presence of this enzyme which catalyzes the following reaction in higher plants was indicated indirectly in several ways.

$$2\,\text{ADP} \rightleftarrows \text{ATP} + \text{AMP}.$$

STUMPF (1950) observed that 5'-AMP met the requirements of pyruvic acid kinase as mentioned above. AXELROD, SALTMAN, BANDURSKI and BAKER (1952) reported the presence of adenyl kinase in peas when they found admittedly impure preparation of phosphofructokinase obtained from this source utilized ADP to one half the extent of ATP as a phosphate donor. CAMPBELL and BANDURSKI (1952) partially purified this enzyme from pea seed meal. Strong evidence pointing to adenyl kinase in plants was provided by the observation of MILLERD, BONNER, AXELROD and BANDURSKI (1951) that mitochondria prepared from *Phaseolus aureus* seedlings utilized AMP as an acceptor for the phosphorylation accompanying the oxidation of the KREBS cycle acids. STUMPF, LOOMIS and MICHELSON (1951) also noted that the glutamo-transferase system from pumpkin seedlings could utilize either ADP or ATP in meeting its requirements for a nucleotide polyphosphate in the exchange reaction. While this was not in itself unequivocal evidence for adenyl kinase LOOMIS and STUMPF (1952) showed by chromatographic means that the preparation did form AMP and ATP from ADP. They also found a similar enzyme in wild cucumber leaves. They also identified ATP among the products by its ability to phosphorylate glucose in the presence of yeast hexokinase. The adenyl kinase obtained from plants in contrast to the muscle enzyme is readily inactivated at 55⁰.

e) Nucleotide diphosphokinase.

The so-called "nudiki" enzymes are very definitely established in yeast and mammalian tissue (BERG and JOKLIK 1953, KREBS and HEMS 1953, SANADI, GIBSON and AYENGAR 1954). In the case of higher plants the inference is strong that they are present but a study of the isolated enzymes remains to be made. These reactions, to cite one example:

$$\text{ATP} + \text{uridine diphosphate} \rightleftarrows \text{uridine triphosphate} + \text{ADP},$$

are useful in permitting one nucleotide diphosphate or triphosphate to act as a generator of another. The ability of plant enzymes to utilize nucleotide poly-phosphates other than adenylic acid derivatives is established. Thus, inosine triphosphate replaced ATP in the phosphoglyceryl kinase reaction, but gave a much lower rate (AXELROD and BANDURSKI 1953). LEVINTOW, MEISTER, HOGE-BOOM and KUFF (1955) showed that highly purified pea glutamo-transferase utilized inosine, cytidine and uridine triphosphates although less rapidly than ATP. On the other hand, the succinyl-CoA phosphorylation reaction in spinach leaves was shown to specifically require adenine nucleotides by KAUFMAN (1955):

$$P_i + \text{succinyl-CoA} + \text{ADP} \rightarrow \text{succinic acid} + \text{CoA} + \text{ATP}.$$

This is especially interesting because the heart muscle enzyme requires guanosine or inosine diphosphates as the primary acceptor and cannot utilize ADP as SANADI, GIBSON and AYENGAR (1954) have shown.

f) Hexokinase.

This enzyme which promotes the phosphorylation of glucose (and some other hexoses) at the expense of ATP in accord with the following reaction:

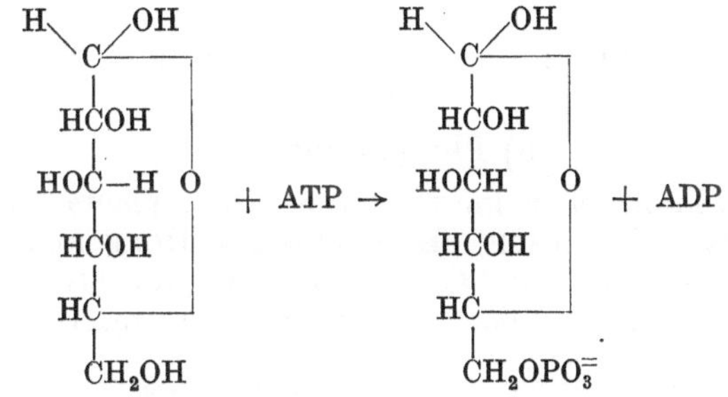

was first demonstrated unequivocally in higher plants by MILLERD, BONNER, AXELROD and BANDURSKI (1951) by the isolation and identification of glucose-6-phosphate formed by the hexokinase associated with mitochondria of *Phaseolus aureus*. The ATP was formed *in situ* by the phosphorylation which accompanied the oxidation of ketoglutarate. The enzyme, however, is not always associated with the particles; at least it does not remain so during the course of its extraction. SALTMAN (1953a) who studied the hexokinase of a number of plant tissues obtained the results shown in Table 1. He found the enzyme to act upon fructose, mannose and glucosamine in addition to glucose. Galactose was not phosphorylated although it can be transformed into sucrose when infiltered into barley leaves as McCREADY and HASSID (1941) have shown. It is hardly likely that

Table 1. *Distribution of hexokinase in soluble and insoluble fraction of several tissues.*
(SALTMAN 1953 b.)

Tissue	Insoluble units[1] per fraction	Per cent activity	Soluble units per fraction	Per cent activity
Wheat germ (commercial) . .	46.8	33	93.6	67
Wheat germ (dissected) . . .	100.8	68	46.8	32
Potato	16.2	100	0.0	0
Pea seed (var. Alaska) . . .	6.1	30	13.8	70
Mung bean seed	16.4	28	42.0	72
Mung bean hypocotyl . . .	14.0	19	61.0	81
Avena seed	3.5	19	14.7	81
Spinach leaf.	1.3	28	3.4	72

[1] 1 unit = 1 mole of glucose phosphorylated per 10 minutes in 1.5 ml. of standard reaction mixture.

galactose could be utilized without phosphorylation. It is possible, however, that it is converted to galactose-1-phosphate by a galacto-kinase such as is found in yeast (WILKINSON 1949).

g) Glucose-1-phosphate kinase.

The enzyme was discovered in jack bean meal by CARDINI (1951). The product. 1,6-diphosphoglucose, is the cofactor for phosphoglucomutase.

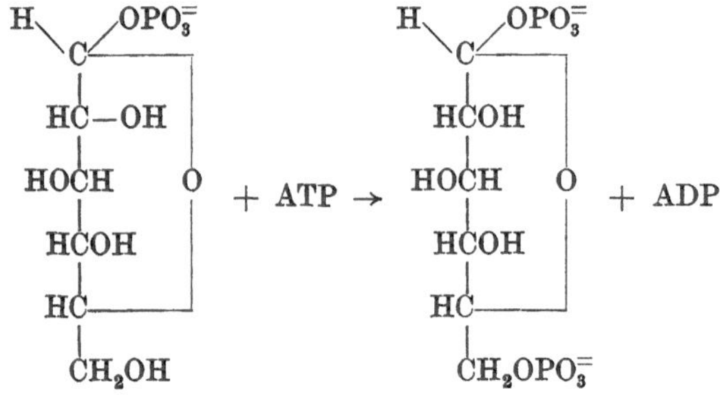

h) Phosphoribokinase.

This enzyme, insofar as is known is confined to plants. It was first isolated from spinach leaves (WEISBACH, SMYRNIOTIS and HORECKER 1954). Because of its key role in photosynthetic CO_2 fixation it is probably present in all green plants (see HASSID, this Handbook, Volume VI). The reaction which is catalyzed is as follows:

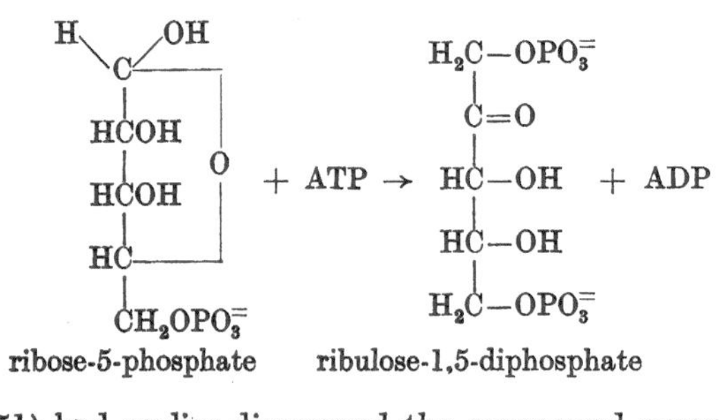

ribose-5-phosphate   ribulose-1,5-diphosphate

BENSON (1951) had earlier discovered the compound among the products of photosynthesis in a number of plants. Since the spinach enzyme was probably contaminated with phosphoriboisomerase it is likely that the actual acceptor was ribulose-5-phosphate. Like all kinases, this enzyme required the addition of Mg^{++}. RACKER (1955) also reported a partial purification of the enzyme and found it to be sensitive to maleimide, a reagent specific for sulfhydryl groups.

C. Phosphomonoesterases as transferases.

a) Citrus phosphatase.

This enzyme obtained from Navel orange juice by salt precipitation and absorption on celite resembles other "non-specific" plant phosphatases (AXELROD 1947). It is of particular interest because it has been studied somewhat intensively and its transferring activity has been investigated in some detail. The enzyme was completely inhibited by 0.02 M NaF and stimulated about 50% by 0.02 M $MgSO_4$. It showed a wide range of specificity, acting upon aromatic and aliphatic monoesters of phosphoric acid, acetyl phosphate, phosphoprotein, nucleotides, inorganic polyphosphates, and, very feebly, phytic acid. It is interesting that it was completely without action on glucose-1-phosphate. It did not hydrolyze diphenyl phosphate. The enzyme also possessed the ability to transfer, under certain conditions, the phosphate of the substrate to acceptor hydroxyl compounds, and with greater efficiency than to water (AXELROD 1948a). This ability to transfer P without the intervention of nucleotides has since been shown to occur with phosphatases from a number of other sources, including animals. With the citrus enzyme the following generalities were observed: Primary alcohols were the best acceptors, secondary alcohols next and tertiary alcohol was entirely unsuitable. Carbohydrates were also unsuited although polyhydroxy alcohols did accept. The ubiquity of phosphorylated and free inositol throughout the plant kingdom makes it interesting that inositol did not serve as an acceptor. Citrus juice, incidentally, is relatively rich in inositol. Aryl phosphates, such as p-nitrophenyl phosphate, phenyl phosphate and phenolphthalein phosphate were satisfactory donors. This may have been a consequence of the fact that these substrates saturate the enzyme at relatively low concentrations while the alkyl phosphates formed have high K_M values and are slowly hydrolyzed. The ease of detection of the liberated phenol makes aryl phosphates ideal for such studies. The likely possibility that alkyl phosphates would be good donors was not tested. Following a suggestion by MEYERHOF and GREEN (1950) that the efficacy of p-nitrophenylphosphate might be due to a high $-\Delta F$ of hydrolysis AXELROD (1951) measured this value, and found it to be only moderately high. If the generalized mechanism discussed below is valid it would seem that any phosphate ester which can be split by this phosphatase should be a donor; i. e., the enzyme-phosphoryl complex is the donor and it does not matter from what substrate it arises. However, if the product is readily hydrolyzed compared to the donor, then the net accumulation of new ester will be low; thus, the donor would be only "apparently" poor.

The influence of p_H on the reaction was complex. With increasing acidity the proportion of P transferred from p-nitrophenyl phosphate to n-propanol to that transferred to water increased constantly, reaching the highest value at slightly over p_H 3.0. Below this p_H the reaction was very slow. The p_H optimum for nitrophenol liberation was 5.0, the same value which was obtained in the absence of acceptor. The p_H optimum for phosphate liberation shifted close to 6 as a result of the less favorable transfer at high p_H values. Similar results were obtained when methanol was the acceptor. It is noteworthy that even such small quantities of methanol as 1.1% produced an increased rate of nitrophenol liberation. The effectiveness of a primary alcohol as an acceptor was clearly greater than that of water; thus 7% of propanol resulted in transfer of 70% of the cleaved P. The superiority of the alcohol appears even greater when it is realized that the molar ratio of water to propanol was about 48:1 in these experiments.

It was further shown that the alcohol phosphate formation could not be explained on the basis of synthesis from inorganic phosphate and methanol. First, the concentrations of reagents were too low to permit a significant synthesis. Secondly, it was demonstrated by using P^{32}-labeled nitrophenyl phosphate in the presence of a pool of unlabeled phosphate that the resultant methyl phosphate had the same molar specific activity as the original nitrophenyl phosphate (Axelrod 1948 b). Conversely, when the label was in the inorganic phosphate instead of the nitrophenyl phosphate, the resulting methyl phosphate was unlabeled. It is thus abundantly clear that the transferred P had no existence as free inorganic phosphate. As will be seen below this is compatible with the proposed mechanism for all "phosphoryl" transferring enzymes of which phosphatases are but a special class.

Phosphatases having "phosphotransferase" activity are probably very generally widespread throughout the plant kingdom. Such enzymes are present in the juice and peel of all citrus fruits, apple, takadiastase (Axelrod 1947) and alfalfa and tobacco (Axelrod, Unpublished). However, while all higher plants have "acid" phosphatases which are superficially similar, these enzymes do not invariably catalyze "phosphotransfer". Onion, pear and sweet potato are examples. The existence of isodynamic enzymes in an extract from a single plant source is a valid possibility (Boroughs 1954) and it is hence conceivable that observed "phosphotransferase" efficiency of a preparation may be an expression of the activities of a mixture of transferring and non-transferring enzymes. Separation attempts, differential inhibition and other procedures did not result in any evidence that the transfer capacity and the phosphatase capacity of the citrus enzyme resided in more than one moiety.

b) Nucleoside phosphotransferase.

Brawerman and Chargaff (1953 a, b) suggested by their findings a metabolic significance for "phosphotransferase". Germinated barley, they found, contained a phosphatase which utilized not only phenyl phosphate, but 5'-mononucleotides as donors for the following acceptors: ribosides and desoxyribosides of hypoxanthine, uracil, cytosine and thymine. It is significant that only the 5'-phosphate esters were formed. These workers (Brawerman and Chargaff 1955) using 5'-inosinic acid as donor and ribocytidine as acceptor detected transfer in the following plant materials: wheat seed, potato tuber, potato sprout, pea seed, pea seedling, spinach leaves and carrot root. In support of their suggestion that the transfer activity was biologically significant they showed that the ratio of transfer to hydrolysis improved in wheat with germination. This, according to them would be in keeping with the need to generate nucleotides in growth. The change in ratio supports the idea that more than one phosphatase is present.

D. Phosphomonoesterases as hydrolytic enzymes.

a) General.

Although phosphatases were first discovered almost fifty years ago by Suzuki, Yoshimura and Takaishi (1906) their role in biological tissue is not always clear. Indeed one of the few instances wherein a logical role could be assigned to phosphatase was indicated by this initial discovery. The enzyme, phytase, acted upon inositol hexaphosphate in bran of rice and wheat liberating inorganic P. Since much of the P within a seed is tied up as inositol hexaphosphate it is important to the developing embryo that an enzyme exist to liberate inorganic P which is required during development. It appears that seed phytase although not strictly specific is in fact an enzyme distinct from the other "non-specific" phosphatases so widely dispersed in Nature (FLEURY and COURTOIS 1947, COURTOIS 1947). It is perhaps possible that these enzymes do have catabolic roles, although normally there are few verified instances in which phosphate hydrolysis is a necessary step in a metabolic pathway in plants. One may speculate that the uptake or transport of carbohydrates may require phosphorylation and then dephosphorylation. Perhaps, if it is true that sucrose is sometimes formed as the phosphate (BUCHANAN 1953, LELOIR and CARDINI 1955), then there would be a need for a hydrolytic enzyme. Another possible function of hydrolysis might be to correct any one-sided accumulations of phosphate in an intermediate, which might deprive the organism of an adequate supply of Pi or the substance esterified to maintain the metabolic processes.

There are countless reports on the presence of phosphomonoesterases in higher plants various tissues, species, varieties, *etc.*, which for the present can be of only limited interest. In general, these so-called acid phosphatases are always to be found in higher plants; they have a wide range of specificity attacking N-phosphates, C—O phosphates and pyrophosphates (both organic and inorganic). Their p_H optimum is usually between 4 and 6 and they are inhibited by fluoride and oxalate. Their response to metal activation is variable, usually showing a slight activation by Mg^{++}. Because almost all of the studies have been made, and unfortunately continue to be made, with very impure preparations and without a systematic study of specificity a compilation of the properties of the enzymes of various plant origins is hardly warranted.

The existence of isodynamic enzymes in the cytoplasmic protein of sugar beet and spinach leaves has been indicated by BOROUGHS (1954). He placed sugar beet leaf cytoplasm on a cation ion exchange resin and obtained a division of the phosphatase into two fractions by eluting first with citrate buffer at p_H 4.9 and then with tartrate buffer at 7.7. The first fraction exhibited an optimum activity at p_H 5.7–5.8. The second fraction was most active at p_H 5.0–5.1.

Another claim for the existence of isodynamic phosphatases was based on the difference in p_H optima exhibited by the phosphatase in spinach cytoplasm which has been dialyzed at different p_H values. The original cytoplasm which had been dialyzed at p_H 7.0 in the course of preparation had an optimum of 5.2; after dialysis at p_H 4.8 it exhibited an optimum of 6.4. The antibody-antigen reaction of sugar beet cytoplasm resulted in a partial precipitation of the enzyme activity. The supernatant was maximally active at p_H 4.9–5.0 while the precipitate was most active at p_H 5.3.

In view of the uncertain status of metal ion activation, the experiments of BOROUGHS with spinach cytoplasmic phosphatase are very interesting. The enzyme was inactivated by dialysis against citrate buffer, p_H 2.0. The addition of Cu^{++} restored 75% of the original activity. Zn^{++}, Mg^{++}, Co^{++}, Mn^{++} and Fe^{++} were also effective but to a much smaller degree. Ca^{++} and Fe^{+++} were without effect.

b) 5′-Nucleotidase.

Although it has been known for some time that mononucleotides can be split by plant extracts (JONO 1930, CONTARDI and RAVAZZONI 1934) until recently there has been no indication that this action was the property of a specific phosphatase. More recently Kornberg and Pricer (1950 a) have claimed a 50-fold purification, on a protein nitrogen basis, of a 5'-AMP hydrolyzing enzyme. During this purification the activity against 3'-AMP increased only five times. At p_H 5.0 the preparation was only twice as effective against 5'-AMP as against 3'-AMP. However, at p_H 9.4 the cleavage of 2'- and 3'-AMP was completely inhibited while only two-thirds of the 5'-AMP activity was lost. A rigorous proof of the existence of a specific 5'-nucleotidase would depend on the isolation of an enzymically homogenous product. The fact that different substrates for the same enzyme have differing values for K_M, differing p_H dependence for K_M and activity and qualitatively different metal requirements always stands as a barrier in the interpretation of such experiments, especially with impure preparations carrying indeterminate amounts of activators and inhibitors. Whatever be the interpretation Kornberg and Pricer have nevertheless provided biochemists with a specific reagent for the cleavage of 5'-AMP.

c) 3'-Nucleotidase.

Shuster and Kaplan (1953) have found this enzyme in a number of plants including barley, rye, wheat, corn, oats, soybeans, and rye grass. The last is the richest source. The enzyme dephosphorylates the following 3'-nucleotides: adenylic, inosinic, cytidylic, uridylic and guanylic. Adenylic acid was the best substrate. The purified preparation showed little or no action against glycerol phosphate and the sugar phosphates. However, these substrates were present at concentrations of one micromole per ml. when tested, a concentration far below the K_M values found for these substrates with ordinary "non-specific" phosphatases. The enzyme acted optimally at p_H 7.5. Metal activation was not observed. The purified preparation has proven to be useful in discriminating between 2'- and 3'-nucleotides. Thus, for example, it has been used to show that the adenylic acid in coenzyme A is phosphorylated in the 3' position.

d) Apyrase.

Following the interest in ATP-splitting enzymes which was aroused by the discovery that muscle contraction was accompanied by ATP hydrolysis, Kalckar (1944) concentrated an enzyme from potato tubers which also cleaved ATP. Unlike the actomyosin reaction, the potato enzyme removed both the terminal and penultimate phosphate. Krishnan (1949) purified this enzyme and verified its apyrase activity. Even after 1000 fold purification it still retained traces of "non-specific" phosphatase. It exhibited a broad p_H optimum in the region of 6.5 and was responsive to Ca^{++}. Lee and Eiler (1951) found at low temperatures a discontinuity in the curve of phosphate release vs. time which was due to the slower release of the middle phosphate. At 7^0 C and below only the terminal phosphate was hydrolyzed.

Eggman, T'so and Vinograd (1955) have examined the actomyosin-like protein of slime mold and found that it reacts with ATP to undergo a reversible decrease in viscosity. On recovery, Pi is liberated, analogously to the behavior of muscle actomyosin. This would appear to be ATP-ase activity.

E. Diesterases.

a) Ribonucleases.

Diesterases which hydrolyze diphenyl phosphate have been detected in higher plants, e. g. rice bran (Uzawa 1932) and wheat (Booth 1944). Enzymes which hydrolyze nucleic acids are also diesterases and these too have been recognized in plants. Whether these are distinct enzymes or actually one has not been adequately examined. WHITFIELD, HEPPEL and MARKHAM (1955) obtained a preparation from rye grass which hydrolyzed the inner diester, cyclic adenosine-2′,3′-phosphate to yield both the 2′- and 3′-adenylic acids. The latter compound was further hydrolyzed to adenosine presumably by a 3′-nucleotidase which was also present. The diesterase is presumably the ribonuclease which SHUSTER and KAPLAN (1953) had already detected in rye grass preparations. BROWN, HEPPEL and HILMOE (1954) found that a similar preparation (as well as one from potatoes) split benzyl alcohol from the diesters, adenosine-3′ benzyl phosphate and cytidine-3′-benzyl phosphate. Ribonucleases of pancreas and spleen can hydrolyze the appropriate diesters of simple nucleotide in addition to nucleic acid and they can also form diesters by transfer to mononucleotides. Plant ribonucleases have not been yet tested for their ability to transfer.

Ribonuclease was first discovered in a variety of plants including soybeans by JONO (1930). SCHLAMOWITZ and GARNER (1946) decided from a partial purification of this enzyme from soybeans that ribonuclease and phosphatase were distinct enzymes. They found the enzyme content of the soybeans rose on germination, reaching a maximum in 96 hours. BERNHEIMER and STEELE (1955) found the ribonuclease in a number of higher plants. Lilac leaves and privet leaves contained in addition an inhibitor of pancreatic and asclepais ribonuclease.

HOLDEN and PIRIE (1955 b) also found ribonuclease in a great number of other higher plants and studied it quite extensively. Pea seeds, initially low in both ribonuclease and "phosphatase", showed a marked gain after 2 to 6 days germination. Upon purification of ribonuclease from pea seedlings there was a decrease in phosphatase activity, but the latter activity was never completely eliminated. The ribonuclease hydrolyzed yeast nucleic acid more completely than the pancreatic enzyme; the latter enzyme always leaves a refractory "core". PIRIE (1950) had already shown that tobacco leaf ribonuclease left no "core" when acting on ribonucleic acid. HOLDEN and PIRIE (1955 a) found that the pea leaf enzyme did not possess the remarkable stability to heat of the pancreatic enzyme which permits the latter to be heated at 100⁰ for 20 minutes with only a slight loss of activity. They noted that the pea enzyme was more stable to heat than most enzymes but that at 88⁰ it was completely inactivated. In contrast, AXELROD (1947) found that the ribonuclease which accompanied citrus phosphatase resisted boiling. Cu^{++}, Fe^{++}, Mn^{++}, Pb^{++}, Zn^{++} and Mg^{++} inhibited the pea seedling enzyme more strongly than the pancreatic enzyme.

b) Desoxyribonuclease.

GREENSTEIN and JENRETTE (1941) demonstrated the existence of desoxyribonuclease in seedlings of maize, wheat, cucurbit, sunflower and lima bean. FRISCH-NIGGEMEYER, KECK, KALJUNEN and HOFMANN-OSTENHOF (1951) reported a similar enzyme in garlic, but they could not demonstrate it in onion. The enzyme, which had a p_H optimum of 6.5, was resistant to brief exposure to 100⁰. BRAWERMAN and CHARGAFF (1954) reported desoxyribonuclease in germinating barley, which is also a source of phosphodiesterase. HOLDEN and PIRIE (1955 b) detected a weak desoxyribonuclease activity in the preparation discussed above. It had a p_H optimum of 5.4 and attacked not only desoxyribonucleic acid prepared from thymus, but it also hydrolyzed the oligonucleotides formed by the action of streptokinase-streptodornase (a bacterial nuclease). They believed the plant desoxyribonuclease to be an enzyme distinct from ribonuclease.

c) Nucleotide pyrophosphatase.

This enzyme, purified 750-fold from potatoes (Kornberg and Pricer 1950a), cleaves a number of pyrophosphate esters which may be regarded as in a special classification of diesters. The substrates include: diphosphopyridine nucleotide (reduced and oxidized), triphosphopyridine nucleotide, flavin adenine dinucleotide, ADP, ATP and thiamin pyrophosphate. The purified enzyme although slightly contaminated with other phosphatases, is in practice a useful reagent for cleaving these diesters. It has a p_H optimum of 7.4 when acting on DPN. Metal activation and F^- inhibition are erratic.

F. Mutases.

Two mutases are known to be normally involved in yeast and animal glycolysis:

(1) Phosphoglucomutase

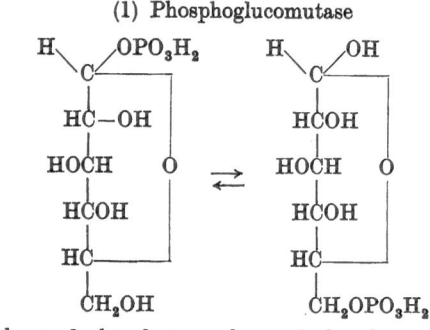

glucose-1-phosphate   glucose-6-phosphate

(2) Phosphoglyceromutase

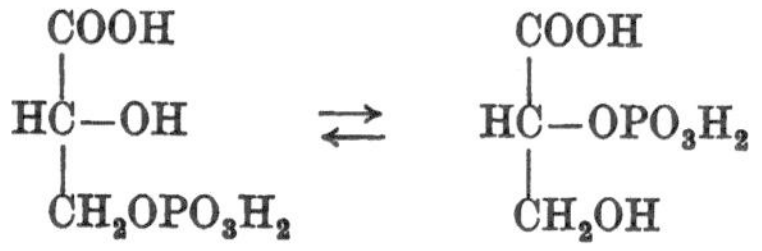

D-3-phosphoglyceric acid   D-2-phosphoglyceric acid

It has been shown that glucose-1,6-diphosphate (Leloir, Trucco, Cardini, Paladini and Caputto 1948) and 2,3-diphosphoglycerate (Sutherland, Poster-nak and Cori 1949) are the coenzymes for these enzymes, respectively.

The first enzyme has long been known in plants (Hanes 1939). Cardini (1951) has established that glucose-1,6-diphosphate is also required, using the enzyme from jackbean. The enzyme, as partially purified from *Phaseolus radiatus* is activated by Mg^{++}, Mn^{++}, Co^{++} and Na_2SO_3 (Ramasarma, Sri Ram and Giri 1954). It is not activated by cysteine, and has a p_H optimum of 7.5. These characteristics are typical of most of the plant preparations of this enzyme.

There is as yet no direct evidence for the presence of phosphoglyceromutase in higher plants although there seems little reason to doubt that it could be readily demonstrated. Stumpf (1950) noting that 3-PGA was converted to pyruvic acid by pea acetone powder extracts assumed the mutase was present.

Both of these mutase reactions may be regarded as special instances of transfer. While formally the reaction appears to be an intramolecular transfer, the discovery of the participation of the coenzyme permitted the following elucidation of the true mechanism.

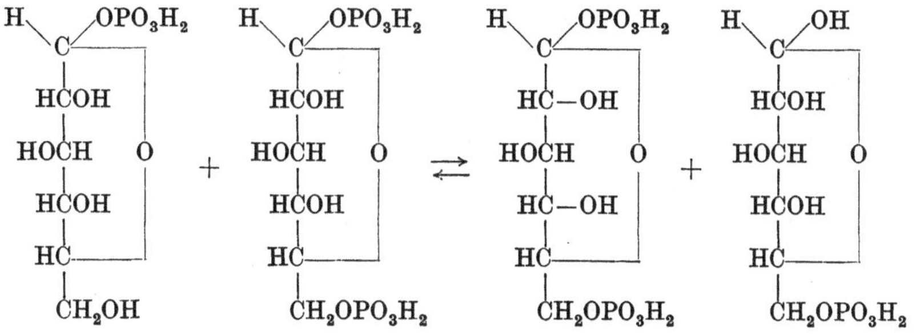

G. Mechanism of phosphate transfer.

The feature which is common to phosphatase activity (transfer to water) "phospho transferase" activity (transfer to an acceptor substrate), mutase and kinase activities is the cleavage of the phosphate bond. In considering the generalized substrate

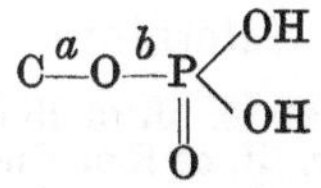

$$C \overset{a}{-} O \overset{b}{-} P \underset{OH}{\overset{OH}{\Big<}}$$
$$\parallel$$
$$O$$

one may ask whether the bond is broken at a or b. This can be decided both directly and indirectly. Thus, where phosphatase hydrolyzes N-phospho compounds or where creatine phosphate reacts with ADP (creatine kinase) the cleavage must occur at the bond connected to P. By extrapolation one could presume that in the case of P—O esters the bond b is cleaved. This inference has been borne out in all cases tested, by using O^{18} as a label. A few cases that may be cited are as follows: alkaline phosphatase (COHN 1949, STEIN and KOSHLAND 1952), acid phosphatase (COHN 1949), acetyl phosphatase (BENTLEY 1949), phosphoglyceryl kinase (COHN 1953, BOYER and HARRISON 1954), hexokinase (COHN 1955). Although in all of these cases animal or yeast enzymes were used, it is not unreasonable to expect the conclusions to apply to enzymes of plant origin.

From the above and from other less direct considerations a reasonable case can be made for the view that all of the reactions considered may be formulated by the following type from reaction:

$$A-PO_3H_2 + \text{Enzyme} \rightarrow \text{Enzyme}-PO_3H_2 + A$$
$$\text{Enzyme}-PO_3H_2 + B \rightarrow B-PO_3H_2 + \text{Enzyme}.$$

Because the cleavage occurs at the bond attached to P, the group which is transferred is a phosphoryl group, one which cannot be expected to have a free existence as such under physiological conditions. Hence, it must be transported by being passed on from donor to enzyme to acceptor.

a) In hydrolysis B is H_2O.

b) When B is an alcohol then one may have an instance of "phosphotransferase".

c) When A is ADP the enzyme will be a phosphatase if B is water, or a kinase if B is a specific acceptor. If $B-PO_3H_2$ and $A-PO_3H_2$ are position isomers, the enzyme is a mutase.

For instance, citrus phosphatase is relatively non-specific with regard to A or B. On the other hand, onion phosphatase is non-specific with regard to A, but apparently requires that B be H_2O. Hexokinase utilizes only ATP and only a limited number of compounds (certain sugars) as acceptors. Of course, it may be discovered that not all kinases are restricted exclusively to ATP.

The following representation of the general mechanism in terms of modern organic mechanistics is due to KOSHLAND (1953, 1954).

$$E: + \quad \underset{O-\delta}{\overset{HO+\delta}{P}}\!\!\overset{OR}{\underset{OH}{}} \quad \rightarrow \quad E:\cdots\cdots\underset{O}{\overset{HO}{P}}\!\!\overset{:OR}{\underset{OH}{}} \quad \rightarrow \quad E\!-\!\underset{O}{\overset{OH}{P}}\!\!\overset{}{\underset{OH}{}} \quad +\; :OR$$

$$E\cdots\cdots\underset{O}{\overset{OH}{P}}\cdots\cdots:x \quad \rightarrow \quad E:\; + \quad \underset{O}{\overset{OH}{P}}\!\!\overset{OH}{\underset{x}{}}$$

A more detailed review of enzymatic phosphate (more properly "phosphoryl") transfer is available (AXELROD 1956).

Literature.

AXELROD, B.: Citrus fruit phosphatase. J. of Biol. Chem. 167, 57–72 (1947). — A new mode of enzymatic phosphate transfer. J. of Biol. Chem. 172, 1–13 (1948a). — A study of the mechanism of "phosphotransferase" activity by the use of radioactive phosphorus. J. of Biol. Chem. 176, 295–298 (1948b). — The free energy of hydrolysis of p-nitrophenyl phosphate. Science (Lancaster, Pa.) 114, 525–526 (1951). — Enzymatic phosphate transfer. Adv. Enzymol. 17, 159–188 (1956). — AXELROD, B., and R. S. BANDURSKI: Phosphoglyceryl kinase in higher plants. J. of Biol. Chem. 204, 939–948 (1953). — AXELROD, B., R. S. BANDURSKI, C. M. GREINER and R. JANG: The metabolism of hexose and pentose phosphates in higher plants. J. of Biol. Chem. 202, 619–634 (1953). — AXELROD, B., P. SALTMAN, R. S. BANDURSKI and R. S. BAKER: Phosphohexokinase in higher plants. J. of Biol. Chem. 197, 89–96 (1952).

BANDURSKI, R. S.: Further studies on the enzymatic synthesis of oxalacetate from phosphorylenolpyruvate and carbon dioxide. J. of Biol. Chem. 217, 137–149 (1955). — BANDURSKI, R. S., and C. M. GREINER: The enzymatic synthesis of oxalacetate from phosphorylenolpyruvate and carbon dioxide. J. of Biol. Chem. 204, 781–786 (1953). — BENSON, A. A.: Identification of ribulose in $C^{14}O_2$ photosynthesis products. J. Amer. Chem. Soc. 73, 2971–2972 (1951). — BENTLEY, R.: The mechanism of hydrolysis of acetyl dihydrogen phosphate. J. Amer. Chem. Soc. 71, 2765–2767 (1949). — BERG, P., and W. K. JOKLIK: Transphosphorylation between nucleoside polyphosphates. Nature (Lond.) 172, 1008–1009 (1953). — BERNHEIMER, A. W., and J. M. STEELE jr.: Ribonuclease and ribonuclease inhibitors among higher plants. Proc. Soc. Exper. Biol. a. Med. 89, 123–126 (1955). — BOOTH. R. C.: Cereal phosphatases. 1. The assay of free wheat phosphomonoesterase and characterization of the free phosphatases of wheat. Biochemic. J. 38, 355–362 (1944). — BOROUGHS, H.: Studies on the acid phosphatase of green leaves. Arch. of Biochem. a. Biophysics 49, 30–42 (1954). — BOYER, P. D., W. H. HARRISON, H. E. ROBERTSON and E. R. ROBBINS: Unpublished work cited in The Mechanism of Enzyme Action, 660–661. Baltimore: Johns Hopkins Press 1954. — BRAWERMAN, G., and E. CHARGAFF: Enzymatic phosphorylation of nucleosides by phosphate transfer. J. Amer. Chem. Soc. 75, 2020–2021 (1953a). — Nucleotide synthesis by malt and prostate phosphatases. J. Amer. Chem. Soc. 75, 4113 (1953b). — On a desoxyribonuclease from germinating barley. J. of Biol. Chem. 210, 445–454 (1954). — On the distribution and biological significance of the nucleoside phototransferases. Biochim. et Biophysics Acta 16, 524–532 (1955). — BROWN, D. M., L. A. HEPPEL and R. J. HILMOE: Nucleotides. Part XXIV. The action of some nucleases on simple esters of monoribonucleotides. J. Chem. Soc. (Lond.) 1954, 40–52. — BUCHANAN, J. G.: The path of carbon in photosynthesis. XIX. The identification of sucrose phosphate in sugar beet leaves. Arch. of Biochem. a. Biophysics 44, 40–52 (1953).

CAMPBELL, J. M., and R. S. BANDURSKI: Adenylate kinase in higher plants. Abstract. Annual Meeting Amer. Soc. Plant Physiologists, Ithaca, New York 1952. — CARDINI, C. E.: Activation of plant phosphoglucomutase by glucose-1,6-diphosphate. Enzymologia (Den Haag) 15, 44–48 (1951). — COHN, M.: Mechanisms of cleavage of glucose-1-phosphate. J. of Biol. Chem. 180, 771–781 (1949). — A study of oxidative phosphorylation with O^{18}-labeled inorganic phosphate. J. of Biol. Chem. 201, 735–777 (1953). — Mechanism of conversion of inorganic phosphate to organic phosphate. (Abstr.) p. 36 C. National Meeting. Amer. Chem. Soc., Cincinnati, Ohio 1955. — CONTARDI, A., and C. RAVAZZONI: Enzymatic cleavage of yeast nucleic acid. Arch. ital. Biol. **92**, 64–75 (1934). — COURTOIS, J.: Investigations on phytase. III. Attempted separation of glycerophosphatase and phytase activities of wheat bran. Biochim. et Biophysics Acta **1**, 270–277 (1947).

EGGMAN, L., P. O. P. T'So and J. VINOGRAD: On the nature of myxomysin—an actomysin-like protein from slime mold plasmodia. Plant Physiol. **30**, xvi (1955).

FLEURY, P., and J. COURTOIS: Investigations on phytase. II. Comparative kinetics of the hydrolysis of glycerophosphate and inositol hexaphosphate by wheat bran. Biochim. et Biophysics Acta **1**, 256–269 (1947). — FRISCH-NIGGEMEYER, W., K. KECK, H. KALJUNEN and O. HOFFMANN-OSTENHOF: A hitherto unknown desoxyribonuclease. Mh. Chem. **82** 758–760 (1951).

GREENSTEIN, J. P., and W. V. JENRETTE: Ribonuclease and thymonucleodepolymerase. J. Nat. Canc. Inst. **2**, 301–303 (1941).

HANES, C. S.: Breakdown and synthesis of starch by an enzyme system from pea seeds. Proc. Roy. Soc. Lond. Ser. B **128**, 421–450 (1940). — HOFFMANN-OSTENHOF, O.: Suggestions for a more rational classification and nomenclature of enzymes. Adv. Enzymol. **14**, 219–260. (1953). — HOLDEN, M., and N. W. PIRIE: A comparison of leaf and pancreatic lipase. Biochemic. J. **60**, 53–62 (1955a). — The partial purification of leaf ribonuclease. Biochemic. J. **60**, 39–46 (1955b).

JONO, Y.: Nucleic acid. I. Enzymes which split nucleic acid. Acta Scholae med. Kioto **13**, 162–175 (1930).

KACHMAR, J. F., and P. D. BOYER: Kinetic analysis of enzyme reactions. II. The potassium activation and calcium inhibition of pyruvic phosphoferase. J. of Biol. Chem. **200**, 669–682 (1953). — KALCKAR, H. M.: Adenyl pyrophosphatase and myokinase. J. of Biol. Chem. **153**, 355–367 (1944). — KAUFMAN, S.: Studies on the mechanism of the reaction catalyzed by the phosphorylating enzyme. J. of Biol. Chem. **216**, 153–164 (1955). — KORNBERG, A., and W. E. PRICER jr.: On the structure of triphosphopyridine nucleotide. J. of Biol. Chem. **186**, 557–567 (1950a). — Nucleotide pyrophosphatase. J. of Biol. Chem. **182**, 763–778 (1950b). — KOSHLAND jr., D. E.: Stereochemistry and the mechanism of enzymic reactions. Biol. Rev. Cambridge Philos. Soc. **28**, 416–436 (1953). — Mechanism of Enzyme Action. Baltimore: Johns Hopkins Press 1954. — KREBS, H. A., and R. HEMS: Some reactions of adenosine and inosine phosphates in animal tissue. Biochim. et Biophysica Acta **12**, 172–180 (1953). — KRISHNAN, P. S.: Studies on apyrase. I. Purification of potato apyrase by fractional precipitation with ammonium sulfate. Arch. of Biochem. **20**, 261–272 (1949).

LARDY, H. A., and J. A. ZIEGLER: The enzymatic synthesis of phosphopyruvate from pyruvate. J. of Biol. Chem. **159**, 343–351 (1945). — LEE, K., and J. J. EILER: Temperature-dependent characteristics of an adenylpyrophosphatase preparation from potatoes. Science (Lancaster, Pa.) **114**, 393–395 (1951). — LELOIR, L. F., and C. E. CARDINI: The biosynthesis of sucrose phosphate. J. of Biol. Chem. **214**, 157–165 (1955). — LELOIR, L. F., R. E. TRUCCO, C. E. CARDINI, A. PALADINI and R. CAPUTTO: The coenzyme of phosphoglucomutase. Arch. of Biochem. **19**, 339–340 (1948). — LEVINTOW, L., A. MEISTER, G. H. HOGEBOOM and E. L. KUFF: Studies on the relationship between the enzymatic synthesis of glutamine and the glutamyl transfer reaction. J. Amer. Chem. Soc. **77**, 5304–5308 (1955). — LIPMANN, F., and L. C. TUTTLE: A specific micromethod for the determination of acyl phosphates. J. of Biol. Chem. **159**, 21–28 (1945). — LOOMIS, W. D., and P. K. STUMPF: Phosphorus Metabolism, vol. 2, p. 29. Baltimore: Johns Hopkins Press 1952.

MCCREADY, R. M., and W. Z. HASSID: Transformation of sugars in excised barley shoots. Plant Physiol. **16**, 599–610 (1941). — MEYERHOF, O., and H. GREEN: Synthetic action of phosphatase. II. Transphosphorylation by alkaline phosphatase in the absence of nucleotides. J. of Biol. Chem. **183**, 377–390 (1950). — MILLERD, A., J. BONNER, B. AXELROD and R. S. BANDURSKI: Oxydative and phosphorylative activity of plant mitochondria. Proc. Nat. Sci. U.S.A. **37**, 855–862 (1951).

NEGELEIN, E.: Cit. in F. KUBOWITZ and P. OTT, Isolation of fermentation enzymes from human muscle. Biochem. Z. **317**, 193–203 (1944).

PIRIE, N. W.: The isolation from normal tobacco leaves of nucleoprotein with some similarity to plant viruses. Biochemic. J. **47**, 614–625 (1950).

RACKER, E.: Synthesis of carbohydrates from carbon dioxide and hydrogen in a cell-free system. Nature (Lond.) **175**, 249–251 (1955). — RAMASARMA, T., J. SRI RAM and K. V. GIRI: Phosphoglucomutase of green gram *(Phaseolus radiatus)*. Arch. of Biochem. a. Biophysics **53**, 167–173 (1954).

SALTMAN, P.: Hexokinase in higher plants. J. of Biol. Chem. **200**, 145–154 (1953a). — Enzymatic phosphate transfer in plant systems. Doctoral thesis. California Institute of Technology, Pasadena, California 1953b. — SANADI, D. R., D. M. GIBSON and P. AYENGAR:

236 BERNARD AXELROD: Transphosphorylation (phosphotransferases, mutases and kinases).

Guanosine triphosphate, the primary product of phosphorylation coupled to the breakdown of succinyl coenzyme A. Biochim. et Biophysica Acta 14, 434–436 (1954). — SCHLAMOWITZ, M., and R. L. GARNER: The ribonucleinase of the soybean. J. of Biol. Chem. 163, 478–497 (1946). — SHUSTER, L., and N. O. KAPLAN: A specific b nucleotidase. J. of Biol. Chem. 201, 535–546 (1953). — STEIN, S. S., and D. E. KOSHLAND jr.: Mechanism of action of alkaline phosphatase. Arch. of Biochem. a. Biophysica 39, 229–230 (1952). — STUMPF, P. K.: Carbohydrate metabolism in higher plants. J. of Biol. Chem. 182, 261–272 (1950). — STUMPF, P. K., W. D. LOOMIS and C. MICHELSON: Amide metabolism in higher plants. 1. Preparation and properties of a glutamyl transphorase from pumpkin seedling. Arch. of Biochem. 30, 126–137 (1951). — SUTHERLAND, E. W., J. POSTERNAK and C. F. CORI: Mechanism of the phosphoglyceric mutase reaction. J. of Biol. Chem. 181, 153 (1949). — SUZUKI, U., Y. YOSHIMURA and M. TAKAISHI: On the occurrence of an enzyme which decomposes anhydroxymethylenephosphoric acid. Tokyo Chem. Soc. 27, 1330–1342 (1906).

TANKO, B.: Hexosephosphates produced in higher plants. Biochemic. J. 30, 692–700 (1936).

UZAWA, T.: The phosphomonoesterases of bran. J. of Biochem. (Tokyo) 15, 1–10 (1932).

WEISBACH, A., P. Z. SMYRNIOTIS and B. L. HORECKER: The enzymatic formation of ribulose diphosphate. J. Amer. Chem. Soc. 76, 5572–5573 (1954). — WHITFIELD, P. R., L. A. HEPPEL and R. MARKHAM: The enzymatic hydrolysis of ribonucleoside-2:3-phosphates. Biochemic. J. 60, 15–19 (1955). — WILKINSON, J. F.: The pathway of the adaptive fermentation of galactose by yeast. Biochemic. J. 44, 460–476 (1949).

The role of the phosphates in energy transfer (adenosine phosphates).

By

Robert S. Bandurski and **Te May Ching.**

With 1 figure.

I. Energy transfer function of adenosine phosphates.

A. Introduction.

It is a remarkable fact that for many years prior to the discovery of the participation of phosphate in carbohydrate metabolism, and even for a time thereafter, that little attention was paid to the mechanism of energy transfer in biological reactions. We recognize today that the respiratory reactions must supply not only the carbon skeletons used by the organism in the synthesis of body substances, but also some provision must be made for the capture and transference of energy liberated during the respiratory oxidation.

Thus, when glucose is oxidized to acetic acid the acetic acid carbons may subsequently be utilized to form the carbon skeleton of fats. Energy liberated during the oxidation is captured as phosphate bond energy in a form in which it may be utilized by energy requiring reactions. It is precisely this role, that of energy capture and transfer, which is filled by the nucleoside phosphates such as adenosine triphosphate (ATP)[1].

The pyrophosphate bonds of the nucleoside phosphate may, in a very real sense, be visualized as an energy storage system. This energy once stored may subsequently be utilized in reactions requiring the input of energy. Thus, the pyrophosphate bonds of nucleoside phosphates are a form of energy "currency". They are made by reactions in which energy is liberated and "spent" in reactions in which energy is required. Since so many of the metabolic reactions of the cell involve ATP, the "currency" is a freely negotiable one analogous to an international money order or bank draft.

In this chapter an attempt will be made to briefly discuss and classify some of the principal ATP making and ATP utilizing reactions. In addition the linkage of other nucleoside phosphate systems to the adenosine nucleotides will be discussed. The reader is also referred to more extensive recent reviews of the nucleotides by CHARGAFF and DAVIDSON and by CARTER.

B. Chemistry of the 5'-nucleoside phosphates.

Purine and pyrimidine glycosides are know as nucleosides. Biologically the nucleosides exist, for the most part, as phosphoric esters know as nucleotides. The nucleotides consist of a pyrimidine or purine base, a sugar (D-ribose or

---

[1] The following abbreviations will be used throughout the present chapter. ATP, adenosine triphosphate; ADP, adenosine diphosphate; AMP, adenosine monophosphate, CoA, coenzyme A; DPN+, diphosphopyridine nucleotide; G-5-P, guanosine-5'-phosphate; GDP, guanosine diphosphate; ITP, inosine triphosphate; IDP, inosine diphosphate; IMP, inosine monophosphate; PP, inorganic pyrophosphate; PPP, inorganic triphosphate.

D-2-deoxyribose), and one or more phosphoric acids. The energy transferring nucleotides, so far as present knowledge extends, are limited to D-ribose and the phosphoric acids are linked to the 5′-position of the ribose[1]. The structures and some useful physical properties of the known energy transferring nucleoside phosphates are summarized below.

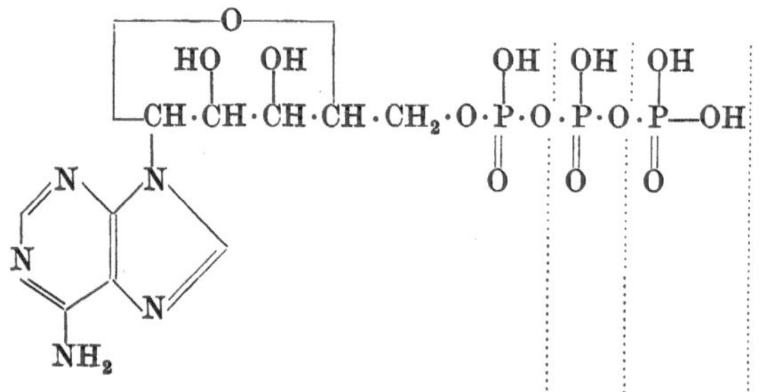

(AMP) adenosine-5′-phosphate (adenylic acid)
(ADP) adenosine-5′-diphosphate
(ATP) adenosine-5′-triphosphate

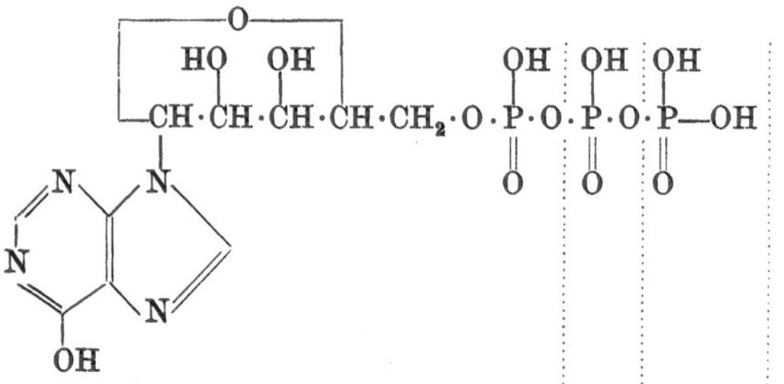

(IMP) inosine-5′-phosphate (inosinic acid)
(IDP) inosine-5′-diphosphate
(ITP) inosine5′-triphosphate

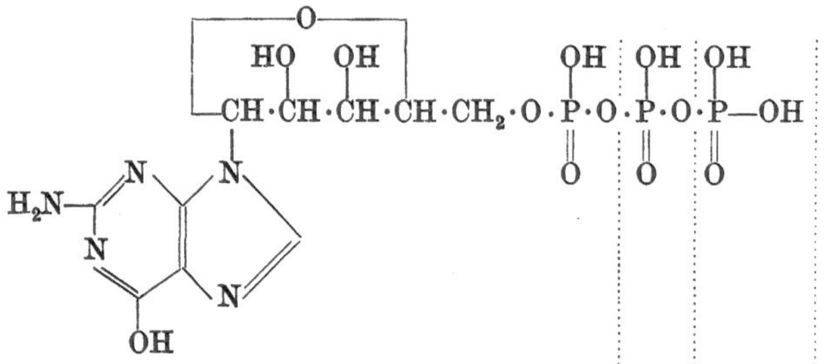

(GMP) guanosine-5′-phosphate
(GDP) guanosine-5′-diphosphate
(GTP) guanosine-5′-triphosphate

In Table 1 some useful physical properties of the nucleotides are summarized. The molecular weights given are for the sodium salts or free acid as indicated in the table and includes the indicated waters of hydration. The wave length of

---

[1] An exception seems to be the adenosine-3′-phosphate-5′-phosphosulfate recently described by Robbins and Lipmann.

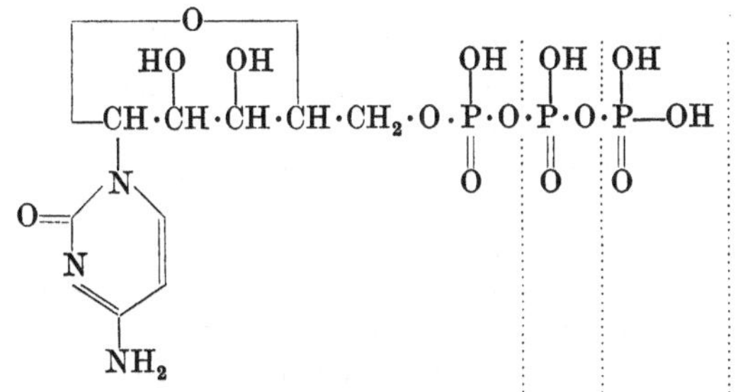

(CMP) cytidine-5′-phosphate
(CDP) cytidine-5′-diphosphate
(CTP) cytidine-5′-triphosphate

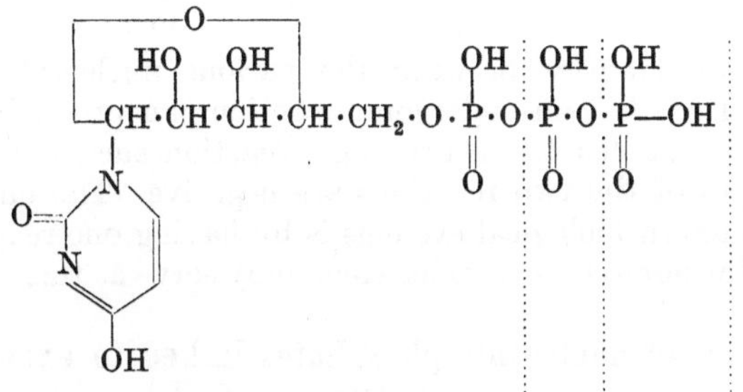

(UMP) uridine-5′-phosphate
(UDP) uridine-5′-diphospahte
(UTP) uridine-5′-triphosphate

maximum absorbancy, at p_H 7, is also given. The molar absorbancy index provides a useful means of measuring nucleotide concentration. The values $\frac{250}{260}$ and $\frac{280}{260}$ represent the ratios of absorbancy at the indicated wavelengths (in mμ) and are helpful in identifying the nucleotides.

Table 1. *Some useful physical properties of the nucleotides*[1].

		Molecular weight	λ max. p_H 7 (mμ)	$\alpha_m \times 10^{-3}$	$\frac{250}{260}$	$\frac{280}{260}$
ATP	Na$_2$H$_2$ATP · 4 H$_2$O . . .	623	259	15.4	0.80	0.15
ADP	NaH$_2$ADP · 2 H$_2$O . . .	485	259	15.4	0.78	0.16
AMP	AMP · 2 H$_2$O.	383	259	15.4	0.79	0.16
GTP	Na$_2$H$_2$GTP · 2 H$_2$O . . .	603	252	13.7	1.17	0.66
GDP	NaH$_2$GDP · 2 H$_2$O . . .	501	252	13.7	1.15	0.66
GMP	Na$_2$GMP · H$_2$O	425	252	13,7	1.16	0.66
CTP	Na$_2$H$_2$CTP · 4 H$_2$O . . .	599	271	9.1	0.84	0.97
CDP	Na$_2$HCDP · 4 H$_2$O . . .	519	271	9.2	0.83	0.98
CMP	Na$_2$CMP · 2 H$_2$O	403	271	9.0	0.84	0.98
UTP	Na$_3$HUTP · 2 H$_2$O . . .	586	262	10.0	0.75	0.38
UDP	Na$_2$HUDP · 3 H$_2$O . . .	502	262	10.0	0.73	0.39
UMP	Na$_2$UMP · 2 H$_2$O	404	262	10.0	0.73	0.39

α_m = molar Absorbancy Index = As/c, As = Absorbancy = $-\log_{10}$T, c = concentration in moles per liter, T = transmission of the sample in a 1 cm. cell with reference to a solvent blank set at T = 1.

[1] Based upon a compilation of data, "Ultraviolet Absorption Spectra of 5′-Ribonucleotides", published by the Pabst Laboratories, Division of Pabst Brewing Company, 1037 W. McKinley Ave., Milwaukee 5, Wisconsin, USA.

C. Transphosphorylation as an energy transfer mechanism.

For convenience, we propose to divide metabolic reactions involving adenosine phosphates, and to a lesser extent other nucleoside phosphates, into three groups: 1. *Exergonic* reactions in which the free energy change of reaction is captured as phosphate bond energy. 2. Reactions in which the free energy of hydrolysis of a pyrophosphate bond of ATP is utilized to permit an otherwise *endergonic* reaction to occur. 3. *Isergonic* reactions involving phosphate transfer from one nucleoside phosphate to another in which little or no overall free energy change results.

The term exergonic was introduced by Coryell to signify a reaction which can produce work (*i. e.* $-\Delta F$ is positive at constant pressure and temperature, ΔF is negative) while endergonic would refer to reactions on which work must be expended to cause the reaction to go ($-\Delta F$ is negative at constant temperature and pressure, ΔF is positive). The term isergonic is introduced here to describe reactions with little of no free energy change and will be restricted to reactions involving phosphate transfer between the various nucleoside phosphates. It should be emphasized that an endergonic reaction can proceed to a measurable extent only if it is coupled to an exergonic reaction such that the sum of the free energy changes of the two reactions are negative. The only way in which such coupling occurs in biological systems is by having one reactant common to both reactions. The adenyl nucleotides commonly serve as this common reactant.

1. Phosphorylation of nucleoside phosphates linked to exergonic reactions.

The best understood example of ATP synthesis coupled to an exergonic reaction is the oxidation of 3-glyceraldehyde phosphate to 3-phosphoglyceric acid with the concomitant formation of ATP. Less well understood is the formation of ATP coupled to the respiratory oxidations of the Krebs cycle. Both of these reactions are discussed in greater detail in Vol. XII[1]. They may in general be formulated as:

$$XH_2 + P_i + ADP + Y \rightleftharpoons X + YH_2 + ATP \tag{1}$$

where: XH_2 is the oxidizable substrate, Y is a hydrogen acceptor, such as DPN^+, YH_2 is the reduced hydrogen acceptor, X is the oxidized substrate, P_i is orthophosphate.

As might be expected this reaction does not occur in one step but may proceed as follows:

$$XH_2 + Y + Enzyme \rightleftharpoons X\text{-}Enzyme + YH_2 \tag{2}$$

$$X\text{-}Enzyme + P_i \rightleftharpoons XP_i + Enzyme \tag{3}$$

$$XP_i + ADP \rightleftharpoons X + ATP \tag{4}$$

Reaction 1 above is simply the sum of reactions 2, 3 and 4.

The essential points concerning this reaction are: 1. the retention of the free energy change of oxidation of XH_2 in a bond between X and enzyme or X and some other acceptor; 2. Phosphorolysis of the bond between X-enzyme or X-acceptor to form X-phosphate; 3. Phosphorylation of ADP by X-phosphate. Thus the free energy change of the oxidation is captured and conserved as phosphate bond energy.

2. Dephosphorylation of nucleoside phosphates linked to endergonic reaction systems.

Synthetic reactions known to be driven by nucleoside phosphate bond energy would include sucrose synthesis (Leloir *et al.* 1953, 1955, Cardini *et al.*, Burma and Mortimer), peptide bond synthesis and amide formation (*cf.* Webster),

---

[1] The group potentials and the role of phosphates in energy transfer, by R. S. Bandurski.

β-carboxylations (UTTER, KURAHASHI and ROSE, BANDURSKI and LIPMANN), fatty acid synthesis (cf. LIPMANN 1954), methylations (CANTONI) and reduction of a carboxyl group to a carbonyl (Vol. XII). Owing to their diverse nature and to the complexity of the reaction sequence it is difficult to write a type reaction. In general, however, these reactions may be formulated as:

$$ATP + X + Y \rightarrow ADP + X\text{-}Y + P_i \tag{5}$$

The synthesis of acetoacetate will serve here as an example. The reaction proceeds as follows (BERG, LIPMANN):

$$2\ ATP + 2\ acetate \rightleftharpoons 2\ acetyl\text{-}AMP + 2\ pyrophosphate \tag{6}$$

$$2\ acetyl\text{-}AMP + 2\ CoA \rightleftharpoons 2\ AMP + 2\ acetyl\text{-}CoA \tag{7}$$

$$2\ acetyl\text{-}CoA \rightleftharpoons acetoacetyl\text{-}CoA + CoA \tag{8}$$

$$acetoacetyl\text{-}CoA + H_2O \rightarrow acetoacetate + CoA \tag{9}$$

$$2\ ATP + 2\ acetate \rightarrow acetoacetate + 2\ AMP + 2\ pyrophosphate \tag{10}$$

In reaction 6 a pyrophosphate is cleaved from ATP, the bond energy however being conserved in the new anhydride type compound, acetyl adenosine monophosphate (acetyl-AMP). The acetyl is transferred in reaction 7 to coenzyme A with the bond energy again being conserved, this time in the thiol ester linkage. In reaction 8 two acetyl CoA are condensed to form acetoacetyl CoA, the new carbon-carbon bond thus being formed at the expense of the thiolester bond. Finally acetoacetyl CoA is hydrolyzed to form free acetoacetate in reaction 9. Alternatively additional molecules of acetyl CoA could be condensed to acetoacetyl CoA to form longer chain fatty acids. The significant points of this type of reaction sequence appears to be, 1. the cleavage of phosphate or pyrophosphate from ATP to form an adenyl-X or enzyme-X intermediate and, 2. the cleavage of the adenyl-X or enzyme-X bond by the acceptor Y to form XY. Thus, the bond energy of ATP has been utilized to form a new bond, as shown in reaction 10.

3. Isergonic phosphate transfer reactions.
a) Nucleotide kinases.

The first enzymatic reaction to be discovered in which phosphate was reversably transferred from ATP to a nucleoside phosphate was the adenyl kinase reaction. COLOWICK and KALCKAR found yeast hexokinase catalyzed reaction 11:

$$ATP + glucose \rightarrow ADP + glucose\text{-}6\text{-}phosphate \tag{11}$$

If a heat stable protein from muscle was added to the yeast reaction mixture it was found that a second molecule of glucose-6-phosphate was formed utilizing the terminal phosphate of ADP. This finding was explained (KALCKAR) by the existence in muscle of a relatively heat stable enzyme, adenyl kinase (myokinase) which catalyzed the reaction shown in 12:

$$ATP + AMP \rightleftharpoons 2\ ADP \tag{12}$$

Equilibrium studies have shown that the reaction proceeds essentially with no change in free energy. Thus a mechanism is provided whereby both anhydride phosphates of ATP may be utilized. As will be discussed later the existence of this enzyme in plants is now well established.

In 1953 KREBS and HEMS and independently BERG and JOKLIK discovered the first of a series of reactions involved in the transference of phosphate from one nucleoside triphosphate to an acceptor nucleoside diphosphate. The enzyme, nucleoside diphosphokinase, catalyzed the reaction shown in Equation 13:

$$NTP + ADP \rightleftharpoons NDP + ATP \tag{13}$$

Where N is either inosine or uridine. Guanosine phosphate was later shown to be active by Sanadi et al.

Three nucleotide kinases analogous to adenyl kinase have now been described. Joklik discovered in yeast an enzyme catalyzing reaction 14:

$$2 IDP \rightleftharpoons ITP + IMP \tag{14}$$

while Lieberman et al. have described an enzyme also from yeast which catalyzes reaction 15:

$$NMP + NTP \rightleftharpoons NDP + NDP \tag{15}$$

in which NMP, NDP and NTP represent respectively the mono-, di- and triphosphates of adenosine, guanosine, or uridine. Ayengar et al., and Strominger et al., have described the reaction shown in Equation 16:

$$ATP + G\text{-}5\text{-}P \rightleftharpoons ADP + GDP \tag{16}$$

ATP and cytidine-5′-phosphate, and cytidine triphosphate and AMP will similarly react.

These enzymes, which may collectively be called nucleotide kinases, together constitute a system whereby the phosphate bond energy of ATP may readily be converted to phosphate bond energy in other nucleotides.

Thus a new metabolic transfer territory has emerged. The "currency" of any of the nucleoside phosphate systems may be converted into the "currency" of any other nucleoside phosphate system. The implications of this isergonic transfer system for regulatory mechanisms of the cell have yet to be investigated. An excellent review of these reactions has recently appeared by Kalckar and Klenow.

b) Inorganic pyrophosphates and metaphosphates.

The existance of enzymes catalyzing the transfer of phosphate from inorganic polyphosphates to adenyl nucleotides seems of great possible significance in the plant kingdom. A reservoir of "high energy" phosphate, as creatine or arginine phosphate, such as occurs in the animal kingdom has so far not been reported for plants. It seems possible, however, that the inorganic polyphosphates may constitute such a reservoir.

The existence of metaphosphate in higher plants and in yeast has been known for over sixty years (Hardin, Lieberman, Ascoli). Two enzymes have now been described which reversibly transfer phosphate from an adenyl nucleotide to inorganic polyphosphate. The first has been termed metaphosphate kinase by Hoffmann-Ostenhof et al. It occurs in yeast and catalyzes reaction 17:

$$ADP + (NaPO_3)_{n+1} \rightharpoonup ATP + (NaPO_3)_n \tag{17}$$

Adenosine monophosphate is inactive in this system. The reaction has so far been shown to proceed only in the direction shown.

Lieberman (1956) has recently discovered that inorganic triphosphate is synthesized by muscle adenylate kinase. The reaction is shown in Equation 18:

$$ADP + PP \rightleftharpoons AMP + PPP \tag{18}$$

ATP is inactive in this system.

WINTERMANS has described an accumulation of polyphosphate by *Chlorella* in light in the absence of carbon dioxide. He observed a disappearance of polyphosphate in the dark and suggested that the polyphosphate might serve as an energy reservoir. SCHMIDT has recently prepared an excellent review of the metabolism of inorganic polyphosphate to which the reader is referred for further discussion.

II. Occurrence and determination of nucleoside phosphates.

A. Occurrence of nucleoside phosphates.

1. Adenosine phosphates.

The development in recent years of the techniques of chromatography and electrophoresis together with the use of radio isotopes has considerably facilitated studies of the biological occurrence of the nucleotides. Nonetheless, our knowledge of the kinds and amounts of nucleotides in plants remains rudimentary. For example, little is at present known of the distribution of nucleotides within the various subcellular structures nor of the variations in kinds and amounts which may characterize stages in the life history of plants or the various organ functions. It may be hoped that with the chromatographic and electrophoretic techniques now available rapid progress will ensue in this field.

The first successful isolation and characterization of ATP from higher plant tissue was that of ALBAUM *et al.* (1950), using mung bean seedlings *(Phaseolus aureus)*. Earlier ALBAUM and OGUR (1947) had isolated an adenosine phosphate from *Avena* seedlings having an adenine to pentose to phosphorous ratio of $1:1:2$, corresponding to ADP. Following hydrolysis in normal HCl at $100^{\circ}C$ for seven minutes, one-half of the phosphorous was liberated as inorganic phosphorous. The adenosine monophosphate remaining seemed not to be identical with authentic muscle adenosine-5'-phosphate as it was not acted on by adenylic acid deaminase nor did it give the characteristic rate of color development for a ribose-5'-phosphate in the orcinol-pentose reaction.

Using mung bean seedlings ALBAUM, OGUR and HIRSHFELD did however succeed in isolating a compound identical with muscle ATP. Only a fraction of the total adenosine polyphosphate fraction was however identical with animal ATP, the remainder differing on the basis of solubility of its barium salt in dilute mineral acid. Thus, although higher plants do contain ATP, the exciting possibility remains that a substantial part of the adenosine polyphosphate of plant tissue differs structurally from ATP. A di-adenine polyphosphate such as earlier suggested by KIESSLING and MEYERHOF would seem a distinct possibility.

Some indirect evidence for the occurrence of adenosine phosphates in the tissues of higher plants is also available. AXELROD, BANDURSKI and SALTMAN observed that homogenates of pea seeds *(Pisum sativum)* incubated with radioactive phosphorous incorporated some of the radioactivity into a compound identified chromatographically as adenosine-5'phosphate. The existence in plant tissue of adenyl kinase which catalyzes reaction 12 above has been demonstrated by several workers (MAZELIS, *cf.* STUMPF). The occurence of this enzyme together with ATP and AMP would of course result in the formation of ADP.

2. Other nucleoside phosphates.

Just as the periodic table of the elements seems now capable of continued expension as techniques for synthesis of the elements are improved, so it seems that the list of metabolically active nucleoside phosphates will grow as analytical techniques improve. Two new adenosine phosphates have recently been described. LIEBERMAN (1955) has described a linear adenosine tetraphosphate[1] while SACKS has found an adenosine pentaphosphate. In recent papers by SCHMITZ, AYENGAR et al., and McLAUGHLIN et al., the isolation and characterization of the mono-, di- and tri-phosphates of adenosine, uridine, cytidine, thymidine and guanosine are described. The reader is referred to SCHMITZ for a partial review. None of these works deal with higher plants so that it must be said that our knowledge of 5'-nucleotides in the tissues of higher plants is still fragmentary[2].

B. The determination of adenosine phosphates.

Several excellent enzymatic assay methods are now available. Only a few of the most recent will be mentioned here since the reader will find in the bibliography of these papers, references to the earlier literature. BOWEN and KERWIN have described a very simple assay for ATP and ADP based upon phosphorous liberation by the combined action of myosin and adenyl kinase. Sufficient adenyl kinase activity is present in the crude myosin preparation so that only a single easily prepared enzyme fraction is required. Infibition of adenyl kinase activity with ethylenediaminetetraacetate makes the method specific for ATP. CHAPPELL and PERRY have recently described an elegant and sensitive assay using creatine phosphokinase. Alternatively the nucleotides may be separated by paper chromatography (cf. WYATT) or electrophoresis (MARKHAM and SMITH) and estimated by absorbancy in the ultra violet region (see I, B above).

III. Regulatory functions of the nucleoside phosphates.

The present chapter is restricted primarily to the role of the nucleoside phosphates in energy transfer. It is however important to mention that nucleoside phosphates also play an important regulatory role in the metabolism of the organism. We have been accustomed in plant physiology to consider mainly the importance of external factors in regulating the rate of metabolic reactions. There must obviously be internal regulatory devices which function to gear the rate of energy liberating reactions to the demands of energy utilizing reactions. The duality of function of the nucleoside phosphates, acting on the one hand as acceptors of energy from energy liberating reactions, and on the other, as energy donors for energy requiring reactions endows them with a unique capacity for regulatory function in the control of respiratory rates.

If, as is usually the case, respiratory oxidations are obligatorily coupled to the formation of ATP from ADP and orthophosphate, then the respiratory rate will be controlled by the concentration of ADP. As ATP is used in energy requiring reactions, ADP and orthophosphate are liberated; thus more ADP becomes available, permitting an increase in the respiratory rate.

Evidence for a role of AMP in the control of glycogen phosphorolysis rates in animal tissue has recently been presented by CORI. This scheme may act in addition to that discussed above. Phosphorylase exists in two forms, an active form, phosphorylase a and an inactive form, phosphorylase b. Phosphorylase b may be activated by the addition of AMP. Thus it would seem that under conditions of high muscle activity where the concentration of AMP might be

---

[1] Adenosine quatrephosphate would be a preferable term, permitting the non-ambiguous abbreviation AQP.

[2] The work of GINSBERG et al., has recently appeared. GINSBERG, V., P. K. STUMPF and W. Z. HASSID: The isolation of uridine diphosphate derivatives of D-glucose, D-galactose, D-xylose and L-arabinose from mung bean seedlings. J. of Biol. Chem. 223, 977—983 (1956).

expected to increase that conversion of phosphorylase b to phosphorylase a would occur. A more rapid rate of phosphorolysis of starch would thus ensue.

The exact elucidation of the role of nucleotides in governing rates of respiratory reactions will undoubtedly provide a field for future research.

IV. Summary.

A highly schematic summary of the main points of this discussion is given in Fig. 1. As can be seen exergonic reactions of the cell are coupled to the esterification of phosphate so that the free energy change of the reaction is not liberated as heat but is rather stored in the new phosphate bond. This phosphate may now be transferred to ADP to form ATP. The phosphate of ATP may be transferred

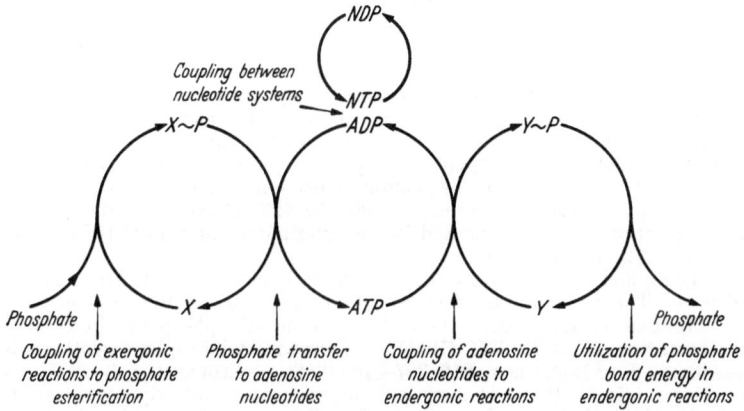

Fig. 1. Diagram of phosphate transfer reactions. X = phosphate acceptor for exergonic reactions. X $\sim$ P = phosphorylated form of acceptor. Y = phosphate acceptor for endergonic reactions. Y $\sim$ P = phosphorylated form of acceptor. NTP = nucleoside triphosphate, other than ATP. NDP = nucleoside diphosphate, other than ADP.

to any of a large number of acceptor molecules. The acceptor molecule, in its phosphorylated form can now enter into the synthetic (endergonic) reactions of the cell. Thus, the free energy change of the exergonic reaction has been coupled to endergonic reaction systems through the intermediary of the adenosine nucleotides. From a stochastic point of view this system provides the organism with an ingenious linkage of the synthetic energy using reactions to the respiratory energy-making reactions. The rate of the respiratory reaction is determined by the rate of the synthetic reactions.

Literature.

ALBAUM, H. G., and M. OGUR: An adenine-pentose-pyrophosphate from plant tissues. Arch. of Biochem. 15, 158–160 (1947). — ALBAUM, H. G., M. OGUR and A. HIRSHFELD: The isolation of adenosine triphosphate from plant tissue. Arch. of Biochem. 27, 130–142 (1950). — ASCOLI, A.: On plasmic acid. Z. physiol. Chem. 28, 426 (1899). — AXELROD, B., R. S. BANDURSKI and P. SALTMAN: Phosphate uptake by pea meal extracts. Federat. Proc. 10, 158 (1951). — AYENGAR, P., D. M. GIBSON, C. H. LEE PANG and D. R. SANADI: Isolation of guanosine di- and triphosphate from yeast. J. of Biol. Chem. 218, 521–533 (1956).

BANDURSKI, R. S., and F. LIPMANN: Studies on an oxalacetic carboxylase from liver mitochondria. J. of Biol. Chem. 219, 741–752 (1956). — BERG, P.: Participation of adenyl-acetate in the acetate activating system. J. Amer. Chem. Soc. 77, 3163–3164 (1955). — BERG, P., and W. K. JOKLIK: Enzymatic phosphorylation of nucleoside diphosphates. J. of Biol. Chem. 210, 657–672 (1954). — BOWEN, W. J., and T. D. KERWIN: A simple method for assaying ATP and ADP in mixtures. J. of Biol. Chem. 220, 9–14 (1956). — BURMA, D. P., and D. C. MORTIMER: The biosynthesis of uridine diphosphate glucose and sucrose in sugar beet leaf. Arch. of Biochem. 62, 16–28 (1956).

CANTONI, G. L.: On the role of high-energy phosphate in transmethylation. In phosphorus metabolism, vol. 1, p. 641–646. Edit. by W. D. McELROY and B. GLASS. Baltimore:

Johns Hopkins Press 1951. — CARDINI, C. E., L. F. LELOIR and J. CHIRIBOGA: The biosynthesis of sucrose. J. of Biol. Chem. 214, 149–155 (1955). — CARTER, C. E.: Metabolism of purines and pyrimidines. Annual. Rev. Biochem. 25, 123–146 (1956). — CHAPPELL, J. B., and S. V. PERRY: Creatine phosphokinase: Assay and application for the micro-determination of the adenine nucleotides. Biochemic. J. 57, 421–427 (1954). — CHARGAFF, E., and J. N DAVIDSON (Editors): The nucleic acids, vol. 2. New York: Academic Press 1955. — COLOWICK, S. P., and H. M. KALCKAR: The role of myokinase in transphosphorylations. I. The enzymatic phosphorylation of hexoses by adenyl pyrophosphate. J. of Biol. Chem. 148, 117–126 (1943).— CORI, C. F.: Regulation of enzyme activity in muscle during work. In Enzymes, Units of Biological Structure and Function, p. 573–583. Edit. by O. H. GAEBLER. New York: Academic Press 1956. — CORYELL, C. D.: The proposed terms "exergonic" and "endergonic" for thermodynamics. Science (Lancaster, Pa.) 92, 380 (1940).

HARDIN, M. B.: The presence of metaphosphoric acid in cottonseed meal. S. Car. Agric. Exper. Stat. Bull. 8, N. S. 10. — HOFFMAN-OSTENHOF, O., J. KENEDY, K. KECK, O. GABRIEL and H. W. SCHÖNFELLINGER: Ein neues phosphat-übertragendes Ferment aus Hefe. Biochim. et Biophysica Acta 14, 285 (1954).

JOKLIK, W. K.: The formation of nucleoside triphosphate from inosine diphosphate in yeast. Biochim. et Biophysica Acta 16, 610–611 (1955).

KALCKAR, H. M.: The role of myokinase in transphosphorylations. II. The enzymatic action of myokinase on adenine nucleotides. J. of Biol. Chem. 148, 127–137 (1943). — KALCKAR, H. M., and H. KLENOW: Non-oxidative and non-proteolytic enzymes. Biosynthesis and metabolism of phosphorus compounds. Annual Rev. Biochem. 23, 527–586 (1954). — KIESSLING, W., u. O. MEYERHOF: Über ein Adenindinucleotid der Hefe: Di-(Adenosin-5'-phosphorsäure). Biochem. Z. 296, 410–425 (1938). — KREBS, H. A., and R. HEMS: Some reactions of adenosine and inosine phosphates in animal tissues. Biochim. et Biophysica Acta 12, 172–180 (1953).

LELOIR, L. F., and C. E. CARDINI: The biosynthesis of sucrose. J. Amer. Chem. Soc. 75, 6084 (1953). — The biosynthesis of sucrose phosphate. J. of Biol. Chem. 214, 157–165 (1955). — LIEBERMAN, I.: Identification of adenosine tetraphosphate from horse muscle. J. Amer. Chem. Soc. 77, 3373–3375 (1955). — Inorganic triphosphate synthesis by muscle adenylate kinase. J. of Biol. Chem. 219, 307–318 (1956). — LIEBERMAN, I., A. KORNBERG and E. SIMMS: Enzymatic synthesis of nucleoside diphosphates and triphosphates. J. of Biol. Chem. 215, 429–440 (1955). — LIEBERMAN, L.: Detection of metaphosphoric acid in the nuclein of yeast. Pflügers Arch. 47, 155–160 (1890). — LIPMANN, F.: Consideration of the role of coenzyme A in some phases of fat metabolism. Fat Metabolism. Edit. by V. A. NAJJAR. Baltimore: Johns Hopkins Press 1954.

MAZELIS, M.: Particulate adenylic kinase in higher plants. Plant Physiol. 31, 37–43 (1956). — MARKHAM, R., and J. D. SMITH: The structure of ribonucleic acid. I. Cyclic nucleotides produced by ribonuclease and by alkaline hydrolysis. Biochemic. J. 52, 552–557 (1952). — McLAUGHLIN, J., G. SCHIFFMAN and A. SZENT-GYÖRGYI: Inosine phosphates in muscle. Biochim. et Biophysica Acta 17, 160 (1955).

ROBBINS, P. W., and F. LIPMANN: Identification of enzymatically active sulfate as adenosine-3'-phosphate-5'-phosphosulfate. J. Amer. Chem. Soc. 78, 2652–2653 (1956).

SACKS, J.: Adenosine pentaphosphate from commercial ATP. Biochim. et Biophysica Acta 16, 436 (1955). — SANADI, D. R., D. M. GIBSON, P. AYENGAR and M. JACOB: α-Ketoglutaric dehydrogenase. V. Guanosine diphosphate in coupled phosphorylation. J. of Biol. Chem. 218, 505–520 (1956). — SCHMIDT, G.: The biochemistry of inorganic phyrophosphates and metaphosphates. In Phosphorus Metabolism, vol. 1. p. 443–475. Edit. by W. B. McELROY, and B. GLASS. Baltimore: Johns Hopkins Press 1951. — SCHMITZ, H.: Isolierung von freien Nucleotiden in verschiedenen Geweben. Vorkommen von Nucleosid-5'-phosphaten im säurelöslichen Extrakt aus Hefe. Biochem. Z. 325, 555–569 (1954). — STROMINGER, J. L., L. A. HEPPEL and E. S. MAXWELL: A new mechanism of nucleoside di- and triphosphate synthesis. I. Transphosphorylation between nucleoside monophosphate and nucleoside triphosphate. Arch. of Biochem. 52, 488–491 (1954). — STUMPF, P. K.: Phosphate assimilation in higher plants. In Phosphorus Metabolism, vol. II, p. 29–67. Edit. by W. D. McELROY and B. GLASS. Baltimore: Johns Hopkins Press 1952.

UTTER, M. F., K. KURAHASHI and I. A. ROSE: Some properties of oxalacetic carboxylase. J. of Biol. Chem. 207, 803–819 (1954).

WEBSTER, G. C.: Nitrogen Metabolism. Annual Rev. Plant Physiol. 6, 43–70 (1955). — WINTERMANS, J. F. G. M.: Polyphosphate formation in Chlorella in relation to photosynthesis. Meded. Landbouwhoogeschool te Wageningen 55, 69–126 (1955). — WYATT, G. R.: Separation of nucleic acid components by chromatography on filter paper. In: The Nucleic Acids, vol. I., p. 243–266. Edit. by E. CHARGAFF and J. N. DAVIDSON. New York: Academic Press 1955.

Namenverzeichnis. — Author Index.

Die *kursiv* gesetzten Seitenzahlen beziehen sich auf die Literatur.
Page numbers in *italics* refer to the bibliography.

Sachverzeichnis.
(Deutsch-Englisch.)

Ä, Ö, Ü sind wie Ae, Oe, Ue eingereiht.
Bei gleicher Schreibweise in beiden Sprachen sind die Stichworte jeweils einfach aufgeführt.

Subject Index.

(English-German.)

Ä, Ö, Ü are taken as Ae, Oe, Ue.
Where English and German spelling of a word is identical the italised (German) entry is omitted.